ADVANCES IN CHEMICAL PHYSICS

VOLUME LXIV

Advances in
CHEMICAL PHYSICS

EDITED BY

I. PRIGOGINE

University of Brussels
Brussels, Belgium
and
University of Texas
Austin, Texas

AND

STUART A. RICE

Department of Chemistry
and
The James Franck Institute
The University of Chicago
Chicago, Illinois

VOLUME LXIV

AN INTERSCIENCE® PUBLICATION
JOHN WILEY & SONS
NEW YORK · CHICHESTER · BRISBANE · TORONTO · SINGAPORE

An Interscience® Publication

Copyright© 1986 by John Wiley & Sons, Inc.

All rights reserved. Published simultaneously in Canada.

Reproduction or translation of any part of this work
beyond that permitted by Section 107 or 108 of the
1976 United States Copyright Act without the permission
of the copyright owner is unlawful. Requests for
permission or further information should be addressed to
the Permissions Department, John Wiley & Sons, Inc.

Library of Congress Catalog Number: 58-9935

ISBN 0-471-82582-4

Printed in the United States of America

10 9 8 7 6 5 4 3 2 1

CONTRIBUTORS TO VOLUME LXIV

TOMAS BAER, Department of Chemistry, University of North Carolina, Chapel Hill, North Carolina, U.S.A.

D. BEDEAUX, Department of Physical and Macromolecular Chemistry, Gorlaeaus Laboratories, University of Leiden, Leiden, The Netherlands

S.-H. CHEN, Nuclear Engineering Department, Massachusetts Institute of Technology, Cambridge, Massachusetts, U.S.A.

JOHN F. GRIFFITHS, Department of Physical Chemistry, The University of Leeds, Leeds, England

J. TEIXEIRA, Laboratoire Leon Brillouin, Comissariat à l'Énergie Atomique and Centre National de la Recherche Scientifique, Saclay, Gif-sur-Yvette, France

INTRODUCTION

Few of us can any longer keep up with the flood of scientific literature, even in specialized subfields. Any attempt to do more and be broadly educated with respect to a large domain of science has the appearance of tilting at windmills. Yet the synthesis of ideas drawn from different subjects into new, powerful, general concepts is as valuable as ever, and the desire to remain educated persists in all scientists. This series, *Advances in Chemical Physics*, is devoted to helping the reader obtain general information about a wide variety of topics in chemical physics, which field we interpret very broadly. Our intent is to have experts present comprehensive analyses of subjects of interest and to encourage the expression of individual points of view. We hope that this approach to the presentation of an overview of a subject will both stimulate new research and serve as a personalized learning text for beginners in a field.

ILYA PRIGOGINE
STUART A. RICE

CONTENTS

ADVANCES IN CHEMICAL PHYSICS

VOLUME LXIV

STRUCTURE AND DYNAMICS
OF LOW-TEMPERATURE WATER
AS STUDIED BY
SCATTERING TECHNIQUES

S.-H. CHEN

Nuclear Engineering Department
Massachusetts Institute of Technology
Cambridge, Massachusetts, U.S.A.

J. TEIXEIRA

Laboratoire Leon Brillouin, Comissariat à l'Energie Atomique
and Centre National de la Recherche Scientifique
Saclay, Gif-sur-Yvette, France

CONTENTS

I. INTRODUCTION

The structure and dynamics of pure water are two of the fundamental problems in chemical physics, because they constitute a cornerstone for understanding all aspects of solution chemistry and the functions of biological macromolecules. In spite of an intense effort by physical scientists for nearly a century, the understanding of these problems is incomplete and in many aspects the physics of water remains an open field.

Historically, the thermodynamic and transport properties of water have attracted the attention of physical chemists not only because of their obvious implications for real-life problems, but because of their unique behavior. The so-called anomalous properties of water are well documented, and a number of good review articles already exist on the subject.[1] Examples of such thermodynamic anomalies are the density maximum above the freezing point (Fig. 1); the minimum in the compressibility of H_2O at 46°C and its rapid rise below 0°C (Fig. 2); and the constant-pressure heat capacity, which increases by about a factor of 2 within the measured supercooled range (0 to −20°C).[1] The transport properties also show anomalous temperature and pressure dependences. For example, the self-diffusion coefficient D and the shear viscosity η are strongly non-Arrhenius and change rapidly with temperature (Fig. 3). It has been pointed out that in the supercooled range the temperature dependences of both the thermodynamic and the transport properties of H_2O can be conveniently fitted by power-law behavior with reference to a temperature $T_s = -45°C$.[2] It is interesting that in spite of their anomalous behavior, the transport coefficients and various measured relaxation times τ satisfy simple relations such as $D\eta/T = \text{const}$ and $D\tau = \text{const}$, where these constants are nearly temperature independent.[3]

Most of the older theories of water tried to explain the anomalous behavior by assuming the presence of two microscopic phases, a solid-like, low-density phase and a liquid-like, higher-density phase, with variable proportions of each depending on external parameters such as temperature or pressure.[4] These models, at least in their primitive forms, are not at all confirmed by experiments, in particular the scattering experiments such as X-ray and neutron scattering. Older reviews on these experimental results exist in the literature.[5]

More recently, the use of effective pair potentials combined with realistic computer molecular-dynamics simulations (CMD) has substantially improved our knowledge of the structure and dynamics of water.[6] Analysis of the CMD data shows the importance of hydrogen bonding and the tetrahedral coordination in the interpretation of thermodynamic and transport properties of water. Moreover, results of such calculations can in general be directly compared with experiments on the dynamics of water.[7] It is clear that a description of the molecular behavior of water by an effective pair potential will never be completely realistic, because of

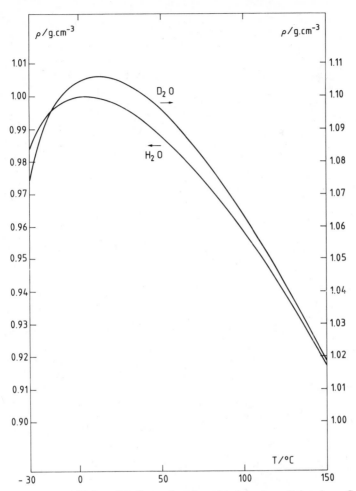

Figure 1.   Densities of H_2O and D_2O as a function of temperature. Taken from G. S. Kell, *J. Chem. Eng. Data* **20**, 97 (1975).

the existence of many-body forces and the complexity of water. Nevertheless, this technique will remain to be a unique and powerful approach for probing microscopic properties of water for some time.

Different scattering experiments such as light, X-ray, and neutron scattering are currently being used to characterize the structure and dynamics of water at the microscopic level. Most of the time these data can be compared directly with the predictions of CMD based on different models of pair potentials. These comparisons serve the dual purpose of checking and improving the model potentials.

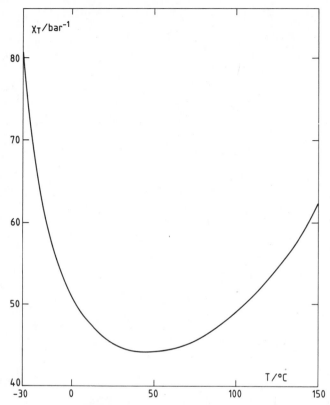

Figure 2.   Isothermal compressibility of H_2O as a function of temperature. Taken from
G. S. Kell, *J. Chem. Eng. Data* **20**, 97 (1975).

Roughly speaking, the scattering experiments indicate that properties
of water appear less anomalous when one probes at the microscopic level.
Thus, a fundamental problem arises in trying to find relations between the
results of microscopic analyses and the macroscopic thermodynamic and
transport properties. Very often, in order to take into account the
temperature dependence of properties of water, the model theories in a
sense assume more structure than is observed in the scattering experiments. That the microscopic origin of the tetrahedral structure of water is
not completely understood explains why the relation of structure to the
thermodynamic anomalies, particularly at low temperatures, remains con-
troversial.[2] In this chapter we summarize some of the more important
recent results obtained by scattering experiments.

X-ray diffraction gives essentially the pair-correlation function of the
oxygen atoms, whereas neutron diffraction gives in principle all the

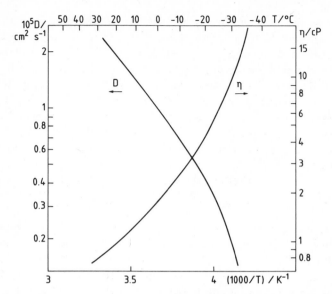

Figure 3. Self-diffusion constant and shear viscosity of H_2O at 1 atm as a function of temperature. Data taken from G. T. Gillen, D. C. Douglass, and M. R. Hoch, *J. Chem. Phys.* **57,** 5117 (1972), and Yu. A. Osipov, B. V. Zheleznyi, and N. F. Bondarenko, *Zh. Fiz. Khim.* **51,** 1264 (1977) (Engl. trans.) Note the non-Arrhenius behavior at low temperatures.

pair-correlation functions between oxygen and hydrogen atoms, and therefore the angular correlations between molecules. Inelastic neutron scattering gives information on dynamical properties associated with both single-particle and collective processes. Light scattering has been used to provide information on density fluctuations in the hydrodynamic limit. Brillouin scattering gives the hypersound velocity and its damping. Depolarized light scattering is sensitive to the rotational motion of the molecules and gives important information on rotational relaxation time. In this article, we limit ourselves to discussion of these topics. Related topics, such as Raman scattering and infrared absorption, have already been reviewed by Walrafen,[8] and Luck,[9] and, more recently, Rice and Sceats.[10]

Although the neutron-scattering technique is not particularly recent, many of the new results we discuss in this article were obtained with it. This is due to two developments: First, there has been a tremendous improvement in the energy resolution and the energy range covered in the new generation of neutron spectrometers, which enables these experiments to cover hitherto inaccessible spectral regions. Second, it was recently realized that the anomalous character of properties of water increases as the temperature is lowered below the freezing point to the supercooled state. This enables the experiment to separate out fast and slow relaxation times and identify their crucial temperature dependence. Many of these new results have just been published or are in the process of being published. The analyses require extensive comparison with CMD treatments, which in many cases have not yet been completed. For this reason, this Chapter can summarize only in a qualitative way the impact of these new experiments on the interpretation of the anomalous behavior of water at low temperature.

II. THE STRUCTURE OF WATER

A. X-Ray Diffraction

The basic information on the equilibrium structure of water is represented by the three partial pair-correlation functions $g_{OO}(r)$, $g_{OH}(r)$ and $g_{HH}(r)$. Figures 4, 5, and 6 illustrate schematically the behavior of g_{OO}, g_{OH} and g_{HH}. These partial pair-correlation functions were generated by a CMD using a model potential proposed by Berendsen and co-workers[11] called a simple-point-charge (SPC) model. The SPC model consists of a pairwise Lennard-Jones interaction between the oxygen atoms together with Coulomb interactions between suitable point charges situated at the oxygen and hydrogen sites. In this simulation, the SPC model was

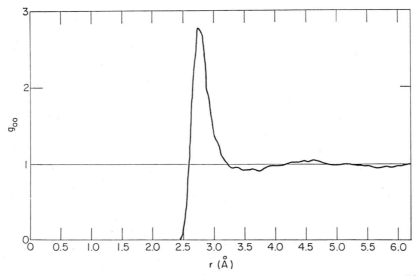

Figure 4.   Partial pair-correlation function $g_{OO}(r)$ generated by CMD from a modified version of the SPC potential. Taken from reference 12.

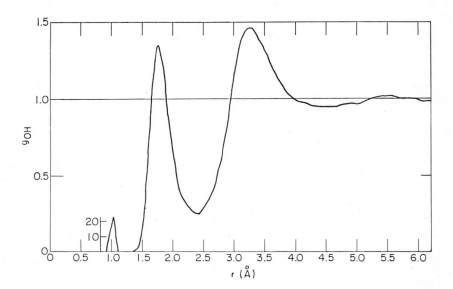

Figure 5. Partial pair-correlation function $g_{OH}(r)$ generated by CMD from a modified version of the SPC potential. Taken from reference 12.

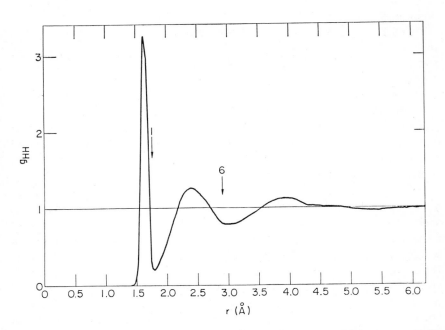

Figure 6. Partial pair correlation function $g_{HH}(r)$ generated by CMD from a modified version of the SPC potential. Taken from reference 12.

TABLE I
Main Peak Positions of g_{OO}, g_{OH}, and g_{HH} in
Angstroms

Peak	g_{OO}	g_{OH}	g_{HH}
1st	2.8	1.035 (intra)	1.66 (intra)
2nd	4.5	1.75	2.28
3rd	—	3.25	3.8

modified to allow for the flexibility of the water molecule by the introduction of the Morse potential representing the intramolecular interaction between the different atoms.[12] We summarize in Table I the important interatomic distances as indicated by the major peaks in the partial pair-correlation functions.

In principle, three independent diffraction measurements can be combined to yield the required partial structure factors but in practice this proves extremely difficult and a definitive set of functions has not yet been obtained even at one temperature. The experimental techniques involved in obtaining these three independent measurements include X-ray, neutron, and electron diffractions[13] or neutron diffraction of different mixtures of H_2O and D_2O.

Among the earlier results of X-ray diffraction were work of Narten and Levy[14] and of Hajdu et al.[15] The neutron-diffraction results were given by Narten.[16] A summary of all diffraction work up to 1982 and a comparison of the results with computer simulations were given by Egelstaff.[17] Most recently, Dore[18] published a short review emphasizing the neutron-diffraction work of his own group.[18] These two review articles emphasized the difficulty of measurement and the uncertainty of the results obtained so far. It is fair to say that the ultimate determination of the three partial pair-correlation functions of water will not be forthcoming for some time.

It is important to realize that while $g_{OO}(r)$ correlations give approximately the distribution of the molecular centers, the other two partial functions, $g_{OH}(r)$ and $g_{HH}(r)$, give the orientational correlations between adjacent molecules. From the point of view of the chemical physics of water, information on the variation of these partials as a function of temperature is more important than the values of these partials at a specific temperature, say room temperature. In this review we shall therefore focus attention on the temperature dependence of the pair-correlation functions over a wide range of temperatures, from room temperature down to −20°C in the supercooled state.

The X-ray structure factor can be shown to reflect mostly the O–O correlations (64%), with some admixture of O–H correlations (32%) and H–H correlations (4%).[19] In an X-ray diffraction experiment, corrections due to multiple scattering, incoherent background, and container contribution reduce the accuracy of the ultimate determination of the structure factor to perhaps a few percent. This degree of accuracy is not sufficient for detection of the weak temperature dependence of the structural rearrangements in water as the temperature is reduced to the supercooled range. It is obvious that study of the structural rearrangement at low temperature is essential for understanding the thermodynamic anomalies in the supercooled regime. To make more evident the variation of the X-ray structure factor $S_X(Q)$ as a function of temperature, Bosio et al.[20] used a differential technique.

The measured X-ray structure factor $S_X(Q)$ can be approximately related to the partial structure factor $S_{OO}(Q)$ by

$$S_X(Q) = \langle F^2(Q) \rangle S_{OO}(Q) \qquad (2.1)$$

where $F(Q)$ is the molecular form factor for X rays. The approximation in Eq. (2.1) involves equating $\langle F^2(Q) \rangle$ and $\langle F(Q) \rangle^2$, which can be done because the electronic distribution in a water molecule is very nearly spherical.[14]

Fortunately, liquid water has the unique property of showing a maximum in the density versus temperature curve, and this feature allows experiments to be performed at two different temperatures but at the same density.[19,31] We shall define the isochoric temperature differential (ITD) of the X-ray structure factor by

$$\Delta S_{OO}(Q, \rho, \Delta T) = \frac{S_X(Q, \rho, T_1) - S_X(Q, \rho, T_2)}{\langle F^2(Q) \rangle} \qquad (2.2)$$

where $\Delta T \equiv T_1 - T_2 > 0$ and T_1 and T_2 are two temperatures on opposite sides of the density maximum for which the density ρ is the same.

In D_2O, the density maximum occurs at 11.2°C. Figure 7 gives $S_{OO}(Q)$ and $g_{OO}(r)$ at this temperature. We note a characteristic double peak of $S_{OO}(Q)$ and a long oscillation extending to $Q > 10 \text{ Å}^{-1}$. It is obviously difficult to obtain an accurate $g_{OO}(r)$ based on finite Q data for $S_{OO}(Q)$. Aside from the main peak in $g_{OO}(r)$, located at 2.9 Å, there is a broad second maximum centered about 4.5 Å. This broad maximum represents the distribution of the second-nearest neighbor in the O–O–O configuration. The temperature dependence of the intensity in this region will represent a completion of the geometrical rearrangement of the tetrahedral coordination. Figure 8 shows the Fourier transform of

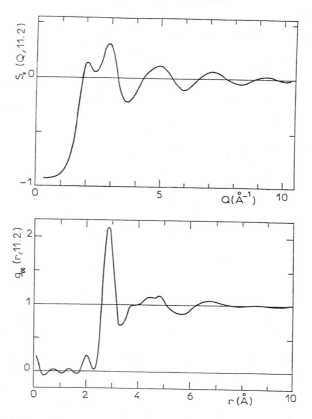

Figure 7. Molecular-center structure factor $S_X(Q)$ and partial pair-correlation function $g_{OO}(r)$ of D_2O at the temperature of maximum density (11.2°C). Taken from reference 20.

$\Delta S_{OO}(Q, \rho, \Delta T)$ corresponding to $\Delta g_{OO}(r, \Delta T)$ at three values of ΔT, namely, $40.0 - (-11.0) = 51.0$, $23.5 - 0.2 = 23.3$, and $15.5 - 7.1 = 8.4°C$, around the maximum-density point, 11.2 °C. From this figure, it is clear that Δg_{OO} is proportional to ΔT and thus the isochoric temperature derivative $\Delta g_{OO}/\Delta T$ is a constant in the temperature range studied (from 40 to -11°C). To interpret physically what is happening to the O–O correlation, Bosio et al.[20] compute from Fig. 8 an integrated quantity ΔN given by

$$\Delta N(R, \Delta T) = \rho \int_0^R \Delta g_{OO}(r, \Delta T) 4\pi r^2 \, dr \qquad (2.3)$$

$\Delta N(R, \Delta T)$ is the temperature variation of the number of neighbors up to

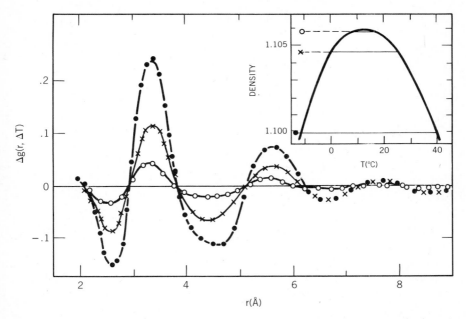

Figure 8.   Plots of $\Delta g(r, \Delta T)$ for the three pairs of temperatures indicated in the insert. Taken from reference 20.

a distance R from a central molecule. The variation of $\Delta N(R, \Delta T)$ with R is depicted in Fig. 9 for $\Delta T = 51.0°C$.

We summarize the conclusions of this study as follows:

1.   Within the accuracy of the experiment, there is no significant change of $S_{OO}(Q)$ for Q beyond $7\ \text{Å}^{-1}$. Therefore, the computation of $\Delta S_{OO}(Q, \rho, \Delta T)$ needs be made only until $Q_{max} \approx 7\ \text{Å}^{-1}$ without losing information in the Fourier transform.

2.   The important rearrangement of the oxygen atoms occurs at the second-nearest-neighbor shell as temperature is varied. It is inferred from Fig. 8 that r-space data indicate that second neighbors have a tendency to move away (toward a more regular tetrahedral coordination) when the temperature is lowered.

3.   As the temperature is lowered, the number of first-nearest neighbors is not significantly affected.

In physical terms, these three conclusions mean that the anomalous increase in molecular volume with temperature can be understood only if one takes into account the positions of the second-nearest neighbors and their variation with temperature, or in other words the O–O–O angular correlations. This angular correlation is connected with the existence of a

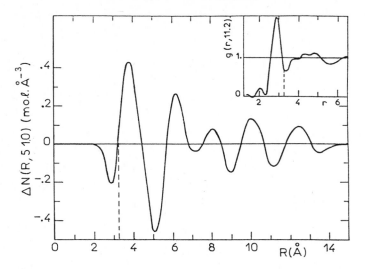

Figure 9. Variation of the number of molecular centers within a distance R from a central molecular for the largest measured temperature interval (51.0°C). The insert shows $g(r)$ at 11.2°C for reference. Taken from reference 20.

finite correlation length, a result obtained by small-angle X-ray measurements, which will be discussed next. Thus, the main conclusion of this work is that the completion of the tetrahedral structure is due to the opening of the O–O–O angles with decreasing temperature.

B. Small-Angle X-Ray Scattering

Small-angle scattering is a way of measuring correlations in an intermediate, semimicroscopic length scale—typically, 10–1000 Å. The peak centered around $Q = 0$ is due to density fluctuations and/or concentration fluctuations. In fact, in an isotropic fluid, $S(Q)$ extrapolated to a zero angle, that is, $S(0)$, is given by

$$S(0) = k_B T \rho \chi_T \qquad (2.4)$$

where k_B is the Boltzmann constant, T the absolute temperature, ρ the number density, and χ_T the isothermal compressibility. In a two-component fluid, or in a two-phase system, extra important contributions to the small-angle scattering come from concentration fluctuations and from individual clusters or particles.

Because of the historical importance of the two-phase models of water,[4] small-angle X-ray scattering by H_2O at room temperature has been used to rule out completely the presence of solid-like clusters in liquid water.[21] The same kind of measurements were extended to the supercooled region by Bosio et al.[22] The results for D_2O are shown in Fig. 10. It is remarkable that even at 75°C the Q dependence of the structure factor at small Q values is not flat. At room temperature, the minimum of $S(Q)$ occurs clearly around $Q = 0.5$ Å^{-1}; the effect is strongly amplified at −20°C. An extrapolation to zero-angle scattering is in agreement with the value predicted by Eq. (2.4). In particular, at −20°C, the value of $S(0)$ is well above the values at higher temperatures and the corresponding $S(Q)$ curve crosses the other curves. This effect is due to the substantial increase of the isothermal compressibility with decreasing temperature[23] (see Fig. 2).

Assuming that the density fluctuation in liquid water can be described by an Ornstein–Zernike form, it is possible to extract a correlation length from the data. This calculation gives a value for the correlation length of ∼8 Å, meaning that the correlations of the density fluctuations do not extend farther than the second neighbors. It will be interesting to determine the temperature dependence of this correlation length and to test its

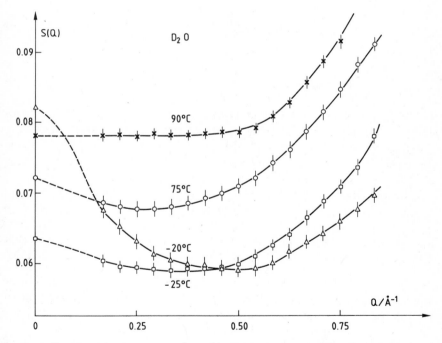

Figure 10. Small-angle X-ray scattering intensity distributions from D_2O at four different temperatures. Dashed lines represent extrapolations. Note the peaks in intensity at $Q = 0$, especially at low temperatures. Taken from reference 22.

possible critical behavior. This type of behavior was suggested by the analysis of the properties of supercooled water by Speedy and Angell.[24]

C. Neutron Diffraction

Neutrons are scattered by oxygen, hydrogen, and deuterium nuclei in different ways. In contrast to X-ray scattering, which is sensitive mainly to the oxygen positions, neutron diffraction can provide the pair correlations between the relative positions of deuterium and oxygen and thus gives information on the orientational correlations of water molecules.

The main use of neutron-diffraction measurements is in the separation of the different partial pair-correlation functions. In principle, this can be achieved in two ways. The first one is to combine data from X-ray, neutron, and electron scatterings. This method has been used by Palinkas et al.,[25] but the low accuracy of electron-scattering data did not allow good separation of the partials. The second possibility, widely used by neutron-diffraction workers, consists of using different isotopes, each with a different scattering length. For example, the composite $g(r)$ from H_2O and D_2O are

$$g(r) = 0.193g_{OO}(r) - 0.492g_{OH}(r) + 0.315g_{HH}(r) \qquad (2.5)$$

and

$$g(r) = 0.092g_{OO}(r) + 0.422g_{OD}(r) + 0.486g_{DD}(r) \qquad (2.6)$$

Different mixtures of H_2O and D_2O provide analogous equations. Because of the negative scattering length of the hydrogen atom, it is even possible to use a mixture of H_2O and D_2O for which the average scattering length of the hydrogen and deuterium atoms is zero and obtain a scattering function analogous to the one determined with X rays.

Unfortunately, this latter technique is plagued with several difficult problems, particularly in the case of water. In spite of very precise and good statistics measurements, it is still impossible to obtain data reliable enough to allow different partials to be extracted. The problems are as follows:

1. The incoherent cross section of hydrogen atoms is very large compared with the coherent cross section. Therefore, for a H_2O sample or for mixtures containing large amounts of H_2O, the coherent part represents only a small fraction of the signal.

2. The effective mass of the water molecule is not much larger than the neutron mass, so the inelasticity correction is very large. Theoretical calculations[17,26] have been done to take this into account, but the model

of the dynamics used is too crude to allow sufficiently accurate corrections to be made.

3.   The effects due to multiple scattering are particularly important in water, and some discrepancies between different sets of data can be attributed to the use of different sample thicknesses in the experiments.

4.   The main problem, however, is that H_2O and D_2O are different liquids in both thermodynamic and dynamic properties. To assume the structure of the two liquids is the same is probably the most misleading approximation in the separation of the partials. This problem is also connected with the central question of the role of hydrogen bonding in the structure and dynamics of water.

A striking feature of the neutron structure factor $S(Q)$ of water is its extremely large temperature dependence. In Fig. 11, we show a typical

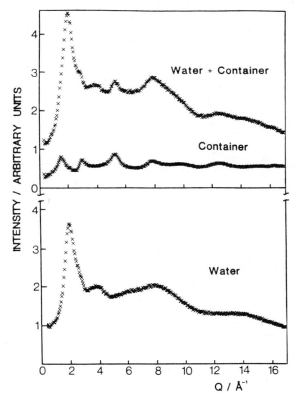

Figure 11.   Diffraction pattern of D_2O at 25°C. From top to bottom, curves are shown for the sample plus the container, the container, and the corrected sample values. Note the drooping of the last curve beyond $Q > 10 \text{ Å}^{-1}$, indicating the size of inelasticity corrections. Taken from reference 27.

structure-factor plot for D_2O at room temperature.[27] Note the drooping of the corrected diffraction pattern in the lower part of the figure for $Q > 10\,\text{Å}^{-1}$. This is due to the inelasticity effect.[17] There is a first diffraction peak at ~$2\,\text{Å}^{-1}$. The position of this peak is strongly temperature dependent, as shown in Fig. 12. An analogous plot from the X-ray diffraction[20] also shows a displacement of the first structural peak toward smaller Q values with decreasing temperature. However, at low temperatures, and in particular in the supercooled region, the temperature dependence of the position of the first neutron-diffraction peak is stronger; one can extrapolate to values as low as $1.7\,\text{Å}^{-1}$ at $-40°C$. The position of the first diffraction peak in amorphous ice is also at $1.7\,\text{Å}^{-1}$.[28] Based on this one may speculate that with decreasing temperature, supercooled water approaches a structure similar to the random-network structure of amorphous ice.[10,18,27] This idea is not incompatible with the fact that the structure of liquid water, which is locally well defined, does not extend to very large distances, even at low temperatures.

Because neutron diffraction is sensitive to both position and angular correlations, the important displacement of the first structural peak with temperature can be attributed to an increase of the angular correlations with decreasing temperature, a conclusion that follows equally from the analysis of precise isochoric differential X-ray scattering measurements, as stated in Section II.A.[20] This means that the increase in the number of

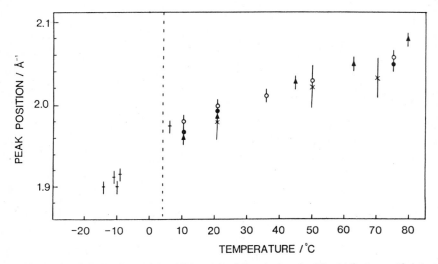

Figure 12.   Shift in the position of the main diffraction peak of liquid D_2O as a function of temperature. Data are from references 27 and 30. Note the strong temperature variation, especially at low temperatures. Taken from reference 27.

intact bonds with decreasing temperature is accompanied by an increase in the angular correlations leading locally to the regular tetrahedral symmetry.

The pronounced temperature dependence of $S(Q)$ suggests the use of a differential technique consisting of taking the differences between two scattered intensities measured at two temperatures T and T_{ref}, where T_{ref} is a reference temperature.[29,30] The Fourier transform of this difference is then very accurate, because the other corrections all cancel out.

Writing the scattering function $S(Q)$ as

$$S(Q) = f_1(Q) + D_M(Q) \qquad (2.7)$$

where $f_1(Q)$ is the molecular-form factor depending on the conformation of the individual molecules and $D_M(Q)$ is the part of $S(Q)$ depending on the molecular positions, the differential technique gives

$$\Delta S_M(Q, T) = S_M(Q, T) - S_M(Q, T_{ref}) = \Delta D_M(Q, T) \qquad (2.8)$$

assuming $f_1(Q)$ to be temperature independent. An example of this treatment is given in Fig. 13. The Fourier transform then gives the changes in the real-space correlation function

$$\Delta d_L(r, T) = \frac{2}{\pi} \int_0^\infty Q \Delta D_M(Q, T) \sin(Qr) \, dQ \qquad (2.9)$$

This method can be used even more accurately if one exploits the maximum in the density of water. If one takes pairs of points at the same density from each side of the density maximum, the procedure cancels the bulk density effects.[20,31]

The main results of the recent experiments can be compared with those of molecular-dynamics simulations from the theoretical point of view. The results are in reasonable agreement, as can be seen from Fig. 14. However, molecular-dynamics simulations seem to always show too much structure. The use of the model potential in CMD constitutes in effect a compromise between the observed structure and the thermodynamic anomalies. This problem is probably a central one—what is the connection between the structural changes on a molecular scale and the thermodynamic properties of water? It is likely that in the near future the role of the hydrogen-bond network per se will be taken into account and that quantum corrections and their effect on the formation of this network will then have to be considered.

So far, the main results obtained by different neutron-diffraction

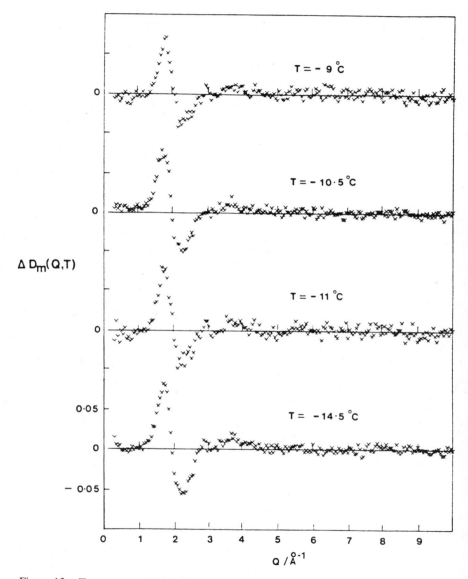

Figure 13. Temperature differential of the neutron-diffraction pattern with respect to the reference temperature $T_{ref} = 21°C$. Taken from reference 27.

18

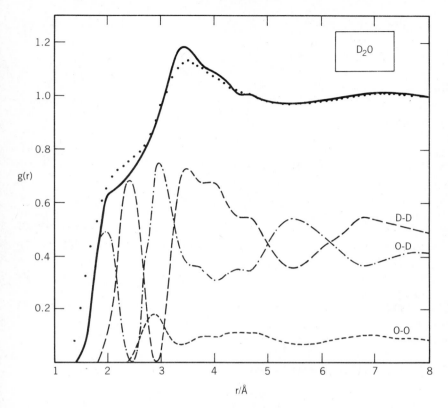

Figure 14.   Comparison between the decomposition into partial correlation functions obtained from reference 13 and neutron-scattering data for D_2O. Taken from reference 30.

experiments still show some discrepancies, essentially because of the difficulties in analyzing the data that we pointed out above. In particular, the pair-correlation function $g_{OH}(r)$ determined by Narten and Thiessen[32] shows a very sharp peak around 1.9 Å corresponding to the distance O $\cdots$ H along the hydrogen bond and there are almost no oscillations beyond 5 Å. In the results of Reed and Dore,[33,18] this first peak of $g_{OH}(r)$ is much less sharp and the oscillations extend up to 8 Å. But in both analyses, the deep minimum at 2.3 Å shows clearly the strong effects on the structure due to hydrogen bonding. The second peak of $g_{OH}(r)$ is at ~3.3 Å and appears sharper in the results of Reed and Dore.[33] A third broader, peak is also present at ~5.5 Å in the analysis of Reed and Dore, but is completely smeared in the work of Narten and Thiessen.[32]

The partial pair-correlation function $g_{HH}(r)$ presents even more problems in the analysis. Essentially, two peaks are present, at 2.4 and 3.8 Å, and again the first appears very sharp in the work of Narten and Thiessen.[32] A new investigation has been done by Soper and Silver,[34] and Soper[35] has reviewed the discrepancies.

From these results, their temperature dependences, and comparisons with molecular-dynamics simulations, it is nontheless possible to draw some conclusions. The local structure of liquid water is close to that of hexagonal ice and the O–O–O angles open toward the ideal value of 108° with decreasing temperature. As a consequence, the local density decreases with decreasing temperature, producing the well-known density maximum. This conclusion is in agreement with a qualitative model that associates the increase in local volume with the degree of bonding.[36] However, this local order does not extend very far—probably not farther than the second-neighbor level. With decreasing temperature, the angular correlations increase rapidly in spite of the fact that there are only small variations at the level of the first neighbors, in agreement with X-ray analysis.[20] The anomalous density variation at low temperatures can be explained only if the positions of the second neighbors are taken into account, and these are, of course, essentially dependent on the O–O–O angle. This observation emphasizes again the importance of hydrogen bonding in the formation of the structure; thus study of such effects will certainly become essential in the near future.

III. THE DYNAMICS OF WATER

A. Quasi-Elastic Incoherent Neutron Scattering

Quasi-elastic neutron scattering (QENS) is generally applied to the study of nonpropagating relaxation modes in liquids. When applied to molecules containing hydrogen atoms, it measures the dominant incoherent scattering from them. In this case, the spectra are dominated by the Fourier transform of the van Hove self-correlation function of the H atoms. In liquid H_2O, the main contributions to the quasi-elastic line are from the self-diffusion of the two equivalent H atoms and from the short-time reorientation of the molecule. Most of the previous studies of water by QENS were performed at temperatures above the freezing point and the analysis was largely done by assuming a single Lorentzian line representing the molecular self-diffusion. Good reviews and summary of the results up to 1972 were given by von Blackenhagen.[37]

A more recent experimental attempt using a three-axis spectrometer at Oak Ridge National Laboratory and done over a wider range of temperatures, from room temperature down to −20°C, has shown that at least two components exist in the QENS spectrum of H_2O.[38] Two series of experiments were performed, with two very different energy resolutions.

The results showed unambiguously a broad line centered around the quasi-elastic position that was distinct from the pronounced QENS due to the self-diffusion line, especially at low temperatures. Figure 15 shows two typical spectra. The top one was obtained at $-20°C$ with a resolution of 97 μeV. The broad line cannot be seen in this graph, because it contributes only a constant background. The bottom one was obtained at 38°C with a resolution of 810 μeV. The width of this peak is dominated by the broad line, because the width of the sharp line at this Q value $(= \sqrt{5}\ Å^{-1})$ is considerably smaller than the resolution. The best estimate of the width of the broad Lorentzian line from this figure is about 2000 μeV. In this work the main conclusion was that the width of the sharp line, Γ_s, at the various temperatures investigated satisfies the scaling relation

$$\Gamma_s(Q, T) = \frac{DQ^2}{1 + Q^2 l_0^2} \qquad (3.1)$$

where D is the macroscopic self-diffusion constant of water at the given temperature T and $l_0 = 0.5 \pm 0.06\ Å$, a characteristic distance that is temperature independent. However, the data were not accurate enough to permit a systematic extraction of the Q dependence of the broad line.

A more accurate experiment was later performed on the high-flux time-of-flight IN6 spectrometer at Grenoble.[39] The high statistical accuracy of the data made it possible to construct a model for simultaneously determining parameters describing the two Lorentzian lines. The Q range covered was $0.25 < Q < 2.0\ Å^{-1}$, and the full width at half maximum of the energy resolution was $\Delta E = 100\ \mu eV$. The transmission of the water sample in a capillary geometry was kept at 95%, with an estimated multiple-scattering contribution of less than 5%. The measurements were done at nine temperatures: 20, 12, 5, -5, -10, -12, -15, -17, and $-20°C$. A separate run was made with the same sample frozen to ice at $-10°C$ in order that comparison with supercooled-water data could be made to ascertain that the sample did not accidentally freeze during the run.

The model analysis was done in the following way. First, one writes the intermediate self-scattering function $F_s(Q, t)$ as

$$F_s(Q, t) = \exp\left(\frac{-Q^2 \langle u^2 \rangle}{3}\right) R(Q, t)\, T(Q, t) \qquad (3.2)$$

where the first factor, the Debye–Waller factor, gives the probability that

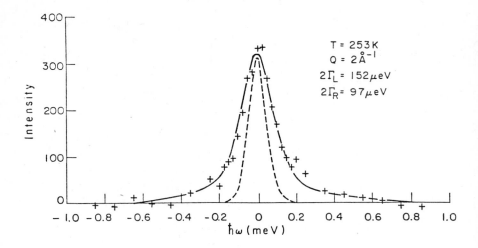

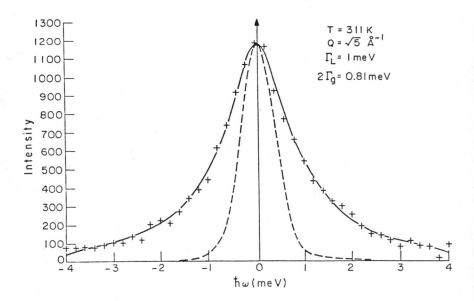

Figure 15. Typical quasi-elastic spectra of water taken with high and low resolutions. The top figure shows a spectrum at $T = -20°C$ and $Q = 2\,\text{Å}^{-1}$. The crosses represent the experimental points, which were fitted with a Lorentzian function convolved with the resolution function (dashed line). The resolution function is a Gaussian function with a width equal to 97 μeV and the extracted Lorentzian function has a width equal to 152 μeV. The bottom figure shows a spectrum at $T = 38°C$ and $Q = \sqrt{5}\,\text{Å}^{-1}$. The dashed line is a Gaussian resolution function with a width equal to 810 μeV. The extracted Lorentzian line has a width equal to 2000 μeV. Taken from reference 38.

22

a neutron is elastically scattered; the second factor, $R(Q, t)$, represents the contribution from the low-frequency rotational motions of the molecule; and the third factor, $T(Q, t)$, represents the contribution from the translational motion. In this model, the translational and rotational motions are decoupled, an assumption that is questionable in water, but has to be made to obtain a numerically tractable analytical model.[40]

The rotational part is further assumed to be given by

$$R(Q, t) = j_0^2(Qa) + 3j_1^2(Qa) \exp\left[\frac{-t}{3\tau_1}\right] + 5j_2^2(Qa) \exp\left[\frac{-t}{\tau_1}\right] \quad (3.3)$$

which are the first three terms of the expansion of the time-correlation function for a rotational diffusion process.[41] Since neutrons probe motions of hydrogen atoms in water, a sensible choice of the parameter a is 0.98 Å, the O–H bond length. With this choice, the product Qa is less than 2.0 and the contributions from higher-order terms in the expansion are negligible. We shall call τ_1 the rotational relaxation time.

The translational part of the correlation function is assumed to be

$$T(Q, t) = \exp[-\Gamma_s(Q)t] \quad (3.4)$$

$$\Gamma_s(Q) = \frac{DQ^2}{1 + DQ^2\tau_0} \quad (3.5)$$

where an explicit assumption of a jump diffusion process is made,[42] and one identifies l_0^2 in Eq. (3.1) with the product $D\tau_0$, where τ_0 is the residence time of the jump diffusion.

Thus, the model contains essentially two parameters, τ_0 and τ_1, since the self-diffusion constant D is known experimentally.[3,43]

Figure 16 displays three typical fits of the model to quasi-elastic peaks at three Q values. The fit is uniformly good to about 1%, consistent with the statistical accuracy of the data. It is rather remarkable that such a simple model can achieve such a good fit, and indicates that the extraction of the narrow translational peak does not depend critically on the model assumed for the broad peak. The main conclusions of this analysis were as follows:

1.   The Debye–Waller factor, which is a measure of the proton delocalization due to vibrations, gives an amplitude of vibration equal to 0.484 Å. This relatively large value can be attributed to the transverse vibrations of the hydrogen atom across the hydrogen bond. It is worth noting that this result is independent of the fitting procedure because it is obtained from the integrated area under the quasi-elastic peak.

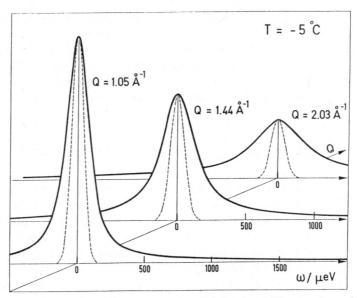

Figure 16.   Results of QENS from water at −5°C for three different values of Q. The experimental points are within the thickness of the solid lines, which represent the best fits. The dashed lines represent the resolution functions. Taken from reference 39.

2. The sharp line can be attributed to the self-diffusion of the hydrogen atoms, which is well described by a jump diffusion mechanism. This point is illustrated in Fig. 17, where the width of the sharp line, Γ_s, is plotted as a function of Q^2. One can see that the solid lines, which were calculated from Eq. (3.5), pass through the data points rather well. Especially noteworthy are the lowest-temperature data, for which $\Gamma_s(Q)$ clearly flattens for $Q \gtrsim 1\,\text{Å}^{-1}$, a striking confirmation of the jump diffusion picture.

3. The broad line is fitted within the conventional rotational diffusion model [Eq. (3.3)]. The width gives the reorientation time τ_1, which has an Arrhenius temperature dependence. We are tempted to think that the molecular reorientation time τ_1 corresponds to the time scale of the breaking of the hydrogen bonds.

The various time scales extracted from different experiments are summarized in Fig. 18. In this figure we display by various dotted lines the temperature dependences of the transport coefficients and of different relaxation times produced from nuclear magnetic resonance (NMR) and dielectric relaxation measurements. A striking feature is that they are all proportional to each other and non-Arrhenius, as pointed out in Section I. This suggests that a single mechanism must explain the temperature

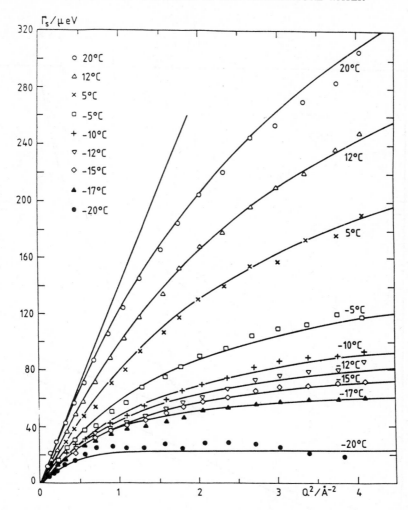

Figure 17.   Line width of the translational component of the spectra plotted versus Q^2. Note that a one-Lorentzian fit, as is commonly done, gives much larger line widths and a different Q dependence. The solid curves represent the best fits from Eq. (3.5), and the straight line corresponds to the self-diffusion constant at 20°C. Taken from reference 39.

dependence of all the dynamic properties of water. Quasi-elastic neutron scattering gives a residence time τ_0, which has the strongest temperature dependence, and a reorientational time τ_1, which has an Arrhenius behavior. Table II summarizes the main quantities obtained by QENS. In this table, $L = \sqrt{6}l_0$ represents the characteristic jump length. Its value is close to that of the intermolecular H–H separation and only slightly

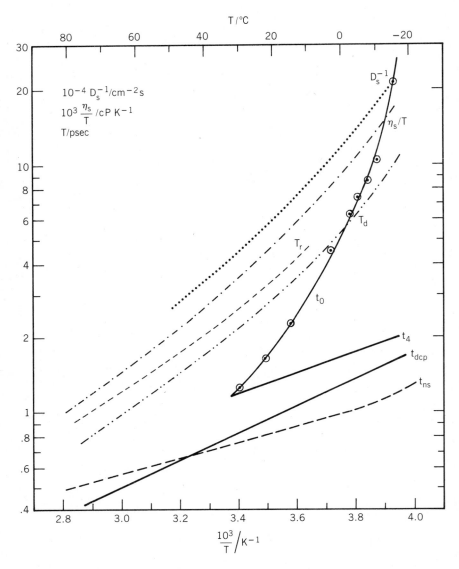

Figure 18.   Transport properties (self-diffusion constant D_s, shear viscosity η_s, dielectric relaxation time τ_d, and orientational relaxation time τ_r) plotted versus temperature. τ_0 and τ_1 are the residence time and the rotational relaxation time obtained by QENS.[39] τ_{dep} is the relaxation time obtained by depolarized Rayleigh scattering,[69] and τ_{HB} is the result of a theoretical calculation of D. Bertolini, M. Cassetari, and G. Salvetti, *J. Chem. Phys.* **76**, 3285 (1982).

26

TABLE II
Parameters Obtained from QENS

T (°C)	τ_0 (ps)	τ_1 (ps)	L (Å)	$\langle u^2 \rangle^{1/2}$ (Å)
20	1.25		1.29	
12	1.66		1.25	
5	2.33	$0.0485 \exp(E_A/k_B T)$	1.32	
−5	4.66		1.54	0.48
−10	6.47	$E_A = 1.85$ kcal/mol	1.65	
−12	7.63		1.70	
−15	8.90		1.73	
−17	10.8		1.80	
−20	22.7		2.39	

temperature dependent. The small increase of L with decreasing temperature is consistent with the opening of the O–O–O angles discussed in Section II.

From all these experimental results, there emerges a possible mechanism for the molecular dynamics of liquid water. The large-amplitude librational movements appear to be the main mechanism of hydrogen-bond breaking. The breaking of an individual hydrogen bond enables a molecular reorientation to occur that is seen experimentally as a molecular rotation. This process is thermally activated, and the associated activation energy (1.85 kcal mol) is comparable to the hydrogen-bond energy. Finally, the molecular diffusion is possible only when several hydrogen bonds are broken simultaneously. Because the number of bonds increases with decreasing temperature, the diffusion process is strongly temperature dependent.

No definitive model for the molecular dynamics of water is yet available. Actually, the extremely strong temperature dependence of the spectra suggests that it may be impossible to apply a single classical model at all temperatures. At least at low temperatures, the main process of diffusion appears to be a rotational jump. When describing the molecular dynamics of water, one needs to pass continuously from the librational movements to self-diffusion through hindered molecular rotations. The coupling of all these motions is temperature dependent. A complete description must take such couplings into account.

B. Inelastic Neutron Scattering

1. High Frequency Scattering

As a result of the existence of spin incoherence in neutron scattering from protons, neutrons are uniquely suited for probing the single-particle motions of protons in hydrogen-containing substances.[40] This feature is particularly valuable for the case of water, because hydrogen motion is sensitive to the formation of hydrogen bonds. Although traditionally Raman scattering has been used extensively to study the hydrogen-bond dynamics in water,[8,44] the quantitative theoretical prediction of the resulting Raman spectra of water has been far from straightforward.[45] This is due to the fact that the intensity of Raman bands is related to the fluctuation spectra of the polarizability tensor of the molecule, which is difficult to calculate from first principles when every molecule is interconnected to its neighbors by the hydrogen-bond network.[46] On the other hand, the incoherent-neutron-scattering spectrum singles out the vibrational and diffusive motions of the hydrogen atoms, which can be simulated with a CMD if a suitable model potential of water is used.[1] Thus, comparisons of neutron-scattering experiments with results of CMD serve the valuable purpose of finding the defects and merits of various proposed intermolecular-potential models and suggesting possible refinements of these models.

Since the vibrational spectrum of water stretches from 0 to ~500 meV (0–4000 cm^{-1}), a wide range that is impossible to cover with a single neutron spectrometer, we have divided this topic between two subsections. This section covers the spectral range of the intramolecular vibrations, and III.B.2 covers that of the intermolecular vibrations.

Very recently an inelastic-neutron-scattering experiment was carried out at the Intense Pulsed Neutron Source (IPNS) at Argonne National Laboratory by Chen, Toukan, Loong, Price, and Teixeira.[47] These authors used a high-resolution medium-energy chopper spectrometer (HRMECS), taking advantage of the availability of epithermal neutrons from the pulse source and small-angle detector banks in the spectrometer. This combination is an absolute necessity for observing high-energy excitations in molecular liquids, because the Doppler broadening of the excitation peaks depends on Q^2. They used an incident energy E_0 of 800 meV and observed the inelastic scattering at an angle $\theta = 8°$. The background-corrected and normalized spectra, which are proportional to $S_s(Q, \omega)$, are displayed in Fig. 19. Starting from the bottom of the figure, the curves successively represent ice at 20 K, supercooled water at −15°C, water at 40°C, and water at 80°C. The notation $E = \hbar\omega$ is used. In Fig. 20, we display a derived function

$$G(Q, E) = \frac{E^2}{Q^2} S_s(Q, E) \qquad (3.6)$$

in a three-dimensional plot. $G(Q, E)$ is a useful function to plot for

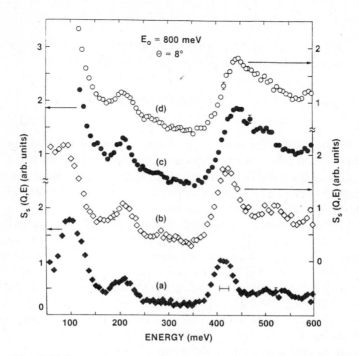

Figure 19. Self-dynamic structure factor at a scattering angle $\theta = 8°$ for ice at 20 K (a), supercooled water at $-15°C$ (b), and water at 40°C (c) and 80°C (d). Note the shift and broadening of the stretch band with temperature variation. Taken from reference 47.

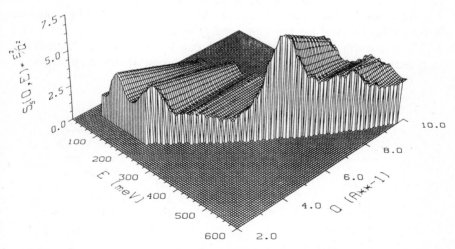

Figure 20. Proton-scattering density function for supercooled water at $-15°C$, plotted in (Q, E) space. The peak at 525 meV is probably due to a mode-coupling process. Taken from reference 47.

liquids, because the small-Q limit of $G(Q, E)$ is related simply to the velocity autocorrelation functions of the hydrogen atom,[48]

$$f_\text{p}(\omega) = \frac{1}{2\pi} \int_{-\infty}^{\infty} dt \, e^{-i\omega t} \langle V_Q^\text{p}(0) V_Q^\text{p}(t) \rangle \qquad (3.7)$$

where $V_Q^\text{p}(t)$ denotes the projection of the velocity of the hydrogen atom in $\hat{Q}$-direction at time t. An advantage of transforming the data to $G(Q, E)$ is that $f_\text{p}(\omega)$ is an easy function to calculate from CMD data. In practice the small-Q limit can be reached approximately by having $Qa < 1$, where a is the O–H distance in water, about 1 Å.[49]

A CMD simulation using an SPC model potential[11] was carried out by Toukan and Rahman[12] to compute $f_\text{p}(\omega)$ and the results were compared with the experiment. The spectra consist of three major bands: the low-frequency librational band E_L centered around 80 meV, which is due to intermolecular coupling; a bending librational band E_B of intramolecular nature around 200 meV; and a stretch vibrational band E_S centered around 425 meV. Table III summarizes the centroids and the widths of the bands at the appropriate Q values. Also listed are the gas-phase Raman frequencies at room temperature.[4]

We can summarize the main features of the experimental spectra as follows:

1. Significant temperature dependence is observed for the bands E_L and E_S, while E_B is nearly temperature independent. The band energy of E_L decreases as the structure changes from ice to water. The opposite behavior is observed for the band energy of E_S.

2. The width of the band E_S, which can be more reliably estimated, increases with temperature. The energy resolution at the stretch band is

TABLE III
Experimental Vibrational Energies in Water and Ice[a]

	T (°C)	E_L (meV)	E_B (meV)	E_S (meV)[b]
Ice	−253	82	207	407 (49)
H$_2$O	−15	74	207	418 (49)
H$_2$O	40	—	207	441 (66)
H$_2$O	80	—	207	443 (70)
Gas (Raman)	25	—	198	454, 465
Q (Å^{-1})		2.6	3.5	6.6

[a] Data are for 800 meV incident neutron energy, except those for E_L, which are given for 500 meV because it gives better resolution.
[b] Values in parentheses are bandwidths.

16 meV, which means that the symmetric and antisymmetric stretch vibrations in the free molecule cannot be resolved in the experiment.

Careful examination of the CMD results[12] both in the gas phase and in liquid phase reveals the significance of the experimental result in the condensed phase. The results of CMD can be summarized as follows:

1.   The intramolecular O–H bond length in the liquid is 2% greater than that in the gas phase. This elongation leads to a weakening of O–H bonds and hence to mode softening of the stretch vibration, as evidenced by the experimental data.

2.   The formation of hydrogen bonds in the liquid phase leads to further softening of the O–H stretch vibrational mode. As the temperature of the system is decreased, more intact hydrogen bonds are formed, leading to a decrease in the stretch-vibrational-band energy. The experimental data, as depicted in Fig. 20, clearly show such behavior. This trend is closely associated with an increase in the librational-band energy. Table IV summarizes the CMD data.

In conclusion, the incoherent-, inelastic-neutron-scattering experiment combined with an improved CMD simulation of water demonstrates the possibility of a systematic study of hydrogen-bond dynamics in water at elevated temperatures as well as in the supercooled state. Comparison of the water data at different temperatures and those of ice clearly indicates that as far as the hydrogen bonding of neighboring water molecules is concerned, supercooled water at −15°C is essentially similar to ice. This is expected, because the number of intact hydrogen bonds is greater than 90% at this temperature.[9] Thus the various thermodynamic and transport anomalies of supercooled water are intimately related to the rapid completion of the hydrogen-bond network at low temperatures. As can be seen from Table III, the main difference between the spectra of supercooled water and that of ice exists at the low-frequency librational band E_L. Since this librational motion of the hydrogen atom is transverse to the direction of hydrogen bonding, the downward shift in E_L in the supercooled state as compared with that in ice would indicate that on the average the hydrogen bond in the liquid state is more distorted from

TABLE IV
Vibrational Energies in Water Derived from CMD
Simulation using the SPC Model[12]

	T (K)	E_L (meV)	E_B (meV)	E_S (meV)
Liquid	325	65	228	429, 448
Gas	—	—	205	475, 490

linearity than it is in the solid state. This trend is more pronounced as the temperature increases in water, as is evidenced also by the broadening of the stretch band E_S. That the main temperature effect is in the librational band proves that a strong temperature dependence of liquid properties is related more to the breaking of bonds due to librational motion than to the number of bonds per se.

2. *Low Frequency Scattering*

The low-frequency inelastic-neutron-scattering spectra of H_2O were determined with great accuracy simultaneously with the quasi-elastic spectra at the IN6 spectrometer of the Institut Laue-Langevin at Grenoble.[50] Data were obtained between 20° and −20°C. The vibrational density of states, $f_p(\omega)$, can in principle be obtained in the usual way by an extrapolation:

$$f_p(\omega) = \lim_{Q \to 0} \frac{\omega^2}{Q^2} S_s(Q, \omega) \tag{3.8}$$

where $S_s(Q, \omega)$ represents the self-dynamic structure factor taken at wavevector transfer Q and energy transfer $\hbar\omega$. An example of the results obtained is given in Fig. 21 for a temperature $T = -17°C$. In this figure, two peaks are clearly seen. The first one, very sharp, is centered around 6 meV. The second peak represents the librational band and has an excitation energy around 70 meV. This band is less well defined because the Q value at this high-energy transfer is larger in an experiment using cold neutrons. The extrapolation $Q \to 0$ in Eq. (3.8) is consequently less reliable. Other, smaller peaks in the intermediate Q range at 22 and 33 meV can also be seen, but their positions are ill defined.

All these peaks correspond to Raman-scattering peaks rather well,[44] as far as their positions are concerned. However, the relative measured intensities are quite different in the two techniques. In Raman scattering the first peak, at 8 meV, is very small. It corresponds to the flexing motions of the O–O–O units and such movement is expected to affect the polarizability of molecules much less than it does the proton displacements. This explains why the peak is dominant in the neutron-scattering data. In contrast, the peak at 22 meV corresponds to hindered translations or stretching movements of the O–O units, and for the same reasons is much more intense in the Raman data. Finally, the peak at 33 meV is very small in both techniques and can be seen only at low temperatures. It was interpreted by Krishnamurthy et al.[44d] as a manifestation of the four-coordinated molecules present in low-temperature water.

The analysis of the temperature dependence of the positions of the main peaks is also in general agreement with the Raman-scattering

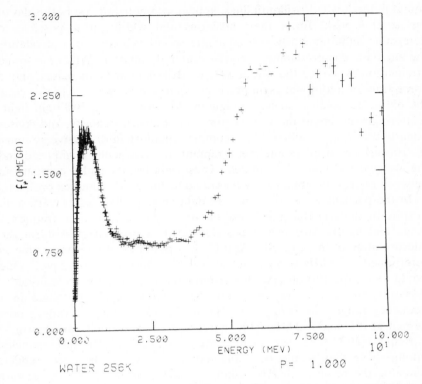

Figure 21.   Proton frequency distribution function $f_p(\omega)$ from H_2O at $-17°C$. Note the two main peaks: one corresponding to the O–O–O bending, centered at 6 meV, and the other to the librational movements, centered around 70 meV. Smaller peaks are seen in the 20–30 meV region.

data.[44] The peak at 8 meV shows almost no variation with temperature and the librational band moves to higher energies with decreasing temperature.

It is important to note that molecular-dynamics simulations using a modification of the SPC potential[12] quite accurately reproduce the neutron spectrum, showing a well-defined line at 9 meV and a bump on the librational band at 25 meV.

The intramolecular librational band is very broad and intense, but experimental limitations do not allow a precise interpretation of the data, as we have pointed out above.

C. Brillouin Light Scattering

Brillouin light scattering is a conventional technique for measuring collective excitations in liquids. Because of Doppler shift, the propagating sound modes, called Brillouin lines, appear symmetrically on both sides of the central peak. From their positions and widths, it is possible to determine simultaneously the adiabatic sound velocity, or equivalently the adiabatic compressibility, and the sound attenuation. When the sound attenuation is large, the width of the Brillouin lines increases and a correction for additional asymmetric peaks has to be made.[51] In this case, the extracted results can be ambiguous. Moreover, in a Brillouin light-scattering experiment the wavevector is an imposed quantity, and consequently the frequency becomes a complex quantity in dispersive systems. In contrast, in an ultrasonic measurement, the frequency is imposed and the wavevector may be complex. These constraints make comparison between the results obtained by these two techniques a complex problem.

In the particular case of water, all data above $-20°C$ agree very well, in spite of the fact that they were obtained over a very large frequency range and by the two techniques. However, a problem persists in the interpretation of these results. At $0°C$, the orientational relaxation time as determined by NMR is 5 ps, but at $-20°C$ it increases to 12 ps.[52] One would then expect to observe dispersion effects in the supercooled region. However, this is not the case and experimental results obtained in a frequency range extending to 5 GHz are identical. An experiment performed at $0°C$ and 7.2 GHz still showed no sound velocity dispersion.[53]

Below $-20°C$ three sets of data are available.[54-56] The discrepancies among these data would be explained partly by the dispersion effect. Actually, the result at 5 GHz[56] shows a minimum at $-24°C$, in contrast with the monotonic temperature variation of the data obtained at 925 MHz.[55] If this interpretation is correct, one finds a structural relaxation time consistent with the orientational relaxation time determined by NMR and dielectric relaxation. Unfortunately, the third set of data, obtained with a levitation technique[54] at very low frequency (54 kHz) falls between the other two sets of data (Fig. 22) and consequently is inconsistent with this interpretation.

The overall analysis of the temperature dependence of the sound velocity has been a source of considerable discussion with regard to the thermodynamic behavior of supercooled water. Because all the properties of water have a very strong temperature dependence, it was suggested by Speedy and Angell[57] that water has a spinodal behavior at $-45°C$. If this is true, then the limiting temperature of a mechanical instability would be related directly to the observed homogeneous nucleation at $-42°C$. In this picture, all the properties of water have a critical behavior as the temperature approaches $-45°C$ and could be fitted by critical laws. This is actually observed down to $-20°C$,[58] and even to $-27°C$[55] for one set of data. Because the measured sound velocity is related only to the adiabatic

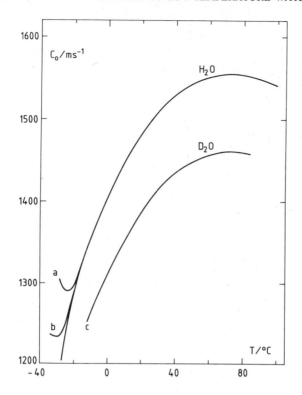

Figure 22.   Temperature dependence of the sound velocity in H_2O and D_2O. Data for D_2O are from O. Conde, J. Teixeira, and P. Papon, *J. Chem. Phys.* **76,** 3747 (1982). Data for H_2O are from references 56 (*a*), 54 (*b*), and 55 (*c*). Note the discrepancy among the data below −20°C.

compressibility, the observation of a minimum does not completely invalidate this type of interpretation.

D.   Coherent Inelastic Neutron Scattering

Coherent scattering in water can in practice be observed only in D_2O. In D_2O, the coherent-scattering cross section is about 80% of the total cross section. We therefore limit our discussion to D_2O.

Short-wavelength collective excitations in dense atomic liquids are in general difficult to observe by coherent inelastic neutron scattering. We may define the evidence of collective excitations as the existence of a peak or a shoulder in the dynamic structure factor $S(Q, \omega)$ on each side of the central line in the Q range of, say, $0\text{--}1.2\,\text{Å}^{-1}$. The neutron-scattering experiments so far indicate that for argon at the triple point, the excitation exists up to $Q\sigma \leq 1.0$[59] and for liquid rubidium and lead near their melting points, up to $Q\sigma \leq 4.0$,[60,61] where σ is the effective hard-core diameter of the atoms. From the CMD of hard spheres by Alley et al.,[62] it was established that the collective excitations exist up to $Q\sigma \leq 0.5$. It was generally believed that the damping of the short-wavelength excitations depends critically on both the steepness of the repulsive part and the depth of the attractive well in the pair potential. For liquid metals, the repulsive part is in general softer and the attractive part deeper. This may be the reason why the excitation is observable up to higher Q.

From the point of view of intermolecular interaction, water is a rather special case. Take, for instance, the ST2 model potential of water.[1] This potential consists of a superposition of a Lennard-Jones potential between molecular centers and directional electrostatic interactions that simulate the hydrogen bonding. A new feature of this potential compared with those of simpler liquids is the presence of a directional strong attraction between molecules. One therefore expects the existence of some kind of short-wavelength excitations persisting up to high Q values.

We performed very recently an experiment on D_2O at room temperature on the IN8 spectrometer of the Institut Laue-Langevin in Grenoble.[63] The result of a constant Q scan at $Q = 0.5\,\text{Å}^{-1}$ is shown in Fig. 23. The strong peak centered at $\omega = 0$ is the quasi-elastic scattering and its width is due entirely to the resolution of the instrument, which is $\Delta\omega = 1.2\,\text{THz}$. If one inspects the figure carefully, one can notice two bands on each side of the central peak. Since the counting statistics are good in this experiment, it is possible to extract the positions and widths of the side bands by fitting the entire spectrum with an analytical function. The resolution function of the instrument is very well-approximated by a Gaussian function. Therefore, one can postulate analytical forms such as a Lorentzian or harmonic-oscillator function for the shape of the side bands. The latter choice is more reasonable based on the generalized hydrodynamics,[64] and in practice it turns out to be the case. The details of the fitting are described in reference 63. The peak positions were found to follow a linear dispersion relation giving a velocity for the collective mode of $c = 3100\,\text{m/s}$. This value is about twice the velocity of ordinary sound in water. The damping of these excitations was also shown to be proportional to Q^2. In reference 63 this new collective excitation was called a "high frequency sound wave" and was conjectured to be an excitation propagation in the patches of hydrogen-bonded regions in water. Among the conclusions drawn from this experi-

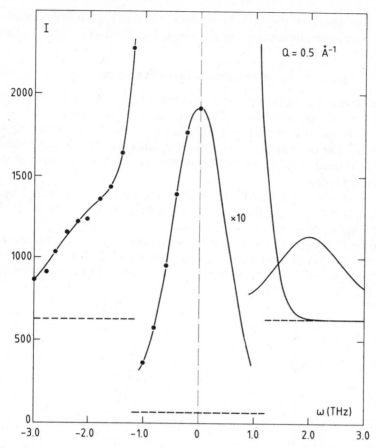

Figure 23.   Coherent, inelastic-neutron-scattering spectrum from D_2O obtained at room temperature. The ordinate is proportional to $S(Q, \omega)$ at $Q = 0.5$ Å^{-1}. Values of ω are expressed in THz (1 THz = 4.13 meV). Notice that the peak centered at zero energy transfer is due to the quasi-elastic scattering, and its width is due entirely to the resolution function of the instrument, which is 1.2 THz. The two side bands are weak but visible, and can be deconvolved from the data by using a generalized hydrodynamic expression. One thus obtained the side peak centered at $\omega_0 = 2.3$ THz with a width equal to 1.8 THz. Taken from reference 63.

ment were the following:

1.   The excitation seems to persist up to Q values of at least 0.6 Å^{-1}, or $Q\sigma \leq 1.8$.

2.   The damping is shown to be much smaller than the value extrapolated to this Q range from the ordinary sound data.

3. The speed of the excitation seems to coincide with that predicted by the computer simulations of Rahman and Stillinger[65] and Impey et al.[66]

E. Depolarized Light Scattering

In depolarized light-scattering experiments the important molecular quantity is the polarizability. The polarizability is a three-dimensional tensor and its antisymmetric part gives rise to the depolarized light scattering. The central line comes from rotation of the molecule, and its width is related directly to the characteristic relaxation times of these rotational processes.

In liquid water the anisotropy of the polarizability is modified by the chemical bonding[67] and the depolarized light scattering is then sensitive to the state of bonding of the molecule with its neighbors.

The first analysis of the temperature dependence of the intensity and the width of the quasi-elastic line was done by Montrose et al.[68] Extensions into the supercooled region and to heavy water have been done more recently.[69]

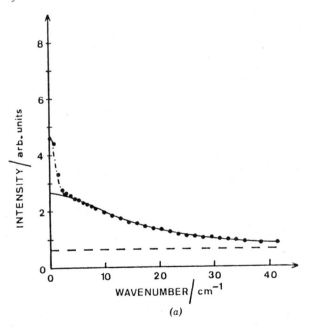

(a)

Figure 24.   Depolarized Rayleigh spectra of water at $-14.7°C$ (a) and $59°C$ (b). The dashed lines are the extrapolations to low frequency of the collision-induced scattering. The dotted and dashed line is the best fit assuming two lines, and the solid line represents the main component related to hydrogen-bond breaking. Taken from reference 69.

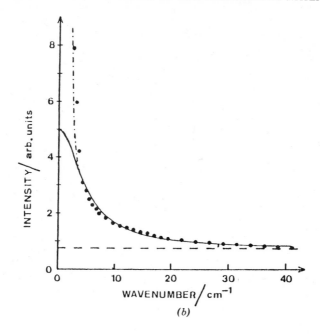

(b)

The analysis of the spectra shows unambiguously the presence of two lines (Fig. 24). The sharp line is less intense and more difficult to analyze, because it is too narrow for resolution by the conventional Raman spectrometer. However, its interpretation is rather clear. It gives a relaxation time identical to the orientational relaxation time as measured by NMR.[52] This result is expected, because the rotational processes contribute to depolarized light scattering through the anisotropy of the polarizability tensor, as said above.

More important is the broader line. Its width is typically 10 times that of the sharp line, and because of its high intensity it can be analyzed easily. It gives a relaxation time shorter than 1 ps (Figs. 18 and 25). The remarkable feature of this line is its Arrhenius temperature dependence. Except for a recent neutron-scattering result,[39] such a temperature dependence is unique in water. The Arrhenius fit extends over the entire normal liquid range and even to the supercooled region. The associated activation energy is ~2.5 kcal/mol and is interpreted to represent the breaking of hydrogen bonds. Actually, librational movements are probably the more efficient mechanism for the breaking of the bonds. Librations contribute to the depolarized spectrum because of their rotational character and large amplitude.[39]

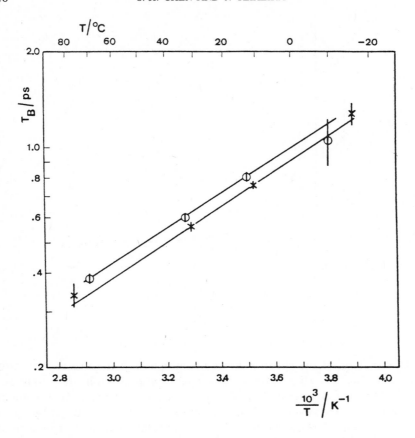

Figure 25.   Arrhenius plot of the hydrogen-bond lifetimes for H_2O (×) and D_2O (○).
Taken from reference 69.

Other arguments can be made in favor of this interpretation. The most important one concerns the intensity of the line. The intensity of the depolarized light scattering can be written as

$$I = \rho\beta^2 \tag{3.9}$$

where β is the anisotropy of polarizability and ρ is the density. For most liquids, this intensity is almost temperature independent, because the thermal dependences of both density and anisotropy are very small. However, in associated liquids, β is sensitive to the chemical bonding, because the deformation of the electron distribution due to the bonding modifies the anisotropy of the polarizability. This effect has been observed in some associated liquids.[67]

In water, the temperature dependence of the intensity I has a quadratic shape with a maximum at $\sim 50°C$ (Fig. 26). It can be fitted by the simple equation

$$I = \text{const } p(1-p) \qquad (3.10)$$

where p represents the number of intact bonds. It is assumed that p has a linear temperature dependence:

$$p = 1.950 - 0.0045T \qquad (3.11)$$

where T is the absolute temperature The values for p from Eq. (3.11) agree very well with results of both theoretical evaluations[36] and experimental determinations using X rays.[70]

To summarize, depolarized light-scattering experiments give a relaxation time that is directly connected with the lifetime of the hydrogen bond and has an Arrhenius temperature dependence. This suggests that the librational motions serve as a mechanism for the breaking of the hydrogen bonds.

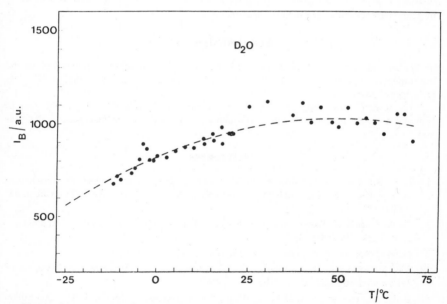

Figure 26. Temperature dependence of the intensity of the broad line of the depolarized Rayleigh spectrum. The dashed line is the best fit using Eqs. (3.10) and (3.11). Taken from reference 69.

IV. PROSPECTS

It is evident from the above discussions that scattering techniques, when performed over a wide range of temperatures including the super-cooled regime, are a potential source of abundant information on both the structure and dynamics of water. Because of the complexity of water, theoretical models for interpreting these results remain to be developed. Meanwhile, CMD can be used to interpolate between the experimental data and simple physical models for understanding them. One can see that there are incentives and potential for developing more accurate dynamic models of water based on the newly available experimental data.

As far as the diffraction experiments are concerned, in both X-ray and neutron diffractions, there are almost insurmountable obstacles to obtaining accurate partial pair-correlation functions. In our opinion, there will be no relief of these difficulties in the near future. Likewise, the various optical spectroscopic techniques have been fully exploited for water and we do not foresee any significant developments in this direction. On the other hand, as far as dynamics experiments are concerned, the new generation of neutron spectrometers are capable of producing even more striking and useful data, especially when used over a wider range of temperatures. We believe it may be fruitful to pursue experiments deeper in the supercooled region of water.

Acknowledgments

We are grateful to M.-C. Bellissent-Funel for her continuing collaboration in the experimental work and also to K. Toukan and A. Rahman for access to their CMD data prior to publication.

This research was supported by the National Science Foundation and the Centre National de la Recherche Scientifique through a grant for United States–French collaboration.

References

1. F. H. Stillinger, *Adv. Chem. Phys.* **31,** 1 (1975); E. W. Lang and H. D. Ludemann, *Angew. Chem. Int. Ed. Engl.* **21,** 315 (1982); C. A. Angell, *Ann. Rev. Phys. Chem.* **34,** 593 (1983).

2. C. A. Angell, in *Water, A Comprehensive Treatise*, Vol. 7, F. Franks, Ed., Plenum, New York, 1982, Ch. 1.

3. H. R. Pruppacher, *J. Chem. Phys.* **56,** 101 (1972).

4. See, for example, D. Eisenberg and W. Kauzmann, *The Structure and Properties of Water*, Oxford Univ. Press, New York, 1969.

5. A. H. Narten and H. A. Levy, in *Water, A Comprehensive Treatise*, Vol. 1, F. Franks, Ed., Plenum, New York, 1972, Ch. 8; D. I. Page, *ibid.*, Ch. 9.

6. A. Rahman and F. H. Stillinger, *J. Chem. Phys.* **55,** 3336 (1971).

7. See the review article of D. W. Wood, in *Water, A Comprehensive Treatise*, Vol. 6, F. Franks, Ed., Plenum, New York, 1979, Ch. 6.

8. G. E. Walrafen, in *Water, A Comprehensive Treatise*, Vol. 1, F. Franks, Ed., Plenum, New York, 1972, Ch. 5.

9. W. A. P. Luck, in *Water, A Comprehensive Treatise*, Vol. 2, F. Franks, Ed., Plenum, New York, 1973, Ch. 4.

10. S. A. Rice and M. G. Sceats, *J. Phys. Chem.* **85**, 1108 (1981).

11. H. J. C. Berendsen, J. P. M. Postma, W. F. van Gunsteren, and J. Hermans, in *Intermolecular Forces*, B. Pullman, Ed., Reidel, Hingham, Massachusetts, 1981.

12. K. Toukan and A. Rahman, "A Molecular Dynamics Study of Atomic Motions in Water," *Phys. Rev.* **B31**, 2643 (1985).

13. E. Kalman, G. Palinkas, and P. Kovacs, *Mol. Phys.* **34**, 505, 525 (1977).

14. A. H. Narten and H. A. Levy, *J. Chem. Phys.* **55**, 2263 (1971).

15. F. Hajdu, S. Lengyel, and G. Palinkas, *J. Phys.* **D12**, 901 (1979).

16. A. H. Narten, *J. Chem. Phys.* **56**, 5681 (1972).

17. P. A. Egelstaff, *Adv. Chem. Phys.* **53**, 1 (1983).

18. J. C. Dore, *J. Physique, Coll. C7*, **45**, 49 (1984).

19. P. A. Egelstaff and J. H. Root, *Chem. Phys.* **76**, 405 (1983).

20. L. Bosio, S.-H. Chen, and J. Teixeira, *Phys. Rev. A* **27**, 1468 (1983).

21. R. W. Hendricks, P. G. Mardon, and L. B. Shaffer, *J. Chem. Phys.* **61**, 319 (1974).

22. L. Bosio, J. Teixeira, and H. E. Stanley, *Phys. Rev. Lett.* **46**, 597 (1981).

23. H. Kanno and C. A. Angell, *J. Chem. Phys.* **70**, 4008 (1979).

24. R. J. Speedy and C. A. Angell, *J. Chem. Phys.* **65**, 851 (1976).

25. G. Palinkas, E. Kalman, and P. Kovacs, *Mol. Phys.* **34**, 525 (1977).

26. J. C. Powles, *Mol. Phys.* **37**, 623 (1979); M. Rovere, L. Blum, and A. H. Narten, *J. Chem. Phys.* **73**, 3729 (1980).

27. L. Bosio, J. Teixeira, J. C. Dore, D. Steytler, and P. Chieux, *Mol. Phys.* **50**, 733 (1983).

28. M. R. Chowdhury, J. C. Dore, and J. T. Wenzel, *J. Non-Cryst. Sol.* **53**, 247 (1982).

29. G. Walford and J. C. Dore, *Mol. Phys.* **34**, 21 (1977).

30. I. P. Gibson and J. C. Dore, *Mol. Phys.* **48**, 1019 (1983).

31. P. A. Egelstaff, J. A. Polo, J. H. Root, L. J. Hahn, and S.-H. Chen, *Phys. Rev. Lett.* **47**, 1733 (1981); J. A. Polo and P. A. Egelstaff, *Phys. Rev.* **A27**, 1508 (1983).

32. A. H. Narten, W. E. Thiessen, and L. Blum, *Science* **217**, 1033 (1982); W. E. Thiessen and A. H. Narten, *J. Chem. Phys.* **77**, 2656 (1982).

33. J. Reed and J. C. Dore, *Mol. Phys.*, in press.

34. A. K. Soper and R. N. Silver, *Phys. Rev. Lett.* **49**, 471 (1982).

35. A. K. Soper, *Chem. Phys.* **88**, 187 (1984).

36. H. E. Stanley and J. Teixeira, *J. Chem. Phys.* **73**, 3404 (1980).

37. P. von Blanckenhagen, *Ber. Bunsen-Ges.* **76**, 891 (1982).

38. S.-H. Chen, J. Teixeira, and R. Nicklow, *Phys. Rev.* **A26**, 3477 (1982).

39. J. Teixeira, M.-C. Bellissent-Funel, S.-H. Chen, and A. J. Dianoux, *Phys. Rev.* **A31**, 1913 (1985).

40. See a review by S. Yip, "Quasi-elastic Scattering in Neutron and Laser Spectroscopy," in *Spectroscopy in Biology and Chemistry—Neutron, X-Ray, Laser,* S.-H. Chen and S. Yip, Eds., Academic, New York, 1974, Ch.2.

41. V. F. Sears, *Can. J. Phys.* **44,** 1299 (1966); **45,** 237 (1966).

42. C. T. Chudley and R. J. Elliot, *Proc. Phys. Soc. (London)* **77,** 353 (1961).

43. K. T. Gillen, D. C. Douglass and M. J. K. Hoch, *J. Chem. Phys.* **57,** 5117 (1972).

44. (a) G. d'Arrigo, G. Maisano, F. Mallamace, P. Migliardo, and F. Wanderlingh, *J. Chem. Phys.* **75,** 4264 (1981); (b) R. Bansil, J. Wiafe-Akenten, and J. L. Taafe, *J. Chem. Phys.* **76,** 2221 (1982); (c) Y. Yek, J. H. Bilgram, and W. Kanzig, *J. Chem. Phys.* **77,** 2317 (1982); (d) S. Krishnamurthy, R. Bansil, and J. Wiafe-Akenten, *J. Chem. Phys.* **79,** 5863 (1983).

45. A. C. Belch and S. A. Rice, *J. Chem. Phys.* **78,** 4817 (1983).

46. F. H. Stillinger, *Science* **209,** 451 (1980).

47. S.-H. Chen, K. Toukan, C. K. Loong, D. L. Price, and J. Teixeira, *Phys. Rev. Lett.* **53,** 1360 (1984).

48. S.-H. Chen and S. Yip, *Physics Today* **29,** 32 (1976).

49. F. H. Stillinger and A. Rahman, in *Molecular Motions in Liquids,* J. Lascomb, Ed., Reidel, 1974, p. 479.

50. J. Teixeira, M.-C. Bellissent-Funel, S.-H. Chen, and A. J. Dianoux, *Phys. Rev.* **A31,** 1913 (1985).

51. B. J. Berne and R. Pecora, *Dynamic Light Scattering,* Wiley, New York, 1976.

52. E. Land and H.-D. Ludemann, *J. Chem. Phys.* **67,** 718 (1977).

53. O. Conde, J. Leblond, and J. Teixeira, *J. Physique* **41,** 997 (1980).

54. E. Trinh and R. E. Apfel, *J. Chem. Phys.* **72,** 6731 (1980).

55. J. C. Bacri and R. Rajaonarison, *J. Physique Lett.* **40,** L404 (1979).

56. G. Maisano, P. Migliardo, F. Aliotta, C. Vasi, F. Wanderlingh, and G. d'Arrigo, *Phys. Rev. Lett.* **52,** 1025 (1984).

57. R. J. Speedy and C. A. Angell, *J. Chem. Phys.* **65,** 851 (1976).

58. J. Rouch, C. C. Lai, and S.-H. Chen, *J. Chem. Phys.* **66,** 5301 (1977); J. Teixeira and J. Leblond, *J. Phys. Lett.* **39,** L83 (1978).

59. K. Skold, J. M. Rowe, G. Ostrowski, and P. D. Randolph, *Phys. Rev.* **A6,** 1107 (1972).

60. J. R. D. Copley and J. M. Rowe, *Phys. Rev. Lett.* **32,** 49 (1974); *Phys. Rev.* **A9,** 1656 (1974).

61. O. Soderstrom, J. R. D. Copley, J.-B. Suck, and B. Dorner, *J. Phys. F: Metal Phys.* **10,** L151 (1980).

62. W. E. Alley, B. J. Alder, and S. Yip, *Phys. Rev.* **A27,** 3158 (1983).

63. J. Teixeira, M.-C. Bellissent-Funel, S.-H. Chen, and B. Dorner, *Phys. Rev. Lett.* **54,** 2681 (1985).

64. L. P. Kadanoff and P. C. Martin, *Ann. of Phys.* **24,** 419 (1963).

65. A. Rahman and F. H. Stillinger, *Phys. Rev.* **A10,** 368 (1974).

66. R. W. Impey, P. A. Madden, and I. R. McDonald, *Mol. Phys.* **46,** 513 (1983).

67. G. Fytas and T. Dorfmuller, *J. Chem. Phys.* **75,** 5232 (1981).

68. G. J. Montrose, J. A. Bucaro, J. J. Marshall-Caokley, and T. A. Litovitz, *J. Chem. Phys.* **60**, 5025 (1974).

69. W. Danninger and G. Zundel, *J. Chem. Phys.* **74**, 2769 (1981); O. Conde and J. Teixeira, *J. Physique* **44**, 525 (1983); O. Conde and J. Teixeira, *Mol. Phys.*, **53**, 951 (1984).

70. P. A. Egelstaff and J. H. Root, *Chem. Phys. Lett.* **91**, 96 (1982).

NONEQUILIBRIUM THERMODYNAMICS AND STATISTICAL PHYSICS OF SURFACES

D. BEDEAUX

Department of Physical and Macromolecular Chemistry
Gorlaeaus Laboratories, University of Leiden
Leiden, The Netherlands

CONTENTS

I. INTRODUCTION

A. Historical Remarks

A consistent phenomenological theory of irreversible processes containing both the Onsager symmetry relations and an explicit expression for the entropy production was formulated by Meixner[1] in 1941 and somewhat later by Prigogine.[2] This was the beginning of the field of nonequilibrium thermodynamics, which developed subsequently in many different directions. In 1962 a book on this subject by de Groot and Mazur[3] was published, which discussed the developments up to that time; it is still the standard text in this field. The method has also been used to describe transport processes through membranes.[4]

This chapter will discuss the use of this general method in the phenomenological description of irreversible processes that take place at a surface of discontinuity between bulk phases. The equilibrium properties of surfaces of discontinuity were discussed extensively by Gibbs[5] in the context of his work on the equilibrium thermodynamics of heterogeneous substances. A general method for the application of nonequilibrium thermodynamics to surfaces of discontinuity consistent with the equilibrium theory for surface thermodynamics formulated by Gibbs was given a hundred years later by Bedeaux, Albano, and Mazur.[6]

One of the difficulties in such an analysis is that the surface of discontinuity not only may move through space but also has a time-dependent curvature. This makes it necessary to use time-dependent orthogonal curvilinear coordinates for the more difficult aspects of the analysis. A shock-wave front, which has no equilibrium analog, may also be described as a moving surface of discontinuity.

The theory gives an explicit expression for the excess production of entropy at the surface of discontinuity. This expression makes it possible to identify the appropriate forces and fluxes. Onsager symmetry relations for the linear constitutive coefficients relating these forces and fluxes are given by the theory. The expression for the fluxes through the surface of discontinuity from one bulk phase to the other lead to the usual boundary conditions containing as constitutive coefficients, for example, the slip coefficient and the temperature-jump coefficient. Other fluxes characterize the flow along the interface and the flow from the bulk regions into the interfacial region and vice versa. The excess entropy production contains the fluxes through and into the surface of discontinuity in the reference frame in which this surface is at rest. This fact gave rise to considerable discussion among workers in the field of membrane transport.[7]

The tensorial nature of the fluxes and forces contributing to the excess entropy production at the surface of discontinuity differs from the corresponding behavior in the bulk regions. The reason for this is that the surface of discontinuity breaks the symmetry; there is still symmetry for translation and rotation along the surface, but not in the direction normal to the surface. As a consequence the force–flux pairs contributing to the excess entropy production at the surface of discontinuity contain 2×2 tensors, two-dimensional vectors, and scalars, rather than 3×3 tensors, three-dimensional vectors, and scalars as in the bulk regions. The number of force–flux pairs is found to be larger than the number of force–flux pairs in the bulk regions, and as a consequence, the number of independent constitutive coefficients needed for a complete description of dynamic processes around the surface of discontinuity is much larger than the number of constitutive coefficients needed in the bulk regions. This large number is needed even though, as a consequence of Curie's symmetry principle, fluxes depend only on forces of the same tensorial nature. The resulting large number of different dynamical phenomena in the neighborhood of a surface of discontinuity is what makes this such an interesting region.

Subsequent work[8–10] extended the original analysis, which was given for a surface of discontinuity between two immiscible one-component fluid phases, to multicomponent systems with mass transport through the surface of discontinuity[8] and including electromagnetic effects.[9] The theory was further extended by Zielinska and Bedeaux[11] to describe spontaneous fluctuations around equilibrium, in which context fluctuation–dissipation theorems were given.

In Sections, II, III, and IV of this chapter we shall discuss, the conservation laws, the entropy balance, and the phenomenological equations for a surface of discontinuity between multicomponent phases in which chemical reactions and mass transport through and into the surface are possible.

In Section V we analyze the equilibrium (equal-time) correlations at a liquid–vapor interface. Expanding the excess entropy of the system to second order in the fluctuations, we find, in addition to the contributions due to the displacement of the interface and usually given in the capillary-wave theory,[12] contributions due to for example, interfacial temperature and velocity fluctuations. These expressions are given in the general case of a nonplanar equilibrium shape of the interface. Some recent work by Bedeaux and Weeks[13] on the behavior of the density–density correlation function and the direct correlation function in the neighborhood of the interface is discussed. For the density–density correlation function, their expressions are generalized to the case of finite bulk compressibilities.

In Section VI the description is extended by the inclusion of random fluxes. Fluctuation–dissipation theorems for these random fluxes are given.[11]

The general method of nonequilibrium thermodynamics is, as we shall discuss in more detail, inherently limited to the description of time-dependent phenomena over distances large compared with the bulk correlation length. We shall therefore not discuss the behavior near and in the surface of discontinuity on a molecular level. The reader is instead referred to the extensive literature on this subject.[14]

B. On the Mathematical Description of Interfaces

We consider here dynamical processes of a system in which two phases coexist. The phases are separated by a moving surface of discontinuity, or interface as we shall often call it, with a time-dependent curvature. The term "surface of discontinuity" does not imply that the discontinuity is sharp, nor that it distinguishes any surface with mathematical precision.[5] It is taken to denote the nonhomogeneous film that separates the two bulk phases. The width of this film is on the order of the bulk correlation length.

In the mathematical description of the dynamical properties of the system, we want to choose a method such that details of the description on length scales smaller than the bulk correlation length do not play a role. In the bulk phases, this implies that one replaces, for example, the molecular density by a continuous field that is obtained after averaging over cells with a diameter of the order of the bulk correlation length. Such a procedure gives an adequate description of the behavior of a bulk phase on a distance scale large compared with the bulk correlation length if the variation of the fields over a bulk correlation length is small. The surface of discontinuity is, in this context, a two-dimensional layer of cells in which the variables change rapidly in one direction over a distance of the order of a bulk correlation length from the value in one phase to the value in the other phase, but change slowly in the other two directions. One now chooses a time-dependent dividing surface in this two-dimensional layer of cells such that the radii of curvature are large compared with the bulk correlation length. Surfaces of discontinuity for which such a choice is impossible are clearly outside the scope of a method meant to describe behavior on length scales large compared with the bulk correlation length. There is clearly a certain amount of freedom in the choice of the dividing surface.[5] It may be shifted over a distance on the order of a bulk correlation length. Because the definition of excess densities and fluxes depends on this choice, it is of some importance, and we shall return to it below.

To describe the time-dependent location of the dividing surface, it is convenient to use a set of time-dependent orthogonal curvilinear coordinates:[15] $\xi_i(\mathbf{r}, t)$, $i = 1, 2, 3$, where $\mathbf{r} = (x, y, z)$ are the Cartesian coordinates and t the time. These curvilinear coordinates are choosen in such a way that the location of the dividing surface at time t is given by

$$\xi_1(\mathbf{r}, t) = 0 \qquad (1.2.1)$$

The dynamical properties of the system are described using balance equations. Consider as an example the balance equation for a variable $d(\mathbf{r}, t)$:

$$\frac{\partial}{\partial t} d(\mathbf{r}, t) + \operatorname{div} \mathbf{J}_d(\mathbf{r}, t) = \sigma_d(\mathbf{r}, t) \qquad (1.2.2)$$

where $\mathbf{J}_d$ is the current of d and σ_d the production of d in the system. In our description d, $\mathbf{J}_d$ and σ_d vary continuously in the bulk regions while the total excess (to be defined precisely below) of d, $\mathbf{J}_d$, and σ_d near the surface of discontinuity is located as a singularity at the dividing surface. We thus write d, $\mathbf{J}_d$, and σ_d in the following form:

$$d(\mathbf{r}, t) = d^-(\mathbf{r}, t)\Theta^-(\mathbf{r}, t) + d^s(\mathbf{r}, t)\delta^s(\mathbf{r}, t) + d^+(\mathbf{r}, t)\Theta^+(\mathbf{r}, t) \qquad (1.2.3)$$

$$\mathbf{J}_d(\mathbf{r}, t) = \mathbf{J}_d^-(\mathbf{r}, t)\Theta^-(\mathbf{r}, t) + \mathbf{J}_d^s(\mathbf{r}, t)\delta^s(\mathbf{r}, t)$$
$$+ \mathbf{J}_d^-(\mathbf{r}, t)\Theta^+(\mathbf{r}, t) \qquad (1.2.4)$$

$$\sigma_d(\mathbf{r}, t) = \sigma_d^-(\mathbf{r}, t)\Theta^-(\mathbf{r}, t) + \sigma_d^s(\mathbf{r}, t)\delta^s(\mathbf{r}, t)$$
$$+ \sigma_d^+(\mathbf{r}, t)\Theta^+(\mathbf{r}, t) \qquad (1.2.5)$$

Here Θ^- and Θ^+ are the time-dependent characteristic functions of the two bulk phases, which are 1 in one phase and zero in the other. Using the time-dependent curvilinear coordinates, one may write these characteristic functions as

$$\Theta^\pm(\mathbf{r}, t) \equiv \Theta(\pm\xi_1(\mathbf{r}, t)) \qquad (1.2.6)$$

where Θ is the Heaviside function. Furthermore, δ^s is, so to speak, the time-dependent "characteristic function" for the surface of discontinuity, which is defined in terms of the curvilinear coordinates as

$$\delta^s(\mathbf{r}, t) \equiv |\text{grad } \xi_1(\mathbf{r}, t)| \, \delta(\xi_1(\mathbf{r}, t)) \qquad (1.2.7)$$

It is clear from this definition that the excess densities in Eqs. (1.2.3)–(1.2.5) can be written as functions of ξ_2 and ξ_3 only; thus one has

$$d^s(\mathbf{r}, t) = d^s(\xi_2(\mathbf{r}, t), \xi_3(\mathbf{r}, t), t) \qquad (1.2.8)$$

and similarly for $\mathbf{J}_d^s$ and σ_d^s. An important consequence of this is that the derivatives of d^s, $\mathbf{J}^s$, and σ^s normal to the dividing surface are zero. For a detailed exposition on the principles of field theories in systems with a surface of discontinuity, we refer the reader to Truesdell and Toupin.[16] Explicit expressions for d^s, $\mathbf{J}_d^s$, and σ_d^s in terms of the corresponding fields describing the system on a more detailed level, that is, a level where variations of these quantities over distances smaller than the bulk correlation length are taken into account, can be derived and will be given below; we shall then also discuss why no contributions to d, $\mathbf{J}_d$, and σ_d are needed proportional to the normal derivative of δ^s.

First we give some definitions and identities:

$$\text{grad } \Theta^{\pm}(\mathbf{r}, t) = \pm \mathbf{n} \delta^s(\mathbf{r}, t) \qquad (1.2.9)$$

where $\mathbf{n}$ is the normal on the dividing surface defined by

$$\mathbf{n}(\xi_2, \xi_3, t) = \mathbf{a}_1(\xi_1 = 0, \xi_2, \xi_3, t) \qquad (1.2.10)$$

and where $\mathbf{a}_i$ is the unit vector in the direction of increasing ξ_i given by

$$\mathbf{a}_i \equiv h_i \text{ grad } \xi_i \qquad \text{with} \qquad h_i \equiv |\text{grad } \xi_i|^{-1} \qquad (1.2.11)$$

These unit vectors defined in each point in space are orthonormal:

$$\mathbf{a}_i \cdot \mathbf{a}_j = \delta_{ij} \qquad (1.2.12)$$

The velocity field describing the time development of the curvilinear coordinate is defined by

$$\mathbf{w}(\xi_1, \xi_2, \xi_3, t) \equiv \frac{\partial}{\partial t} \mathbf{r}(\xi_1, \xi_2, \xi_3, t) \qquad (1.2.13)$$

The velocity of the dividing surface is given in terms of this velocity field by

$$\mathbf{w}^s(\xi_2, \xi_3, t) \equiv \mathbf{w}(\xi_1 = 0, \xi_2, \xi_3, t) \qquad (1.2.14)$$

Using this velocity field, one may show[6] that the time derivative of the characteristic functions for the bulk phases is given by

$$\frac{\partial}{\partial t}\Theta^{\pm}(\mathbf{r}, t) = \mp w_n^s \delta^s(\mathbf{r}, t) \qquad (1.2.15)$$

where the subscript n indicates the normal component. Similarly, one may show[6] that the time derivative of the characteristic function for the surface of discontinuity is given by

$$\frac{\partial}{\partial t}\delta^s(\mathbf{r}, t) = -w_n^s \mathbf{n} \cdot \nabla \delta^s(\mathbf{r}, t) \qquad (1.2.16)$$

where $\nabla \equiv (\partial/\partial x, \partial/\partial y, \partial/\partial z)$ is the Cartesian gradient. One may also show that the gradient of δ^s is normal to the dividing surface:[6]

$$\nabla \delta^s(\mathbf{r}, t) = \mathbf{n}\mathbf{n} \cdot \nabla \delta^s(\mathbf{r}, t) \qquad (1.2.17)$$

These formulas make it possible to analyze the balance equation for d in more detail. In particular, we are interested in the precise form of the balance equation for the excess density d^s. Substitution of the expressions (1.2.3)–(1.2.5) for d, $\mathbf{J}_d$, and σ_d into the general balance equation (1.2.2) and use of the definitions and identities (1.2.6)–(1.2.17) leads to the following more detailed formula for the balance of d:

$$\left[\frac{\partial}{\partial t}d^-(\mathbf{r}, t) + \operatorname{div}\mathbf{J}_d^-(\mathbf{r}, t) - \sigma_d^-(\mathbf{r}, t)\right]\Theta^-(\mathbf{r}, t)$$
$$+ \left[\frac{\partial}{\partial t}d^+(\mathbf{r}, t) + \operatorname{div}\mathbf{J}_d^+(\mathbf{r}, t) - \sigma_d^+(\mathbf{r}, t)\right]\Theta^+(\mathbf{r}, t)$$
$$+ \left[\frac{\partial}{\partial t}d^s(\mathbf{r}, t) + \operatorname{div}\mathbf{J}_d^s(\mathbf{r}, t) - \sigma_d^s(\mathbf{r}, t)\right.$$
$$+ J_{d,n}^+(\mathbf{r}, t) - J_{d,n}^-(\mathbf{r}, t) - w_n^s(\mathbf{r}, t)(d^+(\mathbf{r}, t) - d^-(\mathbf{r}, t))\Big]\delta^s(\mathbf{r}, t)$$
$$+ [J_{d,n}^s(\mathbf{r}, t) - w_n^s(\mathbf{r}, t)d^s(\mathbf{r}, t)]\mathbf{n}(\mathbf{r}, t) \cdot \nabla \delta^s(\mathbf{r}, t) = 0 \qquad (1.2.18)$$

The first two terms in this formula describe the balance in the bulk phases:

$$\frac{\partial}{\partial t} d^{\pm} + \text{div } \mathbf{J}_d^{\pm} = \sigma_d^{\pm} \qquad \text{for} \quad \pm\xi_1(\mathbf{r}, t) > 0 \qquad (1.2.19)$$

The third term in formula (1.2.18) describes the balance of the excess density:

$$\frac{\partial}{\partial t} d^s + \text{div } \mathbf{J}_d^s + J_{d,n,-} - w_n^s d_- = \sigma_d^s \qquad \text{for} \quad \xi_1(\mathbf{r}, t) = 0 \qquad (1.2.20)$$

where the subscript $-$ indicates the difference of the corresponding quantity in the bulk phases from one side of the surface of discontinuity to the other; thus

$$d_-(\mathbf{r}, t) \equiv d^+(\xi_1 = 0, \xi_2(\mathbf{r}, t), \xi_3(\mathbf{r}, t), t) - d^-(\xi_1 = 0, \xi_2(\mathbf{r}, t), \xi_3(\mathbf{r}, t), t)$$
$$(1.2.21)$$

and similarly for $J_{d,n,-}$. We do not follow the more conventional notation, which uses square brackets to indicate this difference.[16,17] The balance equation (1.2.20) for the excess density shows that in addition to the usual contribution occurring also in the balance equation (1.2.19) for the bulk phases, one has a contribution $J_{d,n,-}$, due to flow from the bulk regions into or away from the surface of discontinuity and a contribution $-w_n^s d_-$ due to the fact that the moving surface of discontinuity "scoops up" material on one side and leaves material behind on the other side. The last term in formula (1.2.18) gives

$$J_{d,n}^s - w_n^s d^s = 0 \qquad (1.2.22)$$

This condition expresses the fact that the excess current in a reference frame moving with the surface of discontinuity flows along the dividing surface. Although the validity of this condition is intuitively clear, the above derivation shows that it is also a necessary condition in the context of the above description.

We will now briefly discuss how one may obtain the excess densities and currents from a more detailed description. Crucial to this procedure is the fact that, as we have already elaborated, we are interested only in the temporal behavior of spatial variations over distances long compared with the bulk correlation length. Spatial variations over distances smaller than or comparable to the bulk correlation length are assumed to be in *local* equilibrium. The description given in the context of non-equilibrium thermodynamics is thus, loosely speaking, obtained by averaging the microscopic variables over a "local equilibrium ensemble." After such an averaging procedure, one obtains a description in terms of continuous variables that again satisfy a balance equation and that vary continuously through the surface of discontinuity from the slowly varying value in one phase to the slowly varying value in the other phase. This variation is over the so-called intrinsic width of the surface of discontinuity. The intrinsic width is of the order of the bulk correlation length. As such, this intrinsic width is small compared with the wavelengths of the variations we want to describe. Extrapolating the slowly varying bulk fields to the dividing surface, one may define the following excess fields:

$$d_{ex}(\mathbf{r}, t) \equiv d(\mathbf{r}, t) - d^+(\mathbf{r}, t)\Theta^+(\mathbf{r}, t) - d^-(\mathbf{r}, t)\Theta^-(\mathbf{r}, t)$$

$$\mathbf{J}_{d,ex}(\mathbf{r}, t) \equiv \mathbf{J}_d(\mathbf{r}, t) - \mathbf{J}_d^+(\mathbf{r}, t)\Theta^+(\mathbf{r}, t) - \mathbf{J}_d^-(\mathbf{r}, t)\Theta^-(\mathbf{r}, t) \qquad (1.2.23)$$

$$\sigma_{d,ex}(\mathbf{r}, t) \equiv \sigma_d(\mathbf{r}, t) - \sigma_d^+(\mathbf{r}, t)\Theta^+(\mathbf{r}, t) - \sigma_d^-(\mathbf{r}, t)\Theta^-(\mathbf{r}, t)$$

These excess fields are only unequal to zero in the surface of discontinuity. Albano, Bedeaux, and Vlieger[18] show that if the surface densities and currents are defined as

$$d^s(\xi_2, \xi_3, t) \equiv (h_{2,s}h_{3,s})^{-1} \int_{-\infty}^{\infty} d\xi_1 h_1 h_2 h_3 d_{ex} \qquad (1.2.24)$$

$$J_{d,2}^s - w_2^s d^s \equiv h_{3,s}^{-1} \int_{-\infty}^{\infty} d\xi_1 h_1 h_3 (J_{d,ex,2} - w_2 d_{ex})$$

$$J_{d,3}^s - w_3^s d^s \equiv h_{2,s}^{-1} \int_{-\infty}^{\infty} d\xi_1 h_1 h_2 (J_{d,ex,3} - w_3 d_{ex}) \qquad (1.2.25)$$

$$\sigma_d^s(\xi_2, \xi_3, t) \equiv (h_{2,s}h_{3,s})^{-1} \int_{-\infty}^{\infty} d\xi_1 h_1 h_2 h_3 \sigma_{d,ex} \qquad (1.2.26)$$

then the validity of the balance equation (1.2.20) for d^s follows rigorously from the balance equation for the continuous fields. The subscript s indicates the value of the corresponding quantity at the dividing surface, $\xi_1 = 0$. Using

$$J_{d,1}^s - w_1^s d^s \equiv 0 \qquad (1.2.27)$$

as the definition of the normal current then completes the definition of the fields given in Eqs. (1.2.3)–(1.2.5), which are singular at the dividing surface and the balance equation for which follows rigorously from the balance equation for the continuous fields. If one studies the electromagnetic properties of boundary layers, one may similarly define singular fields, currents, and charge densities.[19] The excess current is usually called *equivalent surface current* and has been a very useful concept for studying the average effect of surface structure on length scales small compared with the wavelength of light.[20]

It follows rigorously from the analysis in references 18 and 19 that no contributions to the fields proportional to normal derivatives of δ^s are needed. The origin of this fact on the one hand is the assumption that averaging the microscopic balance equations over regions with the typical size of the bulk correlation length (the local equilibrium ensemble) again leads to balance equations, but now for continuously varying coarse-grained variables, and, on the other hand, is due to the proper choice of the variables. This last aspect is most apparent for a charge double layer, which may be described not only using a normal derivative of δ^s in the excess charge density, but also using δ^s in the excess polarization density. It is this last choice that is clearly used in reference 19.

An important quantity for a surface of discontinuity is its curvature, which is defined by

$$C(\xi_2, \xi_3, t) \equiv -\left[h_1^{-1} \frac{\partial}{\partial \xi_1} \ln(h_2 h_3) \right]_s = \frac{1}{R_1} + \frac{1}{R_2} \qquad (1.2.28)$$

for the $\xi_1 = 0$ surface.[15] R_1 and R_2 are the so-called radii of curvature. One may derive[6] the following useful identity:

$$C = -(\nabla \cdot \mathbf{n})_s \qquad (1.2.29)$$

In the description we are using the radii of curvature have been assumed to be large compared with the bulk correlation length. As a consequence, the curvature C will be small compared with one divided by the bulk correlation length. For a cell in the surface of discontinuity with a diameter of the order of a bulk correlation length, the surface of discontinuity thus appears to be practically flat. On the basis of this observation it is reasonable to assume that the (equilibrium) relations between the local thermodynamic variables are independent of C.[5]

The velocity field $\mathbf{w}$ gives the motion of the curvilinear coordinate system. In particular, $\mathbf{w}^s$ describes the motion of the dividing surface and thus has a clear-cut physical significance. One may show[19] in particular that the normal on the dividing surface satisfies the following equation of motion:

$$\frac{\partial}{\partial t}\mathbf{n}(\mathbf{r}, t) = -(\mathbf{I} - \mathbf{nn}) \cdot (\nabla w_n^s)_s \qquad (1.2.30)$$

Clearly this equation is needed in addition to the balance equation to give a complete description of time development of the system. The unit tensor is written as $\mathbf{I}$ in Eq. (1.2.30).

II. CONSERVATION LAWS

A. Introduction

In this section we shall discuss the conservation laws for a multicomponent system in which chemical reactions may take place in the bulk regions as well as on the surface of discontinuity. All densities and currents are given by expressions with singular contributions on the dividing surface similar to those given in Eqs. (1.2.3) and (1.2.4). Our emphasis will be on the equations describing the balance of the excess densities. In the bulk regions the description is identical to the one given in a one-phase system, and for this we refer the reader to de Groot and Mazur.[3]

B. Conservation of Mass

Consider a system consisting of n components among which r chemical reactions are possible. The balance equation for the mass density ρ_k of component k is written in the form

$$\frac{\partial}{\partial t}\rho_k + \operatorname{div}\rho_k \mathbf{v}_k = \sum_{j=1}^{r} v_{kj}J_j \qquad (2.2.1)$$

Note that the mass density of component k in the bulk regions $\rho_k^{\pm}$ is given per unit of volume while the excess mass density ρ_k^s of component k is given per unit of surface area. Similarly $\mathbf{v}_k^{\pm}$ is the velocity of component k in the bulk regions and is equal to the bulk current of component k per unit of mass of this component, while $\mathbf{v}_k^s$ is the velocity of the excess of component k at the surface of discontinuity and is equal to the excess current of component k per unit of excess mass of this component. Finally, $v_{kj} J_j^{\pm}$ is the production of component k in the jth chemical reaction per unit of volume in the bulk regions, while $v_{kj} J_j^s$ is the excess production of component k in the jth chemical reaction per unit of surface area. The quantity v_{kj} divided by the molecular mass of component k is proportional to the stoichiometric coefficient with which k appears in the chemical reaction j.

The balance equation for ρ_k is of the general form (1.2.2) discussed in the previous section. Using Eq. (1.2.22) for the normal component of the excess current, we find

$$v^s_{k,n} = w^s_n \tag{2.2.2}$$

The velocity of all the excess densities of the various components normal to the dividing surface is thus identical to the normal velocity of the surface, as is to be expected.

Since mass is conserved in each chemical reaction we have

$$\sum_{k=1}^{n} v_{kj} = 0 \tag{2.2.3}$$

As a consequence, the total mass

$$\rho \equiv \sum_{k=1}^{n} \rho_k \Leftrightarrow \begin{cases} \rho^\pm \equiv \sum_{k=1}^{n} \rho_k^\pm & \text{for the bulk regions} \\ \rho^s \equiv \sum_{k=1}^{n} \rho_k^s & \text{for the interface} \end{cases} \tag{2.2.4}$$

is a conserved quantity, as follows if one sums Eq. (2.2.1) over k:

$$\frac{\partial}{\partial t} \rho + \text{div } \rho\mathbf{v} = 0 \tag{2.2.5}$$

where the barycentric velocity is defined by

$$\rho\mathbf{v} \equiv \sum_{k=1}^{n} \rho_k \mathbf{v}_k \Leftrightarrow \begin{cases} \mathbf{v}^\pm \equiv \sum_{k=1}^{n} \frac{\rho_k^\pm \mathbf{v}_k^\pm}{\rho^\pm} & \text{for the bulk regions} \\ \mathbf{v}^s \equiv \sum_{k=1}^{n} \frac{\rho_k^s \mathbf{v}_k^s}{\rho^s} & \text{for the interface} \end{cases} \tag{2.2.6}$$

It should be stressed that $\rho\mathbf{v}$ is the momentum current *per unit of volume* and as a consequence $\rho\mathbf{v}$ may be written in the singular form given in Eq. (1.2.4); the velocity field $\mathbf{v}$ is the momentum current *per unit of mass* and cannot be written in this form. It follows from the definition of the interfacial velocity and Eq. (2.2.2) that the normal component satisfies

$$w^s_n = v^s_n = v^s_{k,n} \tag{2.2.7}$$

Clearly, the dividing surface moves along with the barycentric velocity of the excess mass in the normal direction. We now use the freedom in the choice of curvilinear coordinates to choose them such that this is also the case in the direction parallel to the dividing surface, so that

$$\mathbf{w}^s = \mathbf{v}^s \qquad (2.2.8)$$

The advantage of this choice is that the motion of the dividing surface then has a direct physical meaning.

Using the general equation (1.2.20) and Eq. (2.2.7), the balance equation for the excess mass of component k becomes

$$\frac{\partial}{\partial t} \rho_k^s + \operatorname{div} \rho_k^s \mathbf{v}_k^s + [\rho_k (v_{k,n} - v_n^s)]_- = \sum_{j=1}^r v_{kj} J_j^s \qquad (2.2.9)$$

Similarly, one finds for the total excess mass

$$\frac{\partial}{\partial t} \rho^s + \operatorname{div} \rho^s \mathbf{v}^s + [\rho(v_n - v_n^s)]_- = 0 \qquad (2.2.10)$$

The balance equation for the excess mass of component k can be written in an alternative form if we define the barycentric time derivative for the dividing surface,

$$\frac{d^s}{dt} \equiv \frac{\partial}{\partial t} + \mathbf{v}^s \cdot \operatorname{grad} \qquad (2.2.11)$$

the bulk and interfacial diffusion flows,

$$\mathbf{J}_k^{\pm} \equiv \rho_k^{\pm}(\mathbf{v}_k^{\pm} - \mathbf{v}^{\pm}) \qquad \text{and} \qquad \mathbf{J}_k^s \equiv \rho_k^s(\mathbf{v}_k^s - \mathbf{v}^s) \qquad (2.2.12)$$

and the bulk and interfacial mass fractions,

$$c_k^{\pm} \equiv \frac{\rho_k^{\pm}}{\rho^{\pm}} \qquad \text{and} \qquad c_k^s \equiv \frac{\rho_k^s}{\rho^s} \qquad (2.2.13)$$

Substitution of these definitions into Eq. (2.2.9) and use of Eq. (2.2.10) then gives

$$\rho^s \frac{d^s}{dt} c_k^s + \boldsymbol{\nabla} \cdot \mathbf{J}_k^s + \mathbf{n} \cdot [(\mathbf{v} - \mathbf{v}^s)\rho(c_k - c_k^s) + \mathbf{J}_k]_- = \sum_{j=1}^r v_{ij} J_j^s \qquad (2.2.14)$$

It follows from Eq. (2.2.7) that the interfacial diffusion current is along the dividing surface:

$$\mathbf{n} \cdot \mathbf{J}_k^s = 0 \qquad (2.2.15)$$

Furthermore, it follows from Eq. (2.2.6) that

$$\sum_{k=1}^{n} \mathbf{J}_k^{\pm} = 0 \qquad \text{and} \qquad \sum_{k=1}^{n} \mathbf{J}_k^s = 0 \qquad (2.2.16)$$

which implies that only $n-1$ diffusion currents are independent.

C. The General Form of Interfacial Balance Equations

In the previous section we discussed how the general balance equation (1.2.2) for a quantity d,

$$\frac{\partial}{\partial t} d + \operatorname{div} \mathbf{J}_d = \sigma_d \qquad (2.3.1)$$

leads to the following balance equation [cf. Eqs. (1.2.20) and (2.2.7)] for the excess of d:

$$\frac{\partial}{\partial t} d^s + \operatorname{div} \mathbf{J}_d^s + \mathbf{n} \cdot [\mathbf{J}_d - \mathbf{v}^s d]_- = \sigma_d^s \qquad (2.3.2)$$

Furthermore, one finds the following equation [cf. Eqs. (1.2.22) and (2.2.7)] for the normal component of the excess current:

$$\mathbf{n} \cdot [\mathbf{J}_d^s - \mathbf{v}^s d^s] = 0 \qquad (2.3.3)$$

It is now convenient to define the density of the quantity d per unit of mass by

$$d \equiv \rho a \Leftrightarrow \begin{cases} a^{\pm} \equiv d^{\pm}/\rho^{\pm} & \text{for the bulk regions} \\ a^s \equiv d^s/\rho^s & \text{for the interface} \end{cases} \qquad (2.3.4)$$

Furthermore, it is convenient to write the current as

$$\mathbf{J}_d \equiv \rho a \mathbf{v} + \mathbf{J}_a \Leftrightarrow \begin{cases} \mathbf{J}^{\pm} \equiv \rho^{\pm} a^{\pm} \mathbf{v}^{\pm} + \mathbf{J}^{\pm} & \text{for the bulk region} \\ \mathbf{J}_d^s \equiv \rho^s a^s \mathbf{v}^s + \mathbf{J}_a^s & \text{for the interface} \end{cases} \qquad (2.3.5)$$

where $\rho a \mathbf{v} = d \mathbf{v}$ is the convective contribution to the current. Substituting these definitions in Eq. (2.3.2) and using the balance equation (2.2.10) for the excess mass then gives

$$\rho^s \frac{d^s}{dt} a^s + \mathbf{\nabla} \cdot \mathbf{J}_a^s + \mathbf{n} \cdot [(\mathbf{v} - \mathbf{v}^s)\rho(a - a^s) + \mathbf{J}_a]_- = \sigma_d^s \qquad (2.3.6)$$

while Eq. (2.3.3) gives

$$\mathbf{n} \cdot \mathbf{J}_a^s = 0 \qquad (2.3.7)$$

These alternative equations, which we shall use often in our further analysis, are, of course, equivalent to the original ones. The results we shall find may also be found using Eqs. (2.3.2) and (2.3.3). In particular, one finds the same results even when the dividing surface can be chosen such that $\rho^s = 0$. We shall come back to this point below.

D. Conservation of Momentum

The equation of motion of the system

$$\frac{\partial}{\partial t} \rho \mathbf{v} + \mathbf{\nabla} \cdot (\rho \mathbf{v} \mathbf{v} + \mathsf{P}) = \sum_{k=1}^{n} \rho_k \mathbf{F}_k \qquad (2.4.1)$$

where

$$\rho \mathbf{v} \mathbf{v} = \rho^- \mathbf{v}^- \mathbf{v}^- \Theta^- + \rho^s \mathbf{v}^s \mathbf{v}^s \delta^s + \rho^+ \mathbf{v}^+ \mathbf{v}^+ \Theta^+ \qquad (2.4.2)$$

is the convective contribution to the momentum flow, P is the pressure tensor that gives the rest of the momentum flow, and $\mathbf{F}_k$ is an external force field acting on component k. The pressure tensor can be written as the sum of the hydrostatic pressures p^- and p^+ in the bulk regions, minus the surface tension γ at the dividing surface and the viscous pressure tensor $\mathbf{\Pi}$ both in the bulk regions and on the dividing surface in the following way:

$$\mathsf{P} = p^- \mathsf{I} \Theta^- + p^+ \mathsf{I} \Theta^+ - \gamma(\mathsf{I} - \mathbf{n}\mathbf{n})\delta^s + \mathbf{\Pi} \qquad (2.4.3)$$

We assume that the system possesses no intrinsic internal angular momentum, so that the pressure tensor and, as a consequence, the viscous contribution are symmetric both in the bulk regions[3] and at the interface.[21] Equation (2.3.7) for the normal component of an interfacial current gives in this case

$$\mathbf{n} \cdot \mathsf{P}^s = \mathsf{P}^s \cdot \mathbf{n} = 0 \quad \text{and} \quad \mathbf{n} \cdot \mathbf{\Pi}^s = \mathbf{\Pi}^s \cdot \mathbf{n} = 0 \qquad (2.4.4)$$

where we have used the symmetry. The excess pressure tensor and its viscous part are thus symmetric 2×2 tensors. The assumption that scalar hydrostatic pressures and a scalar surface tension can be used limits the discussion to inelastic media. It is easy to extend most formulas to the general case. If a fluid flows along a solid wall it is usually sufficient to take the appropriate fluxes in the wall to be equal to zero in the analysis below.

The forces will be assumed to be conservative and can be written in terms of time-independent potentials:

$$\mathbf{F}_k = -\text{grad } \psi_k \qquad (2.4.5)$$

Because of these forces the momentum is not conserved locally, so the equation of motion is a balance equation with $\sum_{k=1}^{n} \rho_k \mathbf{F}_k$ as source of momentum. Usually the system is contained in a finite box and the total momentum is conserved by interaction with the walls.

Using Eq. (2.3.6) for the momentum density, one finds as the equation of motion for the dividing surface

$$\rho^s \frac{d^s}{dt} \mathbf{v}^s + \mathbf{\nabla} \cdot [-\gamma(\mathsf{1} - \mathbf{nn}) + \mathbf{\Pi}^s] + \mathbf{n} \cdot [(\mathbf{v} - \mathbf{v}^s)\rho(\mathbf{v} - \mathbf{v}^s) + \mathsf{P}]_- \equiv \sum_{k=1}^{n} \rho_k^s \mathbf{F}_{k,s}$$

$$(2.4.6)$$

The subscript s indicates, as usual, the value of a field at the dividing surface, obtained by putting $\xi_1 = 0$. The limiting value of the field may be different coming from the plus or the minus phase as, for example, for the hydrostatic pressures p_s^+ and p_s^-, but may also be the same, as, for example, the forces $\mathbf{F}_{k,s}^+ = \mathbf{F}_{k,s}^- = \mathbf{F}_{k,s}$. Furthermore, it should be stressed that after taking the gradient or divergence of a quantity that is defined only on the dividing surface, one should also take the result for $\xi_1 = 0$. Thus it would, for example, be better to write $(\mathbf{\nabla}\gamma)_s$ than $\mathbf{\nabla}\gamma$; however, for ease of notation we will not usually do this.

Using the fact that the gradient of a quantity defined on the dividing surface (i.e., independent of ξ_1) is parallel to this surface, one may write [cf. also Eq. (1.2.29)]

$$\mathbf{\nabla} \cdot \gamma(\mathsf{1} - \mathbf{nn}) = \mathbf{\nabla}\gamma - \gamma\mathbf{n} \text{ div } \mathbf{n} = \mathbf{\nabla}\gamma + C\gamma\mathbf{n} \qquad (2.4.7)$$

In equilibrium, when $\mathbf{v}^s = 0$, $\mathbf{\Pi}^s = 0$, and $\mathsf{P}^{\pm} = p^{\pm}\mathsf{I}$, Eq. (2.4.6) gives as the balance of the normal components of the forces

$$-C\gamma + p_- = -C\gamma + p_s^+ - p_s^- = \sum_{k=1}^n \rho_k^s \mathbf{F}_{k,s} \cdot \mathbf{n} \tag{2.4.8}$$

This is a direct generalization of Laplace's equation for the hydrostatic pressure difference $p_s^+ - p_s^-$ in terms of the surface tension and the curvature.[17] Contraction of Eq. (2.4.6) with the normal gives the generalization of Laplace's equation to the dynamic case. In equilibrium Eq. (2.4.6) gives as the balance of the forces along the dividing surface

$$-\boldsymbol{\nabla}\gamma = \sum_{k=1}^n \rho_k^s \mathbf{F}_{k,s} \cdot (\mathsf{I} - \mathbf{nn}) \tag{2.4.9}$$

In the general case the normal part of Eq. (2.4.6) describes the motion of the interface through space, while the parallel part describes the flow of mass along the interface.

Using Eq. (2.4.6), one may derive an equation for the rate of change of the kinetic energy of the excess mass:

$$\rho^s \frac{d^s}{dt} \tfrac{1}{2}|\mathbf{v}^s|^2 = \rho^s \mathbf{v}^s \cdot \frac{d^s}{dt} \mathbf{v}^s = -\boldsymbol{\nabla} \cdot (\mathsf{P}^s \cdot \mathbf{v}^s) + \mathsf{P}^s : \boldsymbol{\nabla}\mathbf{v}^s$$
$$-\mathbf{n} \cdot [(\mathbf{v}-\mathbf{v}^s)\rho(\mathbf{v}-\mathbf{v}^s) + \mathsf{P}]_- \cdot \mathbf{v}^s$$
$$+ \sum_{k=1}^n \rho_k^s \mathbf{F}_{k,s} \cdot \mathbf{v}^s \tag{2.4.10}$$

The potential energy of the excess mass is defined by

$$\rho^s \psi^s \equiv \sum_{k=1}^n \rho_k^s \psi_{k,s} \tag{2.4.11}$$

Using Eq. (2.2.14), one finds for the rate of change of this potential energy

$$\rho^s \frac{d^s}{dt}\psi^s = \rho^s \frac{d^s}{dt}\sum_{k=1}^n \psi_{k,s}c_k^s = \sum_{k=1}^n \psi_{k,s}\rho^s\frac{d^s}{dt}c_k^s + \rho^s\sum_{k=1}^n c_k^s\mathbf{v}^s \cdot \boldsymbol{\nabla}\psi_{k,s}$$
$$= -\sum_{k=1}^n \psi_{k,s}\boldsymbol{\nabla}\cdot\mathbf{J}_j^s - \mathbf{n}\cdot\left[(\mathbf{v}-\mathbf{v}^s)\rho(\psi-\psi^s) + \sum_{k=1}^n \psi_k\mathbf{J}_k\right]_-$$
$$+ \sum_{k=1}^n\sum_{j=1}^r \psi_{k,s}v_{kj}\mathbf{J}_j^s - \sum_{k=1}^n \rho_k^s\mathbf{F}_{k,s}\cdot\mathbf{v}^s \tag{2.4.12}$$

We shall assume that the potential energy is conserved in a chemical reaction:

$$\sum_{k=1}^{n} \psi_{k,s} v_{kj} = 0 \qquad (2.4.13)$$

This is true, for example, in a gravitational field and for charged particles in an electric field. In this case Eq. (2.4.12) reduces to

$$\rho^s \frac{d^s}{dt} \psi^s = - \sum_{k=1}^{n} \psi_{k,s} \nabla \cdot \mathbf{J}_k^s - \mathbf{n} \cdot \left[(\mathbf{v} - \mathbf{v}^s)\rho(\psi - \psi^s) + \sum_{k=1}^{n} \psi_k \mathbf{J}_k \right]$$
$$- \sum_{k=1}^{n} \rho_k^s \mathbf{F}_{k,s} \cdot \mathbf{v}^s \qquad (2.4.14)$$

Equations (2.4.10) and (2.4.13) make clear that neither the kinetic energy nor the potential energy, nor, for that matter, their sum, is conserved.

E. Conservation of Energy

According to the principle of conservation of energy, one has* for the total specific energy e

$$\frac{\partial}{\partial t} \rho e + \mathrm{div}(\rho e \mathbf{v} + \mathbf{J}_e) = 0 \qquad (2.5.1)$$

The total energy density is the sum of the kinetic energy, the potential energy, and the internal energy u:

$$e^\pm = \tfrac{1}{2}|\mathbf{v}^\pm|^2 + \psi^\pm + u^\pm \qquad \text{for the bulk regions}$$
$$e^s = \tfrac{1}{2}|\mathbf{v}^s|^2 + \psi^s + u^s \qquad \text{for the interface} \qquad (2.5.2)$$

Similarly, the energy current may be written as the sum of a mechanical work term, a potential-energy flux due to diffusion, and a heat flow $\mathbf{J}_q$:

$$\mathbf{J}_e = \mathbf{P} \cdot \mathbf{v} + \sum_{k=1}^{n} \psi_k \mathbf{J}_k + \mathbf{J}_q \qquad (2.5.3)$$

Equations (2.5.2) and (2.5.3) may be considered as definitions of the internal energy and the heat flow.

* In de Groot and Mazur[3] $\mathbf{J}_e$ contains also the convective contribution.

The balance equation for the excess energy density is given by [cf. Eq. (2.3.6)],

$$\rho^s \frac{d^s}{dt} e^s + \mathbf{\nabla} \cdot \mathbf{J}_e^s + \mathbf{n} \cdot [(\mathbf{v} - \mathbf{v}^s)\rho(e - e^s) + \mathbf{J}_e]_- = 0 \qquad (2.5.4)$$

The excess energy flow is along the dividing surface [cf. Eq. (2.3.7)]:

$$\mathbf{n} \cdot \mathbf{J}_e^s = 0 \qquad (2.5.5)$$

Using the balance equations (2.4.10) and (2.4.13) for the kinetic- and potential-energy densities as well as the balance equation (2.5.4) for the total energy, one finds as the balance equation for the excess internal energy

$$\rho^s \frac{d^s}{dt} u^s = -\mathbf{\nabla} \cdot \mathbf{J}_q^s - \mathsf{P}^s : \nabla \mathbf{v}^s + \sum_{k=1}^{n} \mathbf{F}_k \cdot \mathbf{J}_k^s$$
$$- \mathbf{n} \cdot \{(\mathbf{v} - \mathbf{v}^s)\rho[u - u^s - \tfrac{1}{2}|\mathbf{v} - \mathbf{v}^s|^2] + \mathbf{J}_q$$
$$+ [(\mathbf{v} - \mathbf{v}^s)\rho(\mathbf{v} - \mathbf{v}^s) + \mathsf{P}] \cdot (\mathbf{v} - \mathbf{v}^s)\}_- \qquad (2.5.6)$$

It follows from Eqs. (2.2.15), (2.4.4), and (2.5.3) that the excess heat flow is also along the dividing surface:

$$\mathbf{n} \cdot \mathbf{J}_q^s = 0 \qquad (2.5.7)$$

The internal energy of the system is not conserved, because of conversion of kinetic and potential energy into internal energy. The balance equation (2.5.6) for the excess internal energy gives the first law of thermodynamics for the interface.

III. ENTROPY BALANCE

A. The Second Law of Thermodynamics

The balance equation for the entropy density is given by

$$\frac{\partial}{\partial t} \rho s + \mathrm{div}(\rho s \mathbf{v} + \mathbf{J}) = \sigma \qquad (3.1.1)$$

where s is the entropy density per unit of mass, $\mathbf{J}$ is the entropy current,*

---

* The subscript s used in de Groot and Mazur[3] has been dropped because it would in this case clearly be confusing.

and σ is the entropy source. All fields have their usual form (1.2.3)–(1.2.5) containing the excess as a singular contribution at the dividing surface. In the bulk regions, Eq. (3.1.1) gives the usual balance equation for the entropy densities s^+ and s^-:

$$\rho^{\pm}\frac{d^{\pm}}{dt}s^{\pm}=\rho^{\pm}\left(\frac{\partial}{\partial t}+\mathbf{v}^{\pm}\cdot\boldsymbol{\nabla}\right)s^{\pm}=-\operatorname{div}\mathbf{J}^{\pm}+\sigma^{\pm} \qquad (3.1.2)$$

One may now conclude[3] from the second law of thermodynamics that

$$\sigma^{\pm}\geq 0 \qquad (3.1.3)$$

For the interface, one finds the following balance equation [cf. Eq. (2.3.6)]:

$$\rho^{s}\frac{d^{s}}{dt}s^{s}=-\boldsymbol{\nabla}\cdot\mathbf{J}^{s}-\mathbf{n}\cdot[(\mathbf{v}-\mathbf{v}^{s})\rho(s-s^{s})+\mathbf{J}]_{-}+\sigma^{s} \qquad (3.1.4)$$

From the second law of thermodyanics, it now follows that

$$\sigma^{s}\geq 0 \qquad (3.1.5)$$

As is to be expected, not only is entropy produced in the bulk regions, but there is also an excess of this production in the interfacial region, which according to the second law is also positive.

B. The Entropy Production

From thermodynamics we know that the entropy for a system in equilibrium is a well-defined function of the various parameters necessary to define the macroscopic state of the system. As discussed already by Gibbs,[5] this is also the case for a system with two phases separated by a surface of discontinuity. For the system under consideration, we use as parameters the internal energy, the specific volume $v^{\pm}\equiv 1/\rho^{\pm}$ or specific surface area $v^{s}\equiv 1/\rho^{s}$, and the mass fractions. We may then write

$$s^{-}=s^{-}(u^{-},v^{-},c_{k}^{-}), \qquad s^{+}=s^{+}(u^{+},v^{+},c_{k}^{+}), \qquad \text{and} \qquad s^{s}=s^{s}(u^{s},v^{s},c_{k}^{s}) \qquad (3.2.1)$$

Three different functions are needed to give the entropies for the two bulk phases and for the interface. At equilibrium the total differential of the entropy is given by the Gibbs relation. In the bulk regions, this relation is

$$T^{\pm}\,ds^{\pm} = du^{\pm} + p^{\pm}\,dv^{\pm} - \sum_{k=1}^{n} \mu_k^{\pm}\,dc^{\pm} \qquad (3.2.2)$$

where $T^{\pm}$ is the temperature and $\mu_k^{\pm}$ is the chemical potential of component k in the bulk. For the interface, it is given by[5]

$$T^s\,ds^s = du^s - \gamma\,dv^s - \sum_{k=1}^{n} \mu_k^s\,dc_k^s \qquad (3.2.3)$$

where T^s is the interfacial temperature and μ_k^s is the interfacial chemical potential of component k. The Gibbs relation for the interface is in fact similar to the one for the bulk regions if one interprets the surface tension as the negative of the surface pressure, $p^s \equiv -\gamma$.

Though the total system we are describing is not in equilibrium, we assume that the system is in so-called local equilibrium. More particularly, we assume that after averaging the microscopic variables over cells with a diameter of the order of the bulk correlation length, one obtains a description in which the local entropy is given, as at equilibrium, by the functions (3.2.1) in terms of the local values of the internal energy, the specific volume or surface area, and the mass fractions. It is clear that the hypothesis of local equilibrium limits the description to situations where the change of the bulk variables is small over distances of the order of the bulk correlation length in all directions and the change of the excess variables on the dividing surface is small over these distances along the surface. Furthermore, the radii of curvature of the dividing surface must be large compared with the bulk correlation length. This also makes the assumption that s^s is not dependent on the curvature a reasonable one. The local-equilibrium hypothesis implies that the Gibbs relation remains valid in the frame moving with the center of mass. We thus have

$$T^{\pm}\frac{d^{\pm}}{dt}s^{\pm} = \frac{d^{\pm}}{dt}u^{\pm} + p^{\pm}\frac{d^{\pm}}{dt}v^{\pm} - \sum_{k=1}^{n} \mu_k^{\pm}\frac{d^{\pm}}{dt}c_k^{\pm} \qquad (3.2.4)$$

in the bulk phases and

$$T^s\frac{d^s}{dt}s^s = \frac{d^s}{dt}u^s - \gamma\frac{d^s}{dt}v^s - \sum_{k=1}^{n} \mu_k^s\frac{d^s}{dt}c_k^s \qquad (3.2.5)$$

for the interface.

As explained in detail by de Groot and Mazur,[3] one may now find explicit expressions for the entropy current and the entropy production in the bulk phases by substituting the barycentric time derivatives of $u^{\pm}$, $v^{\pm}$, and $c_k^{\pm}$ into Gibbs relation and writing the result in the form of the entropy balance equation (3.1.2). They find in this way

$$\mathbf{J}^{\pm} = \left(\mathbf{J}_q^{\pm} - \sum_{k=1}^{n} \mu_k^{\pm} \mathbf{J}_k^{\pm} \right) \Big/ T^{\pm} \qquad (3.2.6)$$

for the entropy current and

$$\sigma^{\pm} = -(T^{\pm})^{-2} \mathbf{J}_q^{\pm} \cdot \operatorname{grad} T^{\pm} + \sum_{k=1}^{n} \mathbf{J}_k^{\pm} \cdot \left[\left(\frac{\mathbf{F}_k}{T^{\pm}} \right) - \operatorname{grad} \left(\frac{\mu_k^{\pm}}{T^{\pm}} \right) \right]$$

$$- (T^{\pm})^{-1} \mathbf{\Pi}^{\pm} : \operatorname{grad} \mathbf{v}^{\pm} - (T^{\pm})^{-1} \sum_{j=1}^{r} J_j^{\pm} A_j^{\pm} \geq 0 \qquad (3.2.7)$$

for the entropy production. The affinities A_j of the chemical reactions are defined by

$$A_j^{\pm} \equiv \sum_{k=1}^{n} \mu_k^{\pm} v_{kj} \qquad \text{and} \qquad A_j^{s} \equiv \sum_{k=1}^{n} \mu_k^{\pm} v_{kj} \qquad (3.2.8)$$

In a similar way, we find a balance equation for the excess entropy by substitution of Eqs. (2.5.6), (2.2.10), and (2.2.14) for the barycentric derivatives of u^s, $\rho^s = 1/v^s$, and c_k^s, respectively, into the Gibbs relation (3.2.5). Comparing the result with Eq. (3.1.4), we find, after some algebra,

$$\mathbf{J}^{s} = \left(\mathbf{J}_q^{s} - \sum_{k=1}^{n} \mu_k^{s} \mathbf{J}_k^{s} \right) \Big/ T^{s} \qquad (3.2.9)$$

for the excess entropy current along the dividing surface and

$$\sigma^{s} = -(T^{s})^{-2} \mathbf{J}_q^{s} \cdot \operatorname{grad} T^{s} + \sum_{k=1}^{n} \mathbf{J}_k^{s} \cdot \left[\frac{\mathbf{F}_k}{T^{s}} - \operatorname{grad} \left(\frac{\mu_k^{s}}{T^{s}} \right) \right]$$

$$- (T^{s})^{-1} \mathbf{\Pi} : \operatorname{grad} \mathbf{v}^{s} - (T^{s})^{-1} \sum_{j=1}^{r} J_j^{s} A_j^{s} + \left\{ [J_{q,n} + (v_n - v_n^s) T\rho s] \left(\frac{1}{T} - \frac{1}{T^s} \right) \right\}$$

$$- \frac{1}{T^s} \{ [\Pi_{n,\|} + (v_n - v_n^s) \rho \mathbf{v}] \cdot (\mathbf{v} - \mathbf{v}^s) \}_- - \frac{1}{T^s} \sum_{k=1}^{n} \left\{ [J_{k,n} + (v_n - v_n^s) \rho_k] \right.$$

$$\times \left[\mu_k - \mu_k^s - \tfrac{1}{2} |\mathbf{v}_\||^2 + \tfrac{1}{2} |\mathbf{v}_\|^s|^2 + \tfrac{1}{2} (v_n - v_n^s)^2 + \frac{1}{\rho} \Pi_{nn} \right] \right\}$$

$$\geq 0 \qquad (3.2.10)$$

for the excess entropy production. The subscript $\parallel$ indicates the projection of a vector on the dividing surface; for example,

$$\mathbf{v}_\parallel \equiv \mathbf{v} \cdot (1 - \mathbf{nn}) \qquad \text{and} \qquad \Pi_{n,\parallel} \equiv \mathbf{n} \cdot \Pi \cdot (1 - \mathbf{nn})$$

The first four terms in the excess entropy are similar to those found in the bulk regions [cf. Eq. (3.2.7)]. The first arises from excess heat flow along the interface, the second from excess diffusion along the interface, the third from excess viscous flow along the interface and the fourth from the excess chemical reaction rate. The fifth term arises from heat flow through and into the interface, the sixth from flow of momentum through and into the interface, and the seventh from diffusive flow through and into the interface. It is interesting that if $v_{n,s}^\pm \neq v_n^s$, one must use the total flow in the rest frame of the dividing surface, which is obtained by adding the bulk convective flow in the rest frame of the dividing surface to the bulk conductive flow, in the last three terms. This fact, which seems reasonable, has been used to obtain the last two terms in a unique form. In the treatment of membrane transport, this gave rise to considerable discussion.[7] One may write down other forms in which the contributions proportional to the third power of the velocity field are differently distributed over the last two terms. In practical situations contributions proportional to the third power of the velocity field are usually small, and as a consequence these alternative choices will not lead to different results. The choice we made above seems to be the only systematic one and is therefore preferred.

The following thermodynamic identities are often useful:

$$T^\pm s^\pm = u^\pm + \frac{p^\pm}{\rho^\pm} - \sum_{k=1}^{n} \mu_k^\pm c_k^\pm \qquad (3.2.11)$$

in the bulk phases and

$$T^s s^s = u^s - \frac{\gamma}{\rho^s} - \sum_{k=1}^{n} \mu_k^s c_k^s \qquad (3.2.12)$$

for the interface. In fact, the last identity is needed when one derives the expression for σ^s.

In the above analysis we have taken all excess densities to be unequal to zero. It is usually possible by the appropriate choice of the dividing surface in the interfacial region to make one of the excess densities, for example, the excess density of the solvent, equal to zero. The corresponding term in Eq. (3.2.12) may simply be eliminated. If one considers a one-component, two-phase system, one may eliminate ρ^s in this way. In that case it is better to use Eq. (3.2.12) in the form

$$T^s(\rho s)^s - (\rho u)^s = -(\rho f)^s = -\gamma \qquad (3.2.13)$$

It is important to realise that $\rho^s = 0$ does not imply that $(\rho s)^s$, $(\rho u)^s$, or $(\rho f)^s$ is equal to zero. In fact, the excess free energy per unit of surface area $(\rho f)^s$ is equal to the surface tension in a one-component system. In all these cases the expression for the entropy production remains the one given in Eq. (3.2.10). An alternative derivation using densities per unit of surface area may easily be given. Of course certain excess currents may for a given system also be negligible. In that case one should simply eliminate the corresponding terms in the entropy-production equation (3.2.10).

The diffusion currents in the bulk and in the interface are not independent. As one see from Eq. (2.2.16), their sum is zero. Using this property one may eliminate $\mathbf{J}_N$ and $\mathbf{J}_N^s$ from the entropy production (3.2.10).* One then obtains

$$\sigma^s = -(T^s)^{-2}\mathbf{J}_q^s \cdot \operatorname{grad} T^s + \sum_{k=1}^{N-1} \mathbf{J}_k^s \cdot \left[\frac{\mathbf{F}_k - \mathbf{F}_N}{T^s} - \operatorname{grad}\left(\frac{\mu_k^s - \mu_N^s}{T^s} \right) \right]$$

$$-(T^s)^{-1}\mathbf{\Pi}^s : \operatorname{grad} \mathbf{v}^s - (T^s)^{-1} \sum_{j=1}^{r} J_j^s A_j^s$$

$$+ \left\{ [J_{q,n} + (v_n - v_n^s) T\rho s]\left(\frac{1}{T} - \frac{1}{T^s} \right) \right\}_-$$

$$- \frac{1}{T^s} \{ [\Pi_{n,\parallel} + (v_n - v_n^s)\rho \mathbf{v}] \cdot (\mathbf{v} - \mathbf{v}^s) \}_-$$

$$- \frac{1}{T^s} \sum_{k=1}^{N-1} \{ [J_{k,n} + (v_n - v_n^s)\rho_k][(\mu_k - \mu_k^s) - (\mu_N - \mu_N^s)] \}_-$$

$$- \frac{1}{T^s} \{ [\Pi_{nn} + (v_n - v_n^s)\rho v_n + \rho(\mu_N - \mu_N^s - \tfrac{1}{2}|\mathbf{v}|^2$$

$$+ \tfrac{1}{2}|\mathbf{v}^s|^2)](v_n - v_n^s) \}_- \qquad (3.2.14)$$

The last four terms are due, on the one hand, to fluxes through the interface from one bulk phase to the other and, on the other hand, to fluxes into the interface from the bulk regions. This may be seen using the

---

* From here on we write the number of components as N rather than as n to avoid confusion with the subscript n for the normal direction.

identity

$$[ab]_- = a_+b_- + a_-b_+ \qquad (3.2.15)$$

where the average of the bulk fields at the dividing surface, a_+, is defined by

$$a_+(\xi_2, \xi_3, t) \equiv \tfrac{1}{2}[a^-(\xi_1 = 0, \xi_2, \xi_3, t) + a^+(\xi_1 = 0, \xi_2, \xi_3, t)] \qquad (3.2.16)$$

With the above identity one may, for example, write the contribution due to the heat current as the sum of two terms:

$$\left\{ [J_{q,n} + (v_n - v_n^s)T\rho s]\left(\frac{1}{T} - \frac{1}{T^s}\right) \right\}_- = [J_{q,n} + (v_n - v_n^s)T\rho s]_+\left(\frac{1}{T^+} - \frac{1}{T^-}\right)$$

$$+ [J_{q,n} + (v_n - v_n^s)T\rho s]_-\left(\left(\frac{1}{T}\right)_+ - \frac{1}{T^s}\right) \qquad (3.2.17)$$

The first term is due to the heat current between the two bulk phases and the second is due to the heat current into the interface; the conjugate forces are $(1/T^+ + 1/T^-)$ and $[(1/T)_+ - 1/T^s]$, respectively. One may rewrite the last three terms as sums of two contributions in a similar way.

IV THE PHENOMENOLOGICAL EQUATIONS

A. Introduction

If the system is in equilibrium, the entropy production is zero. This is the case if the thermodynamic forces are zero. As a result, the fluxes are also zero. Sufficiently close to equilibrium, the fluxes are linear functions of the thermodynamic forces. For a large class of irreversible phenomena these linear relations are in fact sufficient. The linear constitutive coefficients will in general depend on the local values of the variables. Thus a coefficient like the viscosity depends on the temperature, and if the temperature is not uniform, neither is the viscosity. The resulting dependence on the local variables of the fluxes may in fact be nonlinear. As such there is a crucial difference from the linear dependence of the fluxes on the thermodynamic forces. These thermodynamic forces result from variation of the variables over a distance of a typical bulk correlation length. It is crucial for the hypothesis of local equilibrium that the variation over such distances be small. As a consequence, the linear dependence is already implicit in this hypothesis. The affinities of the chemical reactions are an exception in this context, because they depend

72   D. BEDEAUX

only on the values of variables in the same cell with a diameter of a bulk correlation length; here the hypothesis of local equilibrium implies that the reactions are slow compared with the time the cell needs to reach equilibrium. Though for chemical reactions also the reaction rates will be linear functions of the affinities if the system is sufficiently close to equilibrium, one must usually use nonlinear relations. These nonlinear relations are, as is clear from the discussion above, in agreement with the hypothesis of local equilibrium.

It should be emphasized that linear constitutive relations do not lead to linear equations of motion. Because of convection and the nonlinear nature of the equation of state, the equations of motion are in general highly nonlinear. The fully linearized equations play an important role in some problems, as, for example, in the description of fluctuations around equilibrium.

B. The Curie Symmetry Principle

As is clear from the expression (3.2.14) for the entropy production, there is a rather large number of force–flux pairs. In the most general case, the Cartesian components of the fluxes may in principle depend on all the Cartesian components of the forces. The number of constitutive coefficients needed would then be very large. This situation is greatly simplified if there are symmetries. If one considers a fluid–fluid interface, one may use the isotropy of the system along the interface to show that forces and fluxes of a different tensorial character do not couple. In the bulk, this is referred to as the Curie symmetry principle.[3] In this case one may write the excess entropy production as a sum of contributions from symmetric traceless 2×2 tensorial force–flux pairs, two-dimensional vectorial force–flux pairs, and scalar force–flux pairs:

$$\sigma^s = \sigma^s_{tens} + \sigma^s_{vect} + \sigma^s_{scal} \qquad (4.2.1)$$

The only contribution to σ^s_{tens} is due to the excess viscous pressure tensor. If we define the symmetric traceless part of $\mathbf{\Pi}$ by

$$\overline{\mathbf{\Pi}^s} \equiv \mathbf{\Pi}^s - \tfrac{1}{2}(1 - \mathbf{nn})\Pi^s \qquad \text{with} \quad \Pi^s \equiv \text{Tr}\,\mathbf{\Pi}^s, \qquad (4.2.2)$$

where we note that $\mathbf{\Pi}^s$ is already symmetric, we may write

$$\mathbf{\Pi}^s : \text{grad}\,\mathbf{v}^s = \overline{\mathbf{\Pi}^s} : \overline{\text{grad}\,\mathbf{v}^s_{\parallel}} + \Pi^s\,\text{div}\,\mathbf{v}^s_{\parallel} \qquad (4.2.3)$$

The reason that only the parallel component of $\mathbf{v}^s$ enters the equation is that $\mathbf{n} \cdot \mathbf{\Pi}^s = 0$; the symmetric traceless part of grad $\mathbf{v}_\|^s$ is defined by

$$\overline{(\text{grad } \mathbf{v}_\|^s)}_{\alpha\beta} \equiv \tfrac{1}{2}(\text{grad } \mathbf{v}_\|^s)_{\alpha\beta} + \tfrac{1}{2}(\text{grad } \mathbf{v}_\|^s)_{\beta\alpha} - \tfrac{1}{2}(\delta_{\alpha\beta} - n_\alpha n_\beta) \text{ div } \mathbf{v}_\| \qquad (4.2.4)$$

Note that

$$\mathbf{n} \cdot (\text{grad } \mathbf{v}_\|^s) = (\text{grad } \mathbf{v}_\|^s) \cdot \mathbf{n} = 0 \qquad (4.2.5)$$

so that grad $\mathbf{v}_\|^s$ and, as a consequence, $\overline{\text{grad } \mathbf{v}_\|^s}$ are also 2×2 tensors. Using Eq. (4.2.3) and expression (3.2.14) for the excess entropy production, we find

$$\sigma_{\text{tens}}^s = -(T^s)^{-1}\overline{\mathbf{\Pi}^s} : \overline{\text{grad } \mathbf{v}_\|^s} \qquad (4.2.6)$$

The other term, $\Pi^s \text{ div } \mathbf{v}_\|^s$, gives a contribution to σ_{scal}^s.

For the vectorial force–flux pairs, we find

$$\sigma_{\text{vect}}^s = -(T^s)^{-2}\mathbf{J}_q^s \cdot \text{grad } T^s + \sum_{k=1}^{N-1} \mathbf{J}_k^s \cdot \left[\frac{\mathbf{F}_k - \mathbf{F}_N}{T^s} - \text{grad}\left(\frac{\mu_k^s - \mu_N^s}{T^s} \right) \right]$$

$$- \frac{1}{T^s}[\Pi_{n,\|} + (v_n - v_n^s)\rho\mathbf{v}_\|]_+ \cdot \mathbf{v}_{\|,-}$$

$$- \frac{1}{T^s}[\Pi_{n,\|} + (v_n - v_n^s)\rho\mathbf{v}_\|]_- \cdot (\mathbf{v}_{\|,+} - \mathbf{v}_\|^s) \qquad (4.2.7)$$

where we have also used the identity (3.2.15). There thus are $N+2$ vectorial force–flux pairs.

For the scalar force–flux pairs, we find

$$\sigma_{\text{scal}}^s = -(T^s)^{-1}\Pi^s \text{ div } \mathbf{v}_\|^s - (T^s)^{-1} \sum_{j=1}^r J_j^s A_j^s$$

$$+ [J_{q,n} + (v_n - v_n^s)T\rho s]_+ \left(\frac{1}{T^+} - \frac{1}{T^-} \right) + [J_{q,n} + (v_n - v_n^s)T\rho s]_- \left(\left(\frac{1}{T} \right)_+ - \frac{1}{T^s} \right)$$

$$- \frac{1}{T^s} \sum_{k=1}^{N-1} [J_{k,n} - (v_n - v_n^s)\rho_k]_+ (\mu_{k,-} - \mu_{N,-})$$

$$- \frac{1}{T^s} \sum_{k=1}^{N-1} [J_{k,n} - (v_n - v_n^s)\rho_k]_- ((\mu_{k,+} - \mu_k^s) - (\mu_{N,+} - \mu_N^s))$$

$$- \frac{1}{T^s}[\Pi_{nn} + (v_n - v_n^s)\rho v_n + \rho(\mu_N - \mu_N^s - \tfrac{1}{2}|\mathbf{v}|^2 + \tfrac{1}{2}|\mathbf{v}^s|^2)]_+ v_{n,-}$$

$$- \frac{1}{T^s}[\Pi_{nn} + (v_n - v_n^s)\rho v_n + \rho(\mu_N - \mu_N^s - \tfrac{1}{2}|\mathbf{v}|^2 + \tfrac{1}{2}|\mathbf{v}^s|^2)]_- (v_{n,+} - v_n^s)$$

$$(4.2.8)$$

where we have again used the identity (3.2.15). We thus have $3+r+N$ scalar force–flux pairs.

It is good to realize that isotropy of the interfacial region differs from isotropy in the bulk phases. In the bulk phase the isotropy implies that the properties of the system are not dependent on the direction. For the interface, isotropy implies that the properties do not depend on the directions along the interface. The properties in the direction orthogonal to the interface are clearly very different, however. The isotropy in the bulk regions is three-dimensional. The isotropy of the interface, which is described as a two-dimensional structure, is two-dimensional. The relevant tensors and vectors describing excess flows and forces along the interface are in fact two-dimensional. Because the interface is, of course, embedded in a three-dimensional space, a two-dimensional excess vector flow is written as a three-dimensional vector flow with a zero normal component; similarly one writes the 2×2 dimensional tensors as 3×3 tensors with five elements that are zero. Whereas the situation for the excess fluxes is reasonably self evident, this is less true of the extrapolated values of the bulk fluxes at the dividing surface. These extrapolated values are given by the usual three-dimensional isotropic linear laws in terms of the extrapolated bulk forces. A complete specification of the dynamical behavior of the bulk phase requires specifying the normal component of this extrapolated flux at the dividing surface. It is this normal component that appears in the excess entropy production. In this specification it becomes important to realize the two-dimensional isotropy of the interface. This results, for example, in the extrapolated values of the normal components of the heat flow and the diffusion flow being scalar in this context. As a consequence, diffusion through the interface may be driven by a finite value of one of the affinities. This process is called active transport.[4] Similarly, extrapolated values of the viscous pressure tensor give the viscous forces of the bulk phases on the dividing surface. The parallel component of this viscous force, $\Pi_{n,\parallel}$, is a two-dimensional vector and contributes to σ^s_{vect}, while the normal component of this force, Π_{nn}, is a scalar and contributes to σ^s_{scal}. A consequence of the symmetry breaking in the direction normal to the interface is that many cross effects exist that are impossible in the bulk regions. We shall return to these possibilities in the following sections, where we give the linear constitutive equations.

C. The Onsager Relations

There is one more symmetry property that reduces the number of independent linear constitutive coefficients. This property gives the so-called Onsager relations between the linear coefficients describing cross effects, which are based on the time-reversal invariance of the micro-scopic equations of motion. We refer the reader to de Groot and Mazur[3] for a detailed discussion of the Onsager relations. In the following sections we shall simply give the explicit expressions in the various cases.

D. Symmetric Traceless Tensorial Force–Flux Pairs

It is clear from $\sigma_{\text{tens}}^{\text{s}}$ as given in Eq. (4.2.6) that there is only one tensorial force–flux pair. The linear phenomenological equation for this flux is

$$\overline{\overline{\Pi^{\text{s}}}} = -2\eta^{\text{s}}\,\overline{\overline{\text{grad }\mathbf{v}_{\|}^{\text{s}}}} \tag{4.4.1}$$

The linear coefficient η^{s} will be called "interfacial shear viscosity" and has the dimensionality of a regular viscosity times a length.

E. Vectorial Force–Flux Pairs

The linear laws for the vectorial fluxes that follow from Eq. (4.2.7) are

$$
\mathbf{J}_q^{\text{s}} = -L_{q,q}^{\text{s}}\frac{\text{grad }T^{\text{s}}}{(T^{\text{s}})^2} - \sum_{k=1}^{N-1} L_{q,k}^{\text{s}}\left[\text{grad}\left(\frac{\mu_k^{\text{s}}-\mu_N^{\text{s}}}{T^{\text{s}}}\right) - \frac{\mathbf{F}_k - \mathbf{F}_N}{T^{\text{s}}}\right]
$$
$$
- L_{q,v-}^{\text{s}}\frac{\mathbf{v}_{\|,-}}{T^{\text{s}}} - L_{q,v+}^{\text{s}}\frac{\mathbf{v}_{\|,+}-\mathbf{v}_{\|}^{\text{s}}}{T^{\text{s}}} \tag{4.5.1}
$$

for the excess heat flow along the interface,

$$
\mathbf{J}_l^{\text{s}} = -L_{l,q}^{\text{s}}\frac{\text{grad }T^{\text{s}}}{(T^{\text{s}})^2} - \sum_{k=1}^{N-1} L_{l,k}^{\text{s}}\left[\text{grad}\left(\frac{\mu_k^{\text{s}}-\mu_N^{\text{s}}}{T^{\text{s}}}\right) - \frac{\mathbf{F}_k - \mathbf{F}_N}{T^{\text{s}}}\right]
$$
$$
- L_{l,v-}^{\text{s}}\frac{\mathbf{v}_{\|,-}}{T^{\text{s}}} - L_{l,v+}^{\text{s}}\frac{\mathbf{v}_{\|,+}-\mathbf{v}_{\|}^{\text{s}}}{T^{\text{s}}} \quad \text{for} \quad l=1,\dots,N-1 \tag{4.5.2}
$$

for the excess diffusion flow along the interface, and

$$
[\Pi_{n,\|} + (v_n - v_n^{\text{s}})\rho\mathbf{v}_{\|}]_+
$$
$$
= -L_{v-,q}^{\text{s}}\frac{\text{grad }T^{\text{s}}}{(T^{\text{s}})^2} - \sum_{k=1}^{N-1} L_{v-,k}^{\text{s}}\left[\text{grad}\left(\frac{\mu_k^{\text{s}}-\mu_N^{\text{s}}}{T^{\text{s}}}\right) - \frac{\mathbf{F}_k - \mathbf{F}_N}{T^{\text{s}}}\right]
$$
$$
- L_{v-,v-}^{\text{s}}\frac{\mathbf{v}_{\|,-}}{T^{\text{s}}} - L_{v-,v+}^{\text{s}}\frac{\mathbf{v}_{\|,+}-\mathbf{v}_{\|}^{\text{s}}}{T^{\text{s}}} \tag{4.5.3}
$$

and

$$[\Pi_{n,\parallel} + (v_n - v_n^s)\rho\mathbf{v}_{\parallel}]_-$$

$$= -L_{v+,q}\frac{\text{grad }T^s}{(T^s)^2} - \sum_{k=1}^{N-1}L_{v+,k}^s\left[\text{grad}\left(\frac{\mu_k^s - \mu_N^s}{T^s}\right) - \frac{\mathbf{F}_k - \mathbf{F}_N}{T^s}\right]$$

$$- L_{v+,v-}^s\frac{\mathbf{v}_{\parallel,-}}{T^s} - L_{v+,v+}^s\frac{\mathbf{v}_{\parallel,+} - \mathbf{v}_{\parallel}^s}{T^s} \tag{4.5.4}$$

for the viscous forces along the interface due to the bulk phases and vice versa.

The Onsager relations are

$$L_{q,k}^s = L_{k,q}^s, \qquad L_{q,v-}^s = -L_{v-,q}^s, \qquad L_{q,v+}^s = -L_{v+,q}^s,$$

$$L_{k,v-}^s = -L_{v-,k}^s, \qquad L_{k,v+}^s = -L_{v+,k}^s, \qquad L_{v-,v+}^s = L_{v+,v-}^s \tag{4.5.5}$$

The interfacial heat conductivity is given by $\lambda^s \equiv L_{q,q}^s(T^s)^{-2}$; interfacial diffusion coefficients are related to $L_{l,k}^s$; the coefficients of sliding friction are related to $L_{v\pm,v\pm}^s$; and the coefficients $L_{v\pm,q}^s$ describe thermal slip. It is clear that the above linear relations describe a large number of interfacial phenomena. The number of independent constitutive coefficients is $\frac{1}{2}(N + s)(N+3)$. This number is rather large, reflecting the large number of different phenomena that may take place at an interface. All the above linear expressions for the fluxes may be used in the balance equations to obtain explicit equations describing the temporal behavior of the excess densities. Because it is clear how this should be done and because of the length of the resulting expressions, we shall not give the explicit expressions for the general case. It is more appropriate to do this for every particular case using the available information about which effects are important and which unimportant, in order to simplify the resulting expressions.[22,23]

The linear laws for $\Pi_{n,\parallel}^{\pm}$ also serve as boundary conditions for the temporal behavior of the bulk phases, for which they give the forces exerted by the interfacial region on the bulk phases parallel to the dividing surface. To illustrate this, consider the special case of a one-component fluid flowing along a solid wall (at rest) under isothermal conditions. In this case $v_n^- = v_n^+ = v_n^s = 0$ and $\mathbf{v}_{\parallel}^- = 0$ if we take the solid as the minus phase. The viscous pressure in the solid is zero. One then has

$$[\Pi_{n,\parallel} + (v_n - v_n^s)\rho\mathbf{v}_{\parallel}]_- = 2[\Pi_{n,\parallel} + (v_n - v_n^s)\rho\mathbf{v}_{\parallel}]_+ = \Pi_{n,\parallel}^+ \tag{4.5.6}$$

and

$$\mathbf{v}_{\|,-} = 2\mathbf{v}_{\|,+} = \mathbf{v}_{\|}^{+} \qquad (4.5.7)$$

Using these two relations and Eqs. (4.5.3) and (4.5.4), one may solve $\Pi_{n,\|}^{+}$ and $\mathbf{v}_{\|}^{s}$ in terms of $\mathbf{v}_{\|}^{+}$. The expression of interest to us is of the form

$$\Pi_{n,\|}^{+} = \eta\,(\mathbf{n} \cdot \nabla \mathbf{v}_{\|})^{+} = \beta \mathbf{v}_{\|}^{+} \qquad (4.5.8)$$

which is the slip condition and where β, which can be expressed in $L_{v\pm,v\pm}$ is the coefficient of sliding friction.

F. Scalar Force–Flux Pairs

The linear laws for the scalar fluxes that follow from Eq. (4.2.8) are

$$\Pi^{s} = -L_{v,v}^{s}\frac{\operatorname{div}\mathbf{v}_{\|}^{s}}{T^{s}} - \frac{1}{T^{s}}\sum_{j=1}^{r} R_{v,j}^{s}A_{j}^{s} + L_{v,q-}^{s}\left(\frac{1}{T}\right)_{-} + L_{v,q+}^{s}\left(\left(\frac{1}{T}\right)_{+} - \frac{1}{T^{s}}\right)$$

$$- \frac{1}{T^{s}}\sum_{k=1}^{N-1} L_{v,k-}^{s}(\mu_{k,-} - \mu_{N,-})$$

$$- \frac{1}{T^{s}}\sum_{k=1}^{N-1} L_{v,k+}^{s}[(\mu_{k,+} - \mu_{k}^{s}) - (\mu_{N,+} - \mu_{N}^{s})]$$

$$- L_{v,v-}^{s}\frac{v_{n,-}}{T^{s}} - L_{v,v+}^{s}\frac{v_{n,+} - v_{n}^{s}}{T^{s}} \qquad (4.6.1)$$

for the trace of the excess viscous pressure;

$$J_{l}^{s} = -R_{l,v}^{s}\frac{\operatorname{div}\mathbf{v}_{\|}^{s}}{T^{s}} - \frac{1}{T^{s}}\sum_{j=1}^{r} R_{l,j}^{s}A_{j}^{s} + R_{l,q-}^{s}\left(\frac{1}{T}\right)_{-} + R_{l,q+}^{s}\left(\left(\frac{1}{T}\right)_{+} - \frac{1}{T^{s}}\right)$$

$$- \frac{1}{T^{s}}\sum_{k=1}^{N-1} R_{l,k-}^{s}(\mu_{k,-} - \mu_{N,-})$$

$$- \frac{1}{T^{s}}\sum_{k=1}^{N-1} R_{l,k+}^{s}[(\mu_{k,+} - \mu_{k}^{s}) - (\mu_{N,+} - \mu_{N}^{s})]$$

$$- R_{l,v-}^{s}\frac{v_{n,-}}{T^{s}} - R_{l,v+}^{s}\frac{v_{n,+} - v_{n}^{s}}{T^{s}} \qquad \text{for}\quad l = 1, \ldots, r \qquad (4.6.2)$$

for the excess reaction rates;

$$[J_{q,n} + (v_n - v_n^s)T\rho s]_+ = -L_{q-,v}^s \frac{\operatorname{div} \mathbf{v}_{\parallel}^s}{T^s} - \frac{1}{T^s} \sum_{j=1}^{r} R_{q-,j}^s A_j^s$$

$$+ L_{q-,q-}^s \left(\frac{1}{T}\right)_- + L_{q-,q+}^s \left(\left(\frac{1}{T}\right)_+ - \frac{1}{T^s}\right)$$

$$- \frac{1}{T^s} \sum_{k=1}^{N-1} L_{q-,k-}^s (\mu_{k,-} - \mu_{N,-})$$

$$- \frac{1}{T^s} \sum_{k=1}^{N-1} L_{q-,k+}^s [(\mu_{k,+} - \mu_k^s) - (\mu_{N,+} - \mu_N^s)]$$

$$- L_{q-,v-}^s \frac{v_{n,-}}{T^s} - L_{q-,v+}^s \frac{v_{n,+} - v_n^s}{T^s} \qquad (4.6.3)$$

for the heat flow through the interface;

$$[J_{q,n} + (v_n - v_n^s)T\rho s]_- = -L_{q+,v}^s \frac{\operatorname{div} \mathbf{v}_{\parallel}^s}{T^s} - \frac{1}{T^s} \sum_{j=1}^{r} R_{q+,j}^s A_j^s$$

$$+ L_{q+,q-}^s \left(\frac{1}{T}\right)_- + L_{q+,q+}^s \left(\left(\frac{1}{T}\right)_+ - \frac{1}{T^s}\right)$$

$$- \frac{1}{T^s} \sum_{k=1}^{N-1} L_{q+,k-}^s (\mu_{k,-} - \mu_{N,-})$$

$$- \frac{1}{T^s} \sum_{k=1}^{N-1} L_{q+,k+}^s [(\mu_{k,+} - \mu_k^s) - (\mu_{N,+} - \mu_N^s)]$$

$$- L_{q+,v-}^s \frac{v_{n,-}}{T^s} - L_{q+,v+}^s \frac{v_{n,+} - v_n^s}{T^s} \qquad (4.6.4)$$

for the heat flow into the interfacial region;

$$[J_{l,n} - (v_n - v_n^s)\rho_l]_+ = -L_{l-,v}^s \frac{\operatorname{div} \mathbf{v}_{\parallel}^s}{T^s} - \frac{1}{T^s} \sum_{j=1}^{r} R_{l-,j}^s A_j^s$$

$$+ L_{l-,q-}^s \left(\frac{1}{T}\right)_- + L_{l-,q+}^s \left(\left(\frac{1}{T}\right)_+ - \frac{1}{T^s}\right)$$

$$- \frac{1}{T^s} \sum_{k=1}^{N-1} L_{l-,k-}^s (\mu_{k,-} - \mu_{N,-})$$

$$+ \frac{1}{T^s} \sum_{k=1}^{N-1} L_{l-,k+}^s [(\mu_{k,+} - \mu_k^s) - (\mu_{N,+} - \mu_N^s)]$$

$$- L_{l-,v-}^s \frac{v_{n,-}}{T^s} - L_{l-,v+}^s \frac{v_{n,+} - v_n^s}{T^s} \qquad (4.6.5)$$

for the diffusion flow through the interface

$$[J_{l,n} - (v_n - v_n^s)\rho_l]_- = -L_{l+,v}^s \frac{\text{div } \mathbf{v}_\parallel^s}{T^s} - \frac{1}{T^s} \sum_{j=1}^r R_{l+,j}^s A_j^s$$

$$+ L_{l+,q-}^s \left(\frac{1}{T}\right)_- + L_{l+,q+}^s \left(\left(\frac{1}{T}\right)_+ - \frac{1}{T^s}\right)$$

$$- \frac{1}{T^s} \sum_{k=1}^{N-1} L_{l+,k-}^s (\mu_{k,-} - \mu_{N,-})$$

$$+ \frac{1}{T^s} \sum_{k=1}^{N-1} L_{l+,k+}^s [(\mu_{k,+} - \mu_k^s) - (\mu_{N,+} - \mu_N^s)]$$

$$- L_{l+,v-}^s \frac{v_{n,-}}{T^s} - L_{l+,v+}^s \frac{v_{n,+} - v_n^s}{T^s} \tag{4.6.6}$$

for the diffusion flow into the interfacial region; and

$$[\Pi_{nn} + (v_n - v_n^s)\rho v_n + \rho(\mu_N - \mu_N^s - \tfrac{1}{2}|\mathbf{v}|^2 + \tfrac{1}{2}|\mathbf{v}^s|^2)]_+ = -L_{v-,v}^s \frac{\text{div } \mathbf{v}_\parallel^s}{T^s}$$

$$- \frac{1}{T^s} \sum_{j=1}^r R_{v-,j}^s A_j^s + L_{v-,q-}^s \left(\frac{1}{T}\right)_- + L_{v-,q+}^s \left(\left(\frac{1}{T}\right)_+ - \frac{1}{T^s}\right)$$

$$- \frac{1}{T^s} \sum_{k=1}^{N-1} L_{v-,k-}^s (\mu_{k,-} - \mu_{N,-})$$

$$- \frac{1}{T^s} \sum_{k=1}^{N-1} L_{v-,k+}^s [(\mu_{k,+} - \mu_k^s) - (\mu_{N,+} - \mu_N^s)]$$

$$- L_{v-,v-}^s \frac{v_{n,-}}{T^s} - L_{v-,v+} \frac{v_{n,+} - v_n^s}{T^s} \tag{4.6.7}$$

and

$$[\Pi_{nn} + (v_n - v_n^s)\rho v_n + \rho(\mu_N - \mu_N^s - \tfrac{1}{2}|\mathbf{v}|^2 + \tfrac{1}{2}|\mathbf{v}^s|^2)]_- = -L_{v+,v}^s \frac{\text{div } \mathbf{v}_\parallel^s}{T^s}$$

$$+ \frac{1}{T^s} \sum_{j=1}^r R_{v+,j}^s A_j^s + L_{v+,q-}^s \left(\frac{1}{T}\right)_- + L_{v+,q+}^s \left(\left(\frac{1}{T}\right)_+ - \frac{1}{T^s}\right)$$

$$- \frac{1}{T^s} \sum_{k=1}^{N-1} L_{v+,k-}^s (\mu_{k,-} - \mu_{N,-})$$

$$- \frac{1}{T^s} \sum_{k=1}^{N-1} L_{v+,k+}^s [(\mu_{k,+} - \mu_k^s) - (\mu_{N,+} - \mu_N^s)]$$

$$- L_{v+,v-}^s \frac{v_{n,-}}{T^s} - L_{v+,v+}^s \frac{v_{n,+} - v_n^s}{T^s} \tag{4.6.8}$$

for the normal components of the viscous forces on the interface due to the bulk phases and vice versa.

The Onsager relations are

$$
\begin{array}{lll}
R^{s}_{v,j} = -R^{s}_{j,v}, & L^{s}_{v,q-} = -L^{s}_{q-,v}, & L^{s}_{v,q+} = -L^{s}_{q+,v} \\[4pt]
L^{s}_{v,k-} = -L^{s}_{k-,v}, & L^{s}_{v,k+} = -L^{s}_{k+,v}, & L^{s}_{v,v} = L^{s}_{v-,v} \\[4pt]
L^{s}_{v,v+} = L^{s}_{v+,v}, & R^{s}_{l,j} = R^{s}_{j,l}, & R^{s}_{l,q-} = R^{s}_{q-,l} \\[4pt]
R^{s}_{l,q+} = R^{s}_{q+,l}, & R^{s}_{l,k-} = R^{s}_{k-,l}, & R^{s}_{l,k+} = R^{s}_{k+,l} \\[4pt]
R^{s}_{l,v-} = -R^{s}_{v-,l}, & R^{s}_{l,v+} = -R^{s}_{v+,l}, & L^{s}_{q-,q+} = L^{s}_{q+,q-} \\[4pt]
L^{s}_{q-,k-} = L^{s}_{k-,q-}, & L^{s}_{q-,k+} = L^{s}_{k+,q-}, & L^{s}_{q-,v-} = -L^{s}_{v-,q-} \\[4pt]
L^{s}_{q-,v+} = -L^{s}_{v+,q-}, & L^{s}_{q+,k-} = L^{s}_{k-,q+}, & L^{s}_{q+,v-} = -L^{s}_{v-,q+} \\[4pt]
L^{s}_{q+,v+} = -L^{s}_{v+,q+}, & L^{s}_{k-,k+} = L^{s}_{k+,k-}, & L^{s}_{k+,v+} = -L^{s}_{v-,k-} \\[4pt]
L^{s}_{k-,v+} = -L^{s}_{v+,k-}, & L^{s}_{k+,v-} = -L^{s}_{v-,k+}, & L^{s}_{k+,v+} = -L^{s}_{v+,k+}, \\[4pt]
L^{s}_{v-,v+} = L^{s}_{v+,v-}
\end{array}
$$

$$(4.6.9)$$

The number of independent coefficients is $\frac{1}{2}(3+r+N)(4+r+N)$. The dependence on the affinities of the above fluxes is linear. This implies that all reactions have to be rather close to equilibrium. As we discussed in Section IV.A it is consistent with the general method to use a nonlinear dependence of the fluxes on the affinities. This may be done by choosing the constitutive coefficients to be dependent on the affinities. In that case, however, the Onsager relations are incorrect. Equivalent relations such as detailed balance for chemical rate equations may be formulated.

The scalar force–flux pairs in particular show a lot of cross effects. The excess chemical reactions are, for example, coupled to diffusion and heat flow through as well as into the interface. It is clear that these cross effects describe many interesting processes.[4,22]

The constitutive relations should be used in the balance equations to obtain explicit equations describing the temporal behavior of the excess densities. Equations (4.6.3)–(4.6.8) also give boundary conditions necessary to describe the temporal behavior of the bulk phases.

The coefficients $L^{s}_{q\pm,q\pm}$ are related to the temperature jump coefficient, and $\eta^{s}_{s} \equiv L^{s}_{v,v}/T^{s}$ is the interfacial analog of the bulk viscosity.

G. The Normal Components of the Velocity Field at the Dividing Surface

There is a general observation that can be made about the normal components of the velocities at the interface, $v^{+}_{n,s}$, v^{s}_{n}, and $v^{-}_{n,s}$. At equilibrium these velocities are clearly equal to zero. If the system is not in equilibrium these normal velocities will be finite and in general will no longer be equal to each other. Under most conditions the dissipative fluxes given by the linear laws in the preceding paragraphs are relatively small. This implies that the convective contributions in the reference frame of the dividing surface will usually also be small. Thus, for example. $[(v_n - v_n^s)T\rho s]_\pm$ is usually small. In view of the fact that the entropy density difference (the latent heat) is usually not small, one may conclude that in most cases the normal components of the velocities at the dividing surface are almost equal:[24]

$$v_{n,s}^- \simeq v_n^s \simeq v_{n,s}^+ \tag{4.7.1}$$

The importance of the latent heat in the theory of nucleation,[25] where it leads to growth instabilities and pattern formation,[26] is related to this.

There are, of course, also phenomena in which the velocities differ considerably. One example is a shock wave, in which the latent heat may lead to such a high temperature in the shock front that it starts fires.[27] Another example is explosive crystallization.[28]

Clearly not only the heat flux but also other fluxes of densities that differ sufficiently in the two phases lead to Eq. (4.7.1) in most cases

H. The Liquid–Vapor Interface

To show how one obtains differential equations for the temporal behavior of the excess densities and the location of the interface, we shall discuss this procedure for the special case of a liquid–vapor interface in a one-component system.[11] In the following sections, we shall also consider both the equilibrium and the nonequilibrium fluctuations of this system in more detail.

We shall further simplify this discussion by considering only small deviations from the equilibrium state and by fully linearizing the equations. As the force on the system, we use a gravitational force along the z axis,

$$\mathbf{F} = -\text{grad}\ \psi = -g(0, 0, 1) \Leftrightarrow \psi = gz \tag{4.8.1}$$

The equilibrium dividing surface is assumed to be planar, which is adequate for a large container, and is chosen to coincide with the xy plane. At equilibrium all velocities and other fluxes are zero. Equation (2.4.1) gives for the pressure in the bulk regions

$$\frac{\partial}{\partial z} p_{eq}^+(z) = -g\rho_{eq}^+(z) \quad \text{and} \quad \frac{\partial}{\partial z} p_{eq}^-(z) = -g\rho_{eq}^-(z) \tag{4.8.2}$$

while Eq. (2.4.6) gives as the jump in the pressure at the dividing surface

$$p_{eq,-} = -\rho^s_{eq} g \qquad (4.8.3)$$

The surface tension is clearly constant and equal to its equilibrium value [cf. Eq. (2.4.9)]. Because the thermodynamic forces in the entropy production both in the bulk regions [Eq. (3.2.7)], and at the interface [Eq. (3.2.10)] must all be zero at equilibrium, we have

$$T^+_{eq} = T^s_{eq} = T^-_{eq} \equiv T_0 \qquad (4.8.4)$$
$$\mu^+_{eq} = \mu_0 - gz, \qquad \mu^s_{eq} = \mu_0, \qquad \text{and} \qquad \mu^-_{eq} = \mu_0 - gz$$

where μ_0 is the chemical potential for $z = 0$.

Linearizing Eq. (2.2.10) around equilibrium, we find

$$\frac{\partial}{\partial t} \delta\rho^s = -\rho^s_{eq} \text{div } \mathbf{v}^s - [\rho_{eq}(v_n - v^s_n)]_- \qquad (4.8.5)$$

A deviation from the equilibrium value will be indicated by the prefactor δ. Similarly, we find for the interfacial velocity field, after linearizing Eq. (2.4.6) and using Eqs. (2.4.7) and (4.8.3),

$$\rho^s_{eq}\frac{\partial}{\partial t}\mathbf{v}^s = -\boldsymbol{\nabla}\delta\gamma - (C\gamma_{eq} + \delta p_- + g\delta\rho^s)\mathbf{n}_{eq} = g\rho^s_{eq}\delta\mathbf{n} - \mathbf{n}_{eq} \cdot \boldsymbol{\Pi}_-$$
$$(4.8.6)$$

Here we have used the equilibrium expressions for the curvature and the normal:

$$C_{eq} = 0 \qquad \text{and} \qquad \mathbf{n}_{eq} = (0, 0, 1) \qquad (4.8.7)$$

Note that to linear order, both $\boldsymbol{\nabla}\delta\gamma$ and $\delta\mathbf{n}$ are in the xy plane, so that we find for the normal component v^s_n, by contracting Eq. (4.8.6) with $\mathbf{n}_{eq}$,

$$\rho^s_{eq}\frac{\partial}{\partial t}v^s_n = -C\gamma_{eq} + \delta p_- + g\delta\rho^s - \Pi_{n,n,-} \qquad (4.8.8)$$

where we have used the facts that to linear order $v^s_n = v^s_z$ and $\Pi_{n,n} = \Pi_{z,z}$. We find for the parallel components

$$\rho^s_{eq}\frac{\partial}{\partial t}\mathbf{v}^s_\| = -\boldsymbol{\nabla}\delta\gamma + g\rho^s_{eq}\delta\mathbf{n} - \Pi_{n,\|,-} \qquad (4.8.9)$$

For the excess internal energy per unit surface area, we find on linearization of Eq. (2.5.5) and using Eq. (4.8.5)

$$\frac{\partial}{\partial t} \delta(\rho u)^s = -\text{div } \mathbf{J}_q^s + [\gamma_{eq} - (\rho u)_{eq}^s] \text{div } \mathbf{v}^s$$

$$- [(v_n - v_n^s)(\rho_{eq} u_{eq} + p_{eq})]_- - J_{q,n,-} \tag{4.8.10}$$

To obtain more convenient expressions, we shall now use the freedom in the choice of the dividing surface and use

$$\rho^s \equiv 0 \rightarrow \rho_{eq}^s = 0 \quad \text{and} \quad \delta\rho^s = 0 \tag{4.8.11}$$

One usually refers to this choice as that made by Gibbs. In fact, he also mentioned the other possible choices.[5] Substitution of this definition in the above differential equations gives

$$[\rho_{eq}(v_n - v_n^s)]_- = 0 \rightarrow v_n^s = \frac{\rho_{eq}^+ v_n^+ - \rho_{eq}^- v_n^-}{\rho_{eq}^+ - \rho_{eq}^-} \tag{4.8.12}$$

$$\delta p_- + \Pi_{n,n,-} = C\gamma_{eq} \quad \text{(linearized Laplace equation)} \tag{4.8.13}$$

$$\Pi_{n,\parallel,-} = -\boldsymbol{\nabla}\delta\gamma \tag{4.8.14}$$

The equation for the internal energy density remains the same. Notice that in view of the fact that we use Eq. (4.8.9), we have implicitly assumed that the excess mass flow is equal to zero, $(\rho\mathbf{v})^s = 0$.

If we define the interfacial specific heat by

$$c^s \equiv \frac{\delta(\rho u)^s}{\delta T^s} \tag{4.8.15}$$

and use the fact that $(\rho u)^s$ is now a function of the temperature alone, we find the following equation for the interfacial temperature:[11]

$$c_{eq}^s \frac{\partial}{\partial t} \delta T^s = -\text{div } \mathbf{J}_q^s - T_0(\rho s)_{eq}^s \text{ div } \mathbf{v}^s$$

$$- T_0 s_{eq,-} [\rho_{eq}(v_n - v_n^s)]_+ - J_{q,n-} \tag{4.8.16}$$

To obtain this expression, we also used the thermodynamic relation

$$T^s(\rho s)^s = (u\rho)^s - \gamma \tag{4.8.17}$$

which follows from Eq. (3.2.12) for this case. Furthermore, we used the fact that Eqs. (3.2.12), (4.8.12), and (4.8.4) give

$$
\begin{aligned}
[(v_n - v_n^s)(\rho_{eq}u_{eq} + p_{eq})]_- &= [(v_n - v_n^s)\rho_{eq}(T_0 s_{eq} + \mu_{eq})]_- \\
&= [(v_n - v_n^s)\rho_{eq}]_-(T_0 s_{eq,+} + \mu_{eq,+}) \\
&\quad + [(v_n - v_n^s)\rho_{eq}]_+(T_0 s_{eq,-} + \mu_{eq,-}) \\
&= [(v_n - v_n^s)\rho_{eq}]_+ T_0 s_{eq,-}
\end{aligned}
$$

Note that it follows from Eq. (4.8.12) that

$$
\begin{aligned}
[\rho_{eq}(v_n - v_n^s)]_+ &= \rho_{eq,s}^-(v_{n,s}^- - v_n^s) = \rho_{eq,s}^+(v_{n,s}^+ - v_n^s) \\
&= -\rho_{eq,s}^- \rho_{eq,s}^+ \frac{v_{n,s}^+ - v_{n,s}^-}{\rho_{eq,s}^+ - \rho_{eq,s}^-}
\end{aligned}
\tag{4.8.18}
$$

In view of the fact that the latent heat $T_0 s_{eq,-}$ is generally large, the rate of condensation or evaporation will be small and as a consequence the normal velocities will be practically equal at the dividing surface.[24] That this is usually the case we already discussed in a more general context in the previous section.

For a complete description of the behavior of the interface, we also need the linearized equation of motion for the normal on the dividing surface [cf. Eqs. (1.2.30) and (2.2.7)], for which one finds

$$
\frac{\partial}{\partial t}\delta\mathbf{n} = -(1 - \mathbf{n}_{eq}\mathbf{n}_{eq}) \cdot \nabla v_n^s = -\left(\frac{\partial}{\partial x}, \frac{\partial}{\partial y}, 0\right)v_n^s
\tag{4.8.19}
$$

The above equations, the analogous equations in the bulk regions, and the constitutive equations both in the bulk regions and on the interface give a complete description of the dynamical behavior of the liquid–vapor system. It is clear that even for this relatively simple, fully linearized example the general solution in terms of normal modes is extremely complicated. Further simplifications have to be made and in such simplified cases one may study phenomena like the reflection and transmission of sound by the interface or capillary waves along the interface.[23] The main source of the complexity of the general problem is that at the interface bulk excitations couple that are independent in the bulk regions. In this way bulk shear modes couple with, for example, sound modes at the interface. Rather than discussing these further simplifications, for which we refer the reader to the literature,[17,22,23] we shall proceed in the following sections to consider the general case of a one-component liquid–vapor interface and to discuss the equilibrium and the nonequilibrium fluctuations.

V EQUILIBRIUM FLUCTUATIONS OF A LIQUID–VAPOR INTERFACE

A. Introduction

The probability of thermal fluctuations around equilibrium in a closed system is given in terms of the total entropy S of the system by

$$P_{eq} \approx \exp\left(\frac{S}{k_B}\right) \tag{5.1.1}$$

where k_B is Boltzmann's constant. The total entropy is obtained by integrating the entropy density over the volume of the system.

$$S = \int d\mathbf{r}\rho s = \int d\mathbf{r}[\rho^- s^- \Theta^- + \rho^s s^s \delta^s + \rho^+ s^+ \Theta^+] \tag{5.1.2}$$

One may clearly write the total entropy as the sum of the bulk phase and the interfacial contributions

$$S = S^- + S^s + S^+ \tag{5.1.3}$$

which are defined for the bulk phases by

$$S^- = \int d\mathbf{r}\rho^- s^- \Theta^- = \int_{\xi_1<0} d\mathbf{r}\rho^- s^- \quad \text{and} \quad S^+ \equiv \int d\mathbf{r}\rho^+ s^+ \Theta^+$$
$$= \int_{\xi_1>0} d\mathbf{r}\rho^+ s^+ \tag{5.1.4}$$

and for the interface by

$$S^s \equiv \int d\mathbf{r}\rho^s s^s \delta^s = \int d\xi_1\, d\xi_2\, d\xi_3 h_1 h_2 h_3 \rho^s(\xi_2, \xi_3)s^s(\xi_2, \xi_3)h_1^{-1}\delta(\xi_1)$$
$$= \int d\xi_2\, d\xi_3\, h_{2,s}h_{3,s}\rho^s(\xi_2, \xi_3)s^s(\xi_2, \xi_3) \tag{5.1.5}$$

where we have used Eq. (1.2.7) together with Eq. (1.2.11) for δ^s, and the standard conversion of the integration to curvilinear coordinates.[15]

The fluctuations δS of the total entropy are due on the one hand to fluctuations of $(\rho s)^-$, $(\rho s)^s$, and $(\rho s)^+$ around their equilibrium values and on the other hand to fluctuations in the location of the interface around its equilibrium position. The fluctuations due to $(\rho s)^-$ and $(\rho s)^+$ that keep the interface located at its equilibrium position give contributions to δS that may be expressed in terms of $\delta \rho^\pm$, $\delta T^\pm$, and $\mathbf{v}^\pm$ in the same way as in a one-phase system; one finds (cf. reference 11) the usual formula*

$$\int d\mathbf{r}\delta(\rho s)^\pm \Theta_{eq}^\pm = -\frac{1}{2T_0}\int d\mathbf{r}\left[\frac{c_v^\pm}{T_0}(\delta T^\pm)^2 + (\rho_{eq}^\pm)^{-2}(\kappa_T^\pm)^{-1}(\delta\rho^\pm)^2 + \rho_{eq}^\pm|\mathbf{v}^\pm|^2\right]\Theta_{eq}^\pm$$

$$(5.1.6)$$

where $c_v^\pm$ are the equilibrium values of the specific heat at constant volume and $\kappa_T^\pm$ are the equilibrium values of the isothermal compressibility. The analysis of the contribution due to fluctuations of $(\rho s)^s$ that keep the interface located at its equilibrium position proceeds along similar lines to that for the bulk phases, and as is shown in reference 11, one finds a very similar result:*

$$\int d\mathbf{r}\delta(\rho s)^s \delta_{eq}^s = -\frac{1}{2T_0}\int d\mathbf{r}\left[\frac{c_v^s}{T_0}(\delta T^s)^2 + (\rho_{eq}^s)^{-2}(\kappa_T^s)^{-1}(\delta\rho^s)^2 + \rho_{eq}^s|\mathbf{v}^s|^2\right]\delta_{eq}^s$$

$$(5.1.7)$$

where

$$c_v^s \equiv \left(\frac{\partial(\rho u)^s}{\partial T^s}\right)_{\rho^s} \quad \text{and} \quad \kappa_T^s \equiv -\frac{1}{\rho^s}\left(\frac{\partial\rho^s}{\partial\gamma}\right)_{T^s} \qquad (5.1.8)$$

are the interfacial specific heat for constant surface area and the interfacial isothermal compressibility; the equilibrium values of c_v^s and $\kappa_T^\pm$ are used in Eq. (5.1.7).

In fact, one may easily obtain the above contributions (5.1.6) and (5.1.7) to δS in the more general case of a two-phase multicomponent fluid, because one simply finds the usual expressions found also in a one-phase multicomponent fluid.

---

* The equilibrium characteristic functions $\Theta_{eq}^\pm$, δ_{eq}^s are defined as the characteristic functions corresponding to a state of maximum entropy; they should not be confused with $\langle\Theta^\pm\rangle$ and $\langle\delta^s\rangle$, which will be found to be very different. While the same is true in principle for $\rho_{eq}^\pm$, ρ_{eq}^s, and so on, one may in practice neglect the differences from $\langle\rho^\pm\rangle$, $\langle\rho^s\rangle$, and so on.

If one uses Gibbs's definition of the interface, Eq. (5.1.7) reduces to

$$\int d\mathbf{r}\delta(\rho s)^s\delta_{eq}^s = -\frac{1}{2T_0}\int d\mathbf{r}\,\frac{c^s}{T_0}(\delta T^s)^2\delta_{eq}^s \qquad (5.1.9)$$

For this choice, ρ^s does not fluctuate, while $\mathbf{v}^s$ is not an independently fluctuating quantity. We shall discuss the importance of this below.

B. Fluctuations in the Location of the Dividing Surface

The inhomogeneity of the two-phase system leads to contributions to the entropy fluctuations due to fluctuations in the location of the interface. Such fluctuations do not occur in a homogeneous one-phase system and are characteristic of the existence of an interface. In calculating these contributions one must expand $\delta\Theta^{\pm}$ and $\delta(\delta^s)$ to second order in the fluctuations of the curvilinear coordinates around their equilibrium value $\delta\xi_i \equiv \xi_i - \xi_{i,eq}$. It is clear that such an analysis is most easily done using curvilinear coordinates; we refer the reader to reference 11 for the details. We will merely give the resulting expression. First we give some definitions:

$$d(\xi_2, \xi_3) \equiv -h_{1,eq}(0, \xi_{2,eq}, \xi_{3,eq})\delta\xi_1(0, \xi_{2,eq}, \xi_{3,eq}) \qquad (5.2.1)$$

The distance d is simply the distance between the location of the fluctuating dividing surface and the equilibrium dividing surface measured along the $\xi_{2,eq}$ and $\xi_{3,eq}$ constant lines. If the equilibrium interface is the xy plane as in Section IV.G, the distance d is simply the height of the fluctuating dividing surface above or below this plane along the z axis. We also need the length

$$R_c \equiv \left\{-h_1^{-1}\frac{\partial}{\partial\xi_1}h_1^{-1}\frac{\partial}{\partial\xi_1}\ln(h_2h_3)\right\}_{s,eq}^{-1/2} = (R_1^{-2}+R_2^{-2})_{eq}^{-1/2} \qquad (5.2.2)$$

In reference 11 it is shown that the fluctuation of δS due to fluctuations in the location of the interface is

$$\int d\mathbf{r}\{(\rho s)_{eq}^-\delta\Theta^- + (\rho s)_{eq}^s\delta(\delta^s) + (\rho s)_{eq}^+\delta\Theta^+\}$$

$$= -\frac{1}{2T_0}\int d\mathbf{r}\left\{\gamma_{eq}\left[|\delta\mathbf{n}|^2-\left(\frac{d}{R_c}\right)^2\right] - g[n_z\rho_-\right.$$

$$\left. - (\mathbf{n}\cdot\nabla a_{1,z}-Cn_z)_s\rho^s]_{eq}d^2\right\}\delta_{eq}^s \qquad (5.2.3)$$

In this expression the gravitational acceleration is again given by

$-g(0, 0, 1)$. The equilibrium dividing surface may furthermore be curved. In the special case that the equilibrium dividing surface is the xy plane, Eq. (5.2.3) reduces to

$$\int d\mathbf{r}\{(\rho s)_{eq}^{-}\delta\Theta^{-} + (\rho s)_{eq}^{s}\delta(\delta^{s}) - (\rho s)^{+}\delta\Theta^{+}\}$$

$$= -\frac{1}{2T_{0}}\int dx\,dy\,\{\gamma_{eq}|\delta\mathbf{n}|^{2} - g\rho_{eq,-}d^{2}\} \quad (5.2.4)$$

The term proportional to the surface tension is due to an increase of the interfacial energy resulting from an increase in surface area; $(d/R_{c})^{2}$ cancels $\delta\mathbf{n}^{2}$ integration for a parallel displacement of the dividing surface as a whole. The term proportional to g is due to the change of the gravitational energy due to a displacement of the interface. For the above case of a planar interface, the variation of the normal is related to $d(x, y)$ by

$$\mathbf{n}(x, y) = \left(-\frac{\partial d(x, y)}{\partial x}, -\frac{\partial d(x, y)}{\partial y}, 1\right) \bigg/ \left[1 + \left(\frac{\partial d(x, y)}{\partial x}\right)^{2} + \left(\frac{\partial d(x, y)}{\partial y}\right)^{2}\right]^{1/2}$$

$$(5.2.5)$$

To linear order this gives

$$\partial\mathbf{n}(x, y) = -\nabla d(x, y) \quad (5.2.6)$$

Substituting this in Eq. (5.2.4), one obtains for the variation of the entropy due to a fluctuation of the location of the interface to a height $d(x, y)$ above or below the xy plane,

$$\int d\mathbf{r}\{(\rho s)_{eq}^{-}\delta\Theta^{-} + (\rho s)_{eq}\delta(\delta^{s}) + (\rho s)_{eq}^{+}\delta\Theta^{+}\}$$

$$= -\frac{1}{2T_{0}}\int dx\,dy\{\gamma_{eq}|\nabla d|^{2} - g\rho_{eq,-}d^{2}\} \quad (5.2.7)$$

This contribution governs the fluctuations in the location of the interface and is used in the *capillary-wave model* to calculate the average equilibrium profile. As we discuss below, one may also calculate the density–density correlation function and the direct correlation function in the context of this model.

C. The Equilibrium Distribution

In Sections V.A and V.B we have shown that the probability distribution describing the fluctuations of the one-component two-phase system is given by

$$P_{eq} \approx \exp\left(\frac{\delta S}{k_B}\right) \qquad (5.3.1)$$

where

$$
\begin{aligned}
\delta S = -\frac{1}{2T_0} \int d\mathbf{r} \Bigg(&\left[\frac{c_v^-}{T_0}(\delta T^-)^2 + (\rho_{eq}^-)^{-2}(\kappa_T^-)^{-1}(\delta\rho_-)^2 + \rho_{eq}^-|\mathbf{v}^-|^2\right]\Theta_{eq}^- \\
&+ \left[\frac{c_v^+}{T_0}(\delta T^+)^2 + (\rho_{eq}^+)^{-2}(\kappa_T^+)^{-1}(\delta\rho^+)^2 + \rho_{eq}^+|\mathbf{v}^+|^2\right]\Theta_{eq}^+ \\
&+ \left\{\frac{c_v^s}{T_0}(\delta T^s)^2 + (\rho_{eq}^s)^{-2}(\kappa_T^s)^{-1}(\delta\rho^s)^2 + \rho_{eq}^s|\mathbf{v}^s|^2\right. \\
&\left. + \gamma_{eq}\left[|\delta\mathbf{n}|^2 - \left(\frac{d}{R_c}\right)^2\right] - g[n_z\rho_- - (\mathbf{n}\cdot\boldsymbol{\nabla}a_{1,z} - Cn_z)_s\,\rho^s]_{eq}d^2\right\}\delta_{eq}^s\Bigg)
\end{aligned}
$$

$$(5.3.2)$$

An interesting and useful consequence is that the equilibrium fluctuations of δT^-, $\delta\rho^-$, $\mathbf{v}^-$, δT^+, $\delta\rho^+$, $\mathbf{v}^+$, δT^s, $\delta\rho^s$, $\mathbf{v}^s$, and d or $\delta\mathbf{n}$ are all independent. In other words, the equilibrium cross correlations of these variables are all zero. For the self correlations of these variables, we find

$$\langle\delta T^-(\mathbf{r})\delta T^-(\mathbf{r}')\rangle = (k_B T_0^2/c_v^-)\delta(\mathbf{r}-\mathbf{r}')$$

$$\langle\delta\rho^-(\mathbf{r})\delta\rho^-(\mathbf{r}')\rangle = k_B T_0(\rho_{eq}^-)^2\kappa_T^-\delta(\mathbf{r}-\mathbf{r}')$$

$$\langle v_\alpha^-(\mathbf{r})v_\beta^-(\mathbf{r}')\rangle = (k_B T_0/\rho_{eq}^-)\delta_{\alpha,\beta}\delta(\mathbf{r}-\mathbf{r}')$$

$$\langle\delta T^+(\mathbf{r})\delta T^+(\mathbf{r}')\rangle = (k_B T_0^2/c_v^+)\delta(\mathbf{r}-\mathbf{r}')$$

$$\langle\delta\rho^+(\mathbf{r})\delta\rho^+(\mathbf{r}')\rangle = k_B T_0(\rho_{eq}^+)^2\kappa_T^+\delta(\mathbf{r}-\mathbf{r}')$$

$$\langle v_\alpha^+(\mathbf{r})v_\beta^+(\mathbf{r}')\rangle = (k_B T_0/\rho_{eq}^+)\delta_{\alpha,\beta}\delta(\mathbf{r}-\mathbf{r}')$$

$$
\begin{aligned}
&\langle\delta T^s(\xi_{2,eq}, \xi_{3,eq})\delta T^s(\xi_{2,eq}', \xi_{3,eq}')\rangle \\
&\quad = (k_B T_0^2/c_v^s)h_{2,eq}^{-1}\delta(\xi_{2,eq} - \xi_{2,eq}')h_{3,eq}^{-1}\delta(\xi_{3,eq} - \xi_{3,eq}') \\
&\langle\delta\rho^s(\xi_{2,eq}, \xi_{3,eq})\delta\rho^s(\xi_{2,eq}', \xi_{3,eq}')\rangle \\
&\quad = k_B T_0(\rho_{eq}^s)^2\kappa_T^s h_{2,eq}^{-1}\delta(\xi_{2,eq} - \xi_{2,eq}')h_{3,eq}^{-1}\delta(\xi_{3,eq} - \xi_{3,eq}') \\
&\langle v_\alpha^s(\xi_{2,eq}, \xi_{3,eq})v_\beta^s(\xi_{2,eq}', \xi_{3,eq}')\rangle \\
&\quad = (k_B T_0/\rho_{eq}^s)\delta_{\alpha\beta}h_{2,eq}^{-1}\delta(\xi_{2,eq} - \xi_{2,eq}')h_{3,eq}^{-1}\delta(\xi_{3,eq} - \xi_{3,eq}')
\end{aligned}
$$

$$(5.3.3)$$

where α, $\beta = x$, y, or z. The equilibrium correlation function for d or $\delta\mathbf{n}$ is more complicated and will be discussed in the next section. If one uses the Gibbs definition of the dividing surface, $\delta\rho^s = \rho^s_{eq} = 0$. The corresponding equilibrium autocorrelation function in Eq. (5.3.3) may be left out. The same may be done with the equilibrium velocity autocorrelation function if the excess mass transport is negligible $[(\rho\mathbf{v})^s = 0]$, which seems a reasonable assumption for the liquid–vapor interface. In general, one may in this way eliminate excess densities and the corresponding autocorrelation functions if the excess densities are sufficiently small.

It should be realized that the above correlation functions are in a way only a first step if one is calculating the correlation functions for the full density $\rho(\mathbf{r})$, the internal energy density $(\rho u)(\mathbf{r})$, and the momentum density $(\rho\mathbf{v})(\mathbf{r})$. We shall illustrate this for the fluctuations of the density around its average value,

$$
\begin{aligned}
\delta\rho &= \delta(\rho^-\Theta^- + \rho^s\delta^s + \rho^+\Theta^+) \\
&= \rho^-\Theta^- - \langle\rho^-\Theta^-\rangle + \rho^s\delta^s - \langle\rho^s\delta^s\rangle + \rho^+\Theta^+ - \langle\rho^+\Theta^+\rangle \\
&= \rho^-\Theta^- - \langle\rho^-\rangle\langle\Theta^-\rangle + \rho^s\delta^s - \langle\rho^s\rangle\langle\delta^s\rangle + \rho^+\Theta^+ - \langle\rho^+\rangle\langle\Theta^+\rangle \\
&= (\rho^- - \langle\rho^-\rangle)\Theta^- + \langle\rho^-\rangle(\Theta^- - \langle\Theta^-\rangle) + (\rho^s - \langle\rho^s\rangle)\delta^s + \langle\rho^s\rangle(\delta^s - \langle\delta^s\rangle) \\
&\quad + (\rho^+ - \langle\rho^+\rangle)\Theta^+ + \langle\rho^+\rangle(\theta^+ - \langle\Theta^+\rangle)
\end{aligned} \tag{5.3.4}
$$

Now we use the facts that $\langle\rho^\pm\rangle \simeq \rho_{eq}$ and $\langle\rho^s\rangle \simeq \rho^s_{eq}$, so that $\rho^\pm - \langle\rho^\pm\rangle = \delta\rho^\pm$ and $\rho^s - \langle\rho^s\rangle = \delta\rho^s$, the fluctuations of the densities around the densities corresponding to the maximum entropy that we used in the construction of δS (cf. also the footnote in Section V.A). Using $\Theta^- + \Theta^+ = 1$, we may further write

$$
\delta\Theta^- \equiv \Theta^- - \langle\Theta^-\rangle = -(\Theta^+ - \langle\Theta^+\rangle) \equiv -\delta\Theta^+ \tag{5.3.5}
$$

Note that the definition of $\delta\theta^\pm$ and of $\delta(\delta^s) \equiv \delta^s - \langle\delta^s\rangle$ used in this section differs from the one used in section V.B.

The fluctuation of the density around its average value may thus finally be written as

$$
\delta\rho = \delta\rho^-\Theta^- + \delta\rho^s\delta^s + \delta\rho^+\Theta^+ + (\rho^-_{eq} - \rho^+_{eq})\delta\Theta^- + \rho^s_{eq}\delta(\delta^s) \tag{5.3.6}
$$

The density–density correlation function becomes

$$
\begin{aligned}
H(\mathbf{r}, \mathbf{r}') &\equiv \langle\delta\rho(\mathbf{r})\delta\rho(\mathbf{r}')\rangle = \langle\delta\rho^-(\mathbf{r})\delta\rho^-(\mathbf{r}')\rangle\langle\Theta^-(\mathbf{r})\Theta^-(\mathbf{r}')\rangle \\
&\quad + \langle\delta\rho^s(\mathbf{r})\delta\rho^s(\mathbf{r}')\rangle\langle\delta^s(\mathbf{r})\delta^s(\mathbf{r}')\rangle \\
&\quad + (\delta\rho^+(\mathbf{r})\delta\rho^+(\mathbf{r}'))\Theta^+(\mathbf{r})\Theta^+(\mathbf{r}'))
\end{aligned}
$$

$$+ \langle [(\rho_{eq}^-(\mathbf{r}) - \rho_{eq}^+(\mathbf{r}))\delta\Theta^-(\mathbf{r}) + \rho_{eq}^s(\mathbf{r})\delta(\delta^s(\mathbf{r}))]$$
$$\times [(\rho_{eq}^-(\mathbf{r}') - \rho_{eq}^+(\mathbf{r}'))\delta\Theta^-(\mathbf{r}') + \rho_{eq}^s(\mathbf{r}')\delta(\delta^s(\mathbf{r}'))] \rangle$$
$$= k_B T_0 [(\rho_{eq}^-(\mathbf{r}))^2 \langle \Theta^-(\mathbf{r}) \rangle \kappa_T^- + (\rho_{eq}^s(\mathbf{r}))^2 \langle \delta^s(\mathbf{r}) \rangle \kappa_T^s$$
$$+ (\rho_{eq}^+(\mathbf{r}))^2 \langle \Theta^+(\mathbf{r}) \rangle \kappa_T^+]\delta(\mathbf{r} - \mathbf{r}')$$
$$+ \langle [(\rho_{eq}^-(\mathbf{r}) - \rho_{eq}^+(\mathbf{r}))\delta\Theta^-(\mathbf{r}) + \rho_{eq}^s(\mathbf{r})\delta(\delta^s(\mathbf{r}))][(\rho_{eq}^-(\mathbf{r}')$$
$$- \rho_{eq}^+(\mathbf{r}'))\delta\Theta^-(\mathbf{r}') + \rho_{eq}^s(\mathbf{r}')\delta(\delta^s(\mathbf{r}'))] \rangle$$

$$(5.3.7)$$

Note that $\kappa_T^\pm$ and κ_T^s may in fact also depend on $\mathbf{r}$, which has not been explicitly indicated. It is clear that the density–density correlation function can be calculated explicitly also close to the interface by calculating the averages and the correlation functions of the characteristic functions. This will be done in the following sections.

If one uses Gibbs's definition of the dividing surface, as we shall do in the rest of this section, the expression for the density–density correlation function reduces to

$$H(\mathbf{r}, \mathbf{r}') = \langle \delta\rho(\mathbf{r})\delta\rho(\mathbf{r}') \rangle = k_B T_0 [(\rho_{eq}^-(\mathbf{r}))^2 \langle \theta^-(\mathbf{r}) \rangle \kappa_T^-$$
$$+ (\rho_{eq}^+(\mathbf{r}))^2 \langle \theta^+(\mathbf{r}) \rangle \kappa_T^+]\delta(\mathbf{r} - \mathbf{r}')$$
$$+ (\rho_{eq}^-(\mathbf{r}) - \rho_{eq}^+(\mathbf{r}))(\rho_{eq}^-(\mathbf{r}') - \rho_{eq}^+(\mathbf{r}'))\langle \delta\Theta^-(\mathbf{r})\delta\Theta^-(\mathbf{r}') \rangle$$

$$(5.3.8)$$

Other correlation functions may be written out in a similar way.

Usually one replaces $\rho_{eq}^-(\mathbf{r})$ by a constant liquid density ρ_l and $\rho_{eq}^+(\mathbf{r})$ by a constant gas density ρ_g, neglecting the rather small gradients due to the gravitational field in the bulk phases. This gives

$$H(\mathbf{r}, \mathbf{r}') = k_B T_0 [\rho_l \kappa_{T,l} \langle \Theta^-(\mathbf{r}) \rangle + \rho_g \kappa_{T,g} \langle \Theta^+(\mathbf{r}) \rangle]\delta(\mathbf{r} - \mathbf{r}')$$
$$+ (\rho_l - \rho_g)^2 \langle \delta\Theta^-(\mathbf{r})\delta\Theta^-(\mathbf{r}') \rangle$$
$$= k_B T_0 [\rho_l \kappa_{T,l} \langle \Theta^-(\mathbf{r}) \rangle + \rho_g \kappa_{T,g} \langle \Theta^+(\mathbf{r}) \rangle]\delta(\mathbf{r} - \mathbf{r}')$$
$$+ H_{cap}(\mathbf{r}, \mathbf{r}')$$

$$(5.3.9)$$

which is the equation we shall henceforth use for most purposes. H_{cap} is the contribution due to capillary waves.

D. The Height–Height Correlation Function

As a first step toward the calculation of the full density–density correlation function, we calculate the height–height correlation function. We shall restrict ourselves to the case in which the equilibrium dividing surface is the xy plane. Using Eq. (5.3.1) together with Eq. (5.3.2), we have as the probability distribution for a planar dividing surface [cf. also Eq. (5.2.7)] for this case

$$P_{eq}(\{d(x, y)\}) \approx \exp\left\{-\frac{1}{2k_B T_0}\int dx\,dy\,[\gamma_{eq}|\nabla d(x, y)|^2 - g\rho_{eq,-}d^2(x, y)]\right\} \tag{5.4.1}$$

This is the usual probability distribution for the height of the fluctuating interface.

As the interface, we use a square with sides of length L. If we represent the vertical displacement of the distorted surface by the Fourier series

$$d(x, y) = L^{-2}\sum_{k_x,k_y}^{k_{max}}\tilde{d}(k_x, k_y)\,e^{i(k_x x + k_y y)} \tag{5.4.2}$$

then we find as the probability distribution, where $k_{\parallel}^2 \equiv k_x^2 + k_y^2$,

$$P_{eq}(\{\tilde{d}(k_x, k_y)\}) \approx \exp\left\{-\frac{\gamma_{eq}}{2k_B T_0}L^{-2}\sum_{k_x,k_y}^{k_{max}}(k_{\parallel}^2 + L_c^{-2})\tilde{d}(k_x, k_y)\tilde{d}(-k_x, -k_y)\right\} \tag{5.4.3}$$

Here $k_{max} = 2\pi/\xi_B$, where ξ_B is the bulk correlation length, which is consistent with the general restriction of our method to long-wavelength distortions. In Eq. (5.4.3) we have introduced the *capillary length*

$$L_c \equiv \left(\frac{-\gamma_{eq}}{g\rho_{eq,-}}\right)^{1/2} \tag{5.4.4}$$

Note that because of our choice of the coordinate frame, $\rho_{eq,-}$ is negative. As we shall see, L_c is a fundamental length controlling the range of the correlations along the interface. This range approaches infinite in the zero-gravity limit. In our analysis we shall always consider systems large compared with this range, that is, where $L \gg L_c$.

Because of the quadratic nature of Eq. (5.4.3), we have, from Gaussian fluctuation theory,

$$\tilde{S}(k) \equiv L^{-2}\langle\tilde{d}(k_x, k_y)\tilde{d}(-k_x, -k_y)\rangle = \frac{k_B T_0}{\gamma_{eq}}(k_{\parallel}^2 + L_c^{-2})^{-1} \tag{5.4.5}$$

The height–height correlation function as a function of the distance is found by Fourier transformation:

$$S(x - x', y - y') = \langle d(x, y)d(x', y') \rangle$$

$$= \frac{k_B T_0}{\gamma_{eq} L^2} \sum_{k_x, k_y}^{k_{max}} \left[\frac{\exp\{i[k_x(x - x') + k_y(y - y')]\}}{k_{\parallel}^2 + L_c^{-2}} \right] \quad (5.4.6)$$

For a large enough value of L, one may replace the sum by an integral. One then finds,[11,29] if $|\mathbf{r}_{\parallel} - \mathbf{r}'_{\parallel}| \equiv ((x - x')^2 + (y - y')^2)^{1/2}$ is sufficiently large compared with the bulk correlation length,

$$S(|\mathbf{r}_{\parallel} - \mathbf{r}'_{\parallel}|) = \frac{k_B T_0}{\gamma_{eq}(2\pi)^2} \int dk_x \, dk_y \, \frac{\exp\{i[k_x(x - x') + k_y(y - y')]\}}{k_{\parallel}^2 + L_c^{-2}}$$

$$= \frac{k_B T_0}{2\pi\gamma_{eq}} K_0 \left(\frac{|\mathbf{r}_{\parallel} - \mathbf{r}'_{\parallel}|}{L_c} \right) \quad (5.4.7)$$

where K_0 is a Bessel function of the second kind. For distances small compared with the capillary length, one has in good approximation

$$S(r_{\parallel}) = \frac{k_B T_0}{2\gamma_{eq}} \ln \left(\frac{2L_c}{r_{\parallel}} \right) \quad \text{for} \quad r_{\parallel} \ll L_c \quad (5.4.8)$$

whereas for large distances,

$$S(r_{\parallel}) = \frac{k_B T_0}{2\gamma_{eq}} \left(\frac{L_c}{2\pi r_{\parallel}} \right)^{1/2} \exp \left(\frac{-r_{\parallel}}{L_c} \right) \quad \text{for} \quad r_{\parallel} \gg L_c \quad (5.4.9)$$

For $r_{\parallel}$ of the same order as ξ_B or smaller, the integration should not be extended beyond k_{max}. Equations (5.4.7)–(5.4.9) thus show that long-ranged correlations exist along the interface. The correlation length is equal to the capillary length, which is much larger than the bulk correlation length. In the zero-gravity limit the capillary length approaches infinity. In that case $S(r_{\parallel})$ diverges, as Eq. (5.4.8) shows, and the interface is called *rough*.

E. The Average Density Profile

Using the probability distribution given in Eq. (5.3.1) together with Eq. (5.3.2), the average density becomes

$$\langle \rho \rangle = \langle \rho^- \Theta^- + \rho^+ \Theta^+ \rangle = \langle \rho^- \rangle \langle \Theta^- \rangle + \langle \rho^+ \rangle \langle \Theta^+ \rangle \quad (5.5.1)$$

where we use the Gibbs dividing surface. We first consider the average of the characteristic functions,

$$\langle \Theta^- \rangle = 1 - \langle \Theta^+ \rangle = \langle \Theta(d(x, y) - z) \rangle \qquad (5.5.2)$$

The derivative of this average with respect to z gives the singlet height distribution

$$P(z) = \langle \delta(z - d(x, y)) \rangle = -\frac{\partial}{\partial z} \langle \Theta^- \rangle = \frac{\partial}{\partial z} \langle \Theta^+ \rangle \qquad (5.5.3)$$

This probability distribution may easily be calculated using $P_{eq}(\{d(x, y)\})$ as given by Eq. (5.4.1) (see, e.g., reference 30); the result is

$$P(z) = (W\sqrt{2\pi})^{-1} \exp\left[-\tfrac{1}{2}\left(\frac{z}{W}\right)^2 \right] \qquad (5.5.4)$$

where

$$W = \sqrt{S(\xi_B)} = \left[\frac{k_B T_0}{2\gamma_{eq}} \ln\left(\frac{2L_c}{\xi_B}\right) \right]^{1/2} \qquad (5.5.5)$$

The reason that the value of S in $r_\parallel = \xi_B$ should be chosen, rather than that in $r_\parallel = 0$ for systems with a dimensionality lower than 4, is that for these small distances the upper limit of the summation over k_x, k_y becomes important [cf. Eq. (5.4.6)]. To analyze such details, one should use columns of a finite width.[30]

The average characteristic functions may now be found by integration of $P(z)$. One obtains

$$\langle \Theta^- \rangle = 1 - \langle \Theta^+ \rangle = \tfrac{1}{2}\left[1 - \mathrm{erf}\left(\frac{z}{W\sqrt{2}}\right) \right] \qquad (5.5.6)$$

Substituting this result in Eq. (5.5.1) and using $\langle \rho^\pm \rangle = \rho^\pm_{eq}$ (cf. footnote in Section V.A), one finds for the average density profile

$$\langle \rho(\mathbf{r}) \rangle = \tfrac{1}{2}(\rho^-_{eq}(\mathbf{r}) + \rho^+_{eq}(\mathbf{r})) - \tfrac{1}{2}(\rho^-_{eq}(\mathbf{r}) - \rho^+_{eq}(\mathbf{r}))\, \mathrm{erf}\left(\frac{z}{W\sqrt{2}}\right) \qquad (5.5.7)$$

Usually one replaces $\rho^-_{eq}(\mathbf{r})$ by a constant liquid density ρ_l and $\rho^+_{eq}(\mathbf{r})$ by a constant gas density ρ_g, neglecting the rather small gradients in the bulk phases due to the gravitational field. This gives as the average profile

$$\rho_0(z) = \tfrac{1}{2}(\rho_1 + \rho_g) - \tfrac{1}{2}(\rho_1 - \rho_g)\, \text{erf}\left(\frac{z}{W\sqrt{2}}\right) \tag{5.5.8}$$

which is the usual error-function density profile found in the context of capillary-wave theory.[12] As long as z is smaller than or roughly equal to W, one may use $\rho_0(z)$ instead of $\langle\rho(z)\rangle$; for larger values of z, one should use $\langle\rho(z)\rangle$.

An important consequence of the above equations is that the profile width W diverges in the zero-gravity limit proportionally to $[\ln g]^{1/2}$. This divergence, which is not found in the van der Waals theory, has led to considerable discussion. We refer the reader to references 29–37 for more details.

F. The Density–Density Correlation Function

To calculate the density–density correlation function $H(\mathbf{r}, \mathbf{r}')$ given in Eq. (5.3.8) or (5.3.9) explicitly, we must evaluate $H_{\text{cap}}(\mathbf{r}, \mathbf{r}')$, which is proportional to

$$\langle \delta\Theta^-(\mathbf{r})\delta\Theta^-(\mathbf{r}')\rangle = \langle\Theta(d(x, y) - z)\Theta(d(x', y') - z')\rangle - \langle\Theta^-(\mathbf{r})\rangle\langle\Theta^-(\mathbf{r}')\rangle \tag{5.6.1}$$

It is convenient first to calculate the two-point height distribution:

$$P(z_1, z_2, |\mathbf{r}_\| - \mathbf{r}'_\||) \equiv \langle\delta(d(x, y) - z_1)\delta(d(x', y') - z_2)\rangle$$

$$= \frac{\partial^2}{\partial z_1\, \partial z_2}\langle\Theta(d(x, y) - z_1)\Theta(d(x', y') - z_2)\rangle \tag{5.6.2}$$

Using the general results of Wang and Uhlenbeck[38] for a bivariate Gaussian distribution or following the explicit calculation in reference 30, one obtains

$$P(z_1, z_2, r_\|) = \frac{1}{(2\pi)^2} \int_{-\infty}^{\infty} dt_1 \int_{-\infty}^{\infty} dt_2 \exp[i(t_1 z_1 + t_2 z_2)$$

$$- \tfrac{1}{2}W^2(t_1^2 + t_2^2) - S(r_\|)t_1 t_2]$$

$$= \frac{1}{2\pi}[W^4 - S^2(r_\|)]^{-1/2}\exp\left\{\frac{-W^2(z_1^2 + z_2^2) + 2S(r_\|)z_1 z_2}{2[W^4 - S^2(r_\|)]}\right\} \tag{5.6.3}$$

For further analysis, an alternative representation is very important:

$$P(z_1, z_2, r_\parallel) = \exp\left[S(r_\parallel)\frac{\partial^2}{\partial z_1 \partial z_2}\right]\frac{1}{(2\pi)^2}\int_{-\infty}^{\infty}dt_1\int_{-\infty}^{\infty}dt_2$$
$$\times \exp[i(t_1z_1 + t_2z_2) - \tfrac{1}{2}W^2(t_1^2 + t_2^2)]$$
$$= \exp\left[S(r_\parallel)\frac{\partial^2}{\partial z_1 \partial z_2}\right]P(z_1)P(z_2) \qquad (5.6.4)$$

where the exponential operator is defined by its Taylor series expansion.
Using Eqs. (5.6.1), (5.6.2), (5.6.4), and (5.5.3), we find

$$\frac{\partial^2}{\partial z_1 \partial z_2}\langle\delta\Theta^-(\mathbf{r}_1)\delta\Theta^-(\mathbf{r}_2)\rangle = \left\{\exp\left[S(r_\parallel)\frac{\partial^2}{\partial z_1 \partial z_2}\right] - 1\right\}P(z_1)P(z_2)$$
$$(5.6.5)$$

Using further the fact that [cf. Eqs. (5.5.8) and (5.5.4)] $P(z)$ is given in terms of the average profile by

$$P(z) = -(\rho_l - \rho_g)^{-1}\frac{d}{dz}\rho_0(z) \qquad (5.6.6)$$

we may integrate Eq. (5.6.5) and obtain

$$\langle\delta\Theta^-(\mathbf{r})\delta\Theta^-(\mathbf{r}')\rangle = \left\{\exp\left[S(r_\parallel)\frac{\partial^2}{\partial z_1 \partial z_2}\right] - 1\right\}\rho_0(z_1)\rho_0(z_2)(\rho_l - \rho_g)^{-2}$$
$$(5.6.7)$$

for the autocorrelation function of the characteristic function. We may also write the average characteristic functions in terms of the average density profile [cf. Eqs. (5.5.6) and (5.5.8)]:

$$\langle\Theta^-\rangle = \frac{\rho_0(z) - \rho_g}{\rho_l - \rho_g} \quad \text{and} \quad \langle\Theta^+\rangle = \frac{\rho_l - \rho_0(z)}{\rho_l - \rho_g} \qquad (5.6.8)$$

Substitution of Eqs. (5.6.7) and (5.6.8) into Eq. (5.3.9) finally gives the following expression for the density–density correlation function:

$$H(z_1, z_2, \mathbf{r}_\parallel) = k_BT_0\left[\rho_l\kappa_{T,l}\frac{\rho_0(z_1) - \rho_g}{\rho_l - \rho_g}\right.$$
$$\left. + \rho_g\kappa_{T,g}\frac{\rho_l - \rho_0(z_1)}{\rho_l - \rho_g}\right]\delta(\mathbf{r}_\parallel)\delta(z_1 - z_2)$$
$$+ \left\{\exp\left[S(r_\parallel)\frac{\partial^2}{\partial z_1 \partial z_2}\right] - 1\right\}\rho_0(z_1)\rho_0(z_2) \quad (5.6.9)$$

The first term on the right-hand side gives the usual contribution in the liquid if $(-z_1) \gg W$ and gives the usual contribution in the vapor if $z_1 \gg W$. If z_1 is of the same order as W or smaller, these bulk contributions disappear smoothly with a weight function that depends on the average density profile $\rho_0(z_1)$. In this region the behavior of the density–density correlation function is dominated by the second contribution on the right-hand side of Eq. (5.6.9). This contribution is due to the long-wavelength capillary-wave-like fluctuations in the location of the interface and is therefore called H_{cap}.

In the capillary-wave theory one does not consider density fluctuations in the bulk regions. This is equivalent to taking compressibility in both the liquid and the vapor to be equal to zero.[29,30] The density–density correlation function then reduces to

$$H_{cap}(z_1, z_2, r_\parallel) = \left\{ \exp\left[S(r_\parallel) \frac{\partial^2}{\partial z_1 \partial z_2} \right] - 1 \right\} \rho_0(z_1)\rho_0(z_2) \qquad (5.6.10)$$

For considering the novel behavior found only in two-phase systems close to the interface this expression is very useful, and we shall restrict ourselves to this case in Sections V.G. and V.I. The same expression is found for interfaces in systems with a different dimensionality; only $S(r_\parallel)$ and W are different in such systems.

Using the explicit expressions for $S(r_\parallel)$ and $\rho_0(z)$ it follows that H_{cap} may be written as a function of z_1/W, z_2/W and $r_\parallel/L_c$. In this way one may scale the contribution to H due to fluctuations in the location of the interface. A general hypothesis about the scaling behavior of H for distances large compared with ξ_B has been formulated by Weeks[36] for systems of arbitrary dimensionality. For the resulting scaling behavior of $\rho_0(z)$ and the direct correlation function, we also refer the reader to reference 37.

G. Special Representation of the Density–Density Correlation Function in the Capillary-Wave Model

It is possible to write the density–density correlation function H_{cap} given in Eq. (5.6.10) in a spectral form. This will enable us to obtain explicit expressions for the direct correlation functions in the neighborhood of the interface. Consider for this purpose, [cf. Eqs. (5.5.4) and (5.6.6)],

$$(\rho_1 - \rho_g)^{-1} \frac{d^{n+1}}{dz^{n+1}} \rho_0(z) = -\frac{d^n}{dz^n} P(z) = -(W\sqrt{2\pi})^{-1} \frac{d^n}{dz^n} \exp\left[-\tfrac{1}{2}\left(\frac{z}{W} \right)^2 \right]$$

$$= \pi^{-1/2}(W\sqrt{2})^{-(n+1)} \frac{d^n}{d\zeta^n} \exp(-\zeta^2)$$

$$= \pi^{-1/2}(-W\sqrt{2})^{-(n+1)} H_n(\zeta) \exp(-\zeta^2) \qquad (5.7.1)$$

where we have introduced a scaled variable

$$\zeta \equiv \frac{z}{W\sqrt{2}} \qquad (5.7.2)$$

and where H_n is the nth-order Hermite polynomial. Using the eigenfunctions of the quantum-mechanical harmonic oscillators,

$$\psi_n(z) = H_n(\zeta) \exp(-\tfrac{1}{2}\zeta^2)(W2^n n! \sqrt{2\pi})^{-1/2} \qquad (5.7.3)$$

we may write the derivatives of the average density profile in the form

$$\frac{d^{n+1}}{dz^{n+1}} \rho_0(z) = -(\rho_1 - \rho_g)^{1/2}(-W)^{-n}(n!)^{1/2}\psi_n(z)[-\rho_0'(z)]^{1/2} \qquad (5.7.4)$$

where

$$\rho_0'(z) \equiv \frac{d}{dz}\rho_0(z) = -(\rho_1 - \rho_g)(W\sqrt{2\pi})^{-1}\exp(-\zeta^2) = -(\rho_1 - \rho_g)\psi_0^2(z) \qquad (5.7.5)$$

Expanding the exponential operator in Eq. (5.6.10), we obtain as the density–density correlation function

$$H_{\text{cap}}(z_1, z_2, r_{\parallel}) = (\rho_1 - \rho_g)W^2[\rho_0'(z_1)\rho_0'(z_2)]^{1/2} \sum_{n=0}^{\infty} (n+1)^{-1}$$

$$\times \frac{S(r_{\parallel})^{n+1}}{W^2} \psi_n(z_1)\psi_n(z_2)$$

$$= (\rho_1 - \rho_g)W^2[\rho_0'(z_1)\rho_0'(z_2)]^{1/2} \sum_{n=0}^{\infty} (n+1)^{-1}$$

$$\times \left[\frac{K_0(r_{\parallel}/L_c)}{\pi \ln(2L_c/\xi_B)}\right]^{n+1} \psi_n(z_1)\psi_n(z_2) \qquad (5.7.6)$$

where we have used Eqs. (5.4.7) and (5.5.5). It is clear that the long-wavelength capillary-wave-like fluctuations in the position of the interface lead to long-ranged density correlations along the interface. The range L_c diverges in the zero-gravity limit. Such behavior has in fact been verified on the basis of correlation-function identities that follow directly from the microscopic description by Wertheim.[39] The lowest-order ($n = 0$) eigenfunction contribution decays slower than the higher-order ones, as can be verified easily using the $r_\parallel \gg L_c$ behavior of K_0 [cf. Eq. (5.4.9)]. This supports an important assumption with respect to this point made by Wertheim.[39] It shows also, however, that the higher-order eigenfunctions also give contributions that are long-ranged, a fact that one would be tempted to neglect.

Fourier transforming the density–density correlation function with respect to $\mathbf{r}_\parallel$ gives

$$\tilde{H}_{cap}(z_1, z_2, k_\parallel) \equiv \int dx\, dy\, \exp[-i(k_x x + k_y y)] H_{cap}(z_1, z_2, r_\parallel)$$

$$= (\rho_1 - \rho_g) W^2 [\rho_0'(z_1)\rho_0'(z_2)]^{1/2}$$

$$\times \sum_{n=0}^{\infty} (n+1)^{-1} \tilde{H}_n(k_\parallel)\psi_n(z_1)\psi_n(z_2) \qquad (5.7.7)$$

where

$$\tilde{H}_n(k_\parallel) = \int dx\, dy\, \exp[-i(k_x x + k_y y)]\left(\frac{S(r_\parallel)}{W^2}\right)^{n+1} \qquad (5.7.8)$$

For $n = 0$ this gives (cf. Section V.D)

$$\tilde{H}_0(k_\parallel) = \frac{\tilde{S}(k_\parallel)}{W^2} = \frac{k_B T_0}{\gamma_{eq} W^2} (k_\parallel^2 + L_c^{-2})^{-1} \qquad (5.7.9)$$

For larger values of n it is not possible to give an analytic expression.

Finally, we note that the harmonic-oscillator eigenfunctions form a complete orthonormal set:

$$\int \psi_n(z)\psi_m(z)\, dz = \delta_{nm} \quad \text{and} \quad \sum_{n=0}^{\infty} \psi_n(z_1)\psi_n(z_2) = \delta(z_1 - z_2)$$

$$(5.7.10)$$

These properties will be useful in the following sections.

H. A General Identity for the Density–Density Correlation Function

It is interesting to verify that the description in the general context of non-equilibrium thermodynamics, and in particular the elimination of variations and fluctuations of the variables with wavelengths smaller than or equal to the bulk correlation length, does not affect the validity of the following identity for a planar interface:

$$\int dz_2 \int d\mathbf{r}_\parallel H(z_1, z_2, \mathbf{r}_\parallel) = -\frac{k_B T_0}{g} \frac{d}{dz_1} \langle \rho(z_1) \rangle \qquad (5.8.1)$$

This identity was first derived by Wertheim[39] on a microscopic basis. Using the divergences of the right-hand side in the zero-gravity limit, it follows that long-ranged correlations exist along the interface with a correlation length that diverges in the zero-gravity limit.

The derivative of the average density profile is equal to

$$\langle \rho \rangle = \left(\frac{d}{dz} \rho_{eq}^- \right) \langle \Theta^- \rangle + \left(\frac{d}{dz} \rho_{eq}^+ \right) \langle \Theta^+ \rangle + \rho_{eq}^- \frac{d}{dz} \langle \Theta^- \rangle + \rho_{eq}^+ \frac{d}{dz} \langle \Theta^+ \rangle$$

$$(5.8.2)$$

Using Eq. (5.3.8) for the correlation function in Eq. (5.8.1), we have

$$\int H(\mathbf{r}_1, \mathbf{r}_2) \, d\mathbf{r}_2 = k_B T_0 [(\rho_{eq}^-(z_1))^2 \kappa_T^- \langle \Theta^-(\mathbf{r}) \rangle + (\rho_{eq}^+(z_1))^2 \kappa_T^+ \langle \Theta^+(\mathbf{r}) \rangle]$$

$$+ \int H_{cap}(\mathbf{r}_1, \mathbf{r}_2) \, d\mathbf{r}_2 \qquad (5.8.3)$$

The equilibrium profile in the bulk regions is given by

$$(\rho_{eq}^\pm(z))^2 \kappa_T^\pm = -\frac{1}{g} \frac{d}{dz} \rho_{eq}^\pm(z) \qquad (5.8.4)$$

Substitution in Eq. (5.8.3) gives

$$\int H(\mathbf{r}_1, \mathbf{r}_2) \, d\mathbf{r}_2 = -\frac{k_B T_0}{g} \left[\left(\frac{d}{dz_1} \rho_{eq}^-(z_1) \right) \langle \Theta^- \rangle + \left(\frac{d}{dz_1} \rho_{eq}^+(z_1) \right) \langle \Theta^+ \rangle \right]$$

$$+ \int H_{cap}(\mathbf{r}_1, \mathbf{r}_2) \, d\mathbf{r}_2 \qquad (5.8.5)$$

Comparing this relation with the identity and Eq. (5.8.2) shows that the following identity remains to be shown:

$$\int H_{cap}(z_1, z_2, \mathbf{r}_\parallel) \, dz_2 \, d\mathbf{r}_\parallel = -\frac{k_B T_0}{g} (\rho_{eq}^-(z_1) - \rho_{eq}^+(z_1)) \frac{d}{dz_1} \langle \Theta^- \rangle$$

$$(5.8.6)$$

Because the bulk-density variation over a distance W is negligible, it is sufficient to prove

$$\int H_{\text{cap}}(z_1, z_2, \mathbf{r}_{\|}) \, dz_2 \, d\mathbf{r}_{\|} = \int \tilde{H}_{\text{cap}}(z_1, z_2, 0) \, dz_2 = -\frac{k_B T_0}{g} (\rho_1 - \rho_g) \frac{d}{dz_1} \langle \Theta^- \rangle$$

$$= -\frac{k_B T_0}{g} \rho_0'(z_1) \tag{5.8.7}$$

Note how crucial the difference between $\langle \rho(z_1) \rangle$ and $\rho_0(z_1)$ is in this context. Only if $\kappa_{T,l} = \kappa_{T,g} = 0$ may we identify these two functions.

To prove Eq. (5.8.7) we use the spectral representation of H_{cap} as given in Eq. (5.7.7) and the fact that [cf. Eq. (5.7.5)]

$$[-\rho_0'(z_2)]^{1/2} = (\rho_1 - \rho_g)^{-1/2} \psi_0(z_2)$$

As a consequence of this and the orthonormality of the eigenfunctions [cf. Eq. (5.7.10)], we find on integration of Eq. (5.7.7)

$$\int \tilde{H}_{\text{cap}}(z_1, z_2, 0) \, dz_2 = (\rho_1 - \rho_g)^{1/2} W^2 [-\rho_0'(z_1)]^{1/2} \tilde{H}_0(0) \psi_0(z_1)$$

$$= -(\rho_1 - \rho_g) \frac{k_B T_0}{\gamma_{\text{eq}}} L_c^2 \rho_0'(z_1) = -\frac{k_B T_0}{g} \rho_0'(z_1)$$

This finishes the verification of Eq. (5.8.7) and thus of the general identity (5.8.1). Its validity in the context of our analysis shows the consistency of this analysis with results obtained on a microscopic basis.

I. The Direct Correlation Function in the Capillary-Wave Model

The direct correlation function is defined by

$$\int d\mathbf{r}_2 c(\mathbf{r}_1, \mathbf{r}_2) H(\mathbf{r}_2, \mathbf{r}_3) = \delta(\mathbf{r}_1 - \mathbf{r}_3) \tag{5.9.1}$$

as the "inverse" of the density–density correlation function. If we define the Fourier transform of c by

$$\tilde{c}(z_1, z_2, k_{\|}) \equiv \int dx \, dy \, \exp[-i(k_x x + k_y y)] c(z_1, z_2, r_{\|}) \tag{5.9.2}$$

the definition of c may be written as

$$\int dz_2 \tilde{c}(z_1, z_2, k_{\|}) \tilde{H}(z_2, z_3, k_{\|}) = \delta(z_1 - z_3) \tag{5.9.3}$$

Using the general identity in the previous section we may show that $\tilde{c}$

satisfies the following general identity:

$$\int dz_2 \tilde{c}(z_1, z_2, 0) \frac{d}{dz_2} \langle \rho(z_2) \rangle = -\frac{g}{k_B T_0} \qquad (5.9.4)$$

Expanding $\tilde{c}$ in $k_\parallel$

$$\tilde{c}(z_1, z_2, k_\parallel) = \tilde{c}(z_1, z_2, 0) + k_\parallel^2 c_2(z_1, z_2) + \cdots \qquad (5.9.5)$$

we may furthermore verify the Triezenberg–Zwanzig identity[40]

$$\int dz_1 \, dz_2 \left(\frac{d}{dz_1} \langle \rho(z_1) \rangle \right) \tilde{c}_2(z_1, z_2) \left(\frac{d}{dz_2} \langle \rho(z_2) \rangle \right) = \frac{\gamma_{eq}}{k_B T_0} \qquad (5.9.6)$$

using again the general identity in the previous section.

In order to be able to obtain an explicit expression for the direct correlation function, we shall restrict ourselves to the case in which both bulk phases are incompressible, $\kappa_{T,l} = \kappa_{T,g} = 0$, for which case H is given by H_{cap}. In the neighborhood of the interface, that is, for z_1 and z_2 smaller than or roughly equal to W, this should give insight in the behavior of the direct correlation function. Inverting Eq. (5.7.7), we obtain in this way

$$\tilde{c}_{cap}(z_1, z_2, k_\parallel) = (\rho_l - \rho_g)^{-1} W^{-2} [\rho_0'(z_1)\rho_0'(z_2)]^{-1/2}$$
$$\times \sum_{n=0}^{\infty} (n+1) \tilde{H}_n^{-1}(k_\parallel)\psi_n(z_1)\psi_n(z_2) \qquad (5.9.7)$$

Inverse Fourier transformation then gives

$$c_{cap}(z_1, z_2, r_\parallel) = (\rho_l - \rho_g)^{-1} W^{-2} [\rho_0'(z_1)\rho_0'(z_2)]^{-1/2}$$
$$\times \sum_{n=0}^{\infty} (n+1) c_n(r_\parallel)\psi_n(z_1)\psi_n(z_2) \qquad (5.9.8)$$

where

$$c_n(r_\parallel) = \frac{1}{(2\pi)^{-2}} \int dk_x \, dk_y \exp[i(k_x x + k_y y)]\tilde{H}_n^{-1}(k_\parallel) \qquad (5.9.9)$$

One of the properties that would be intersting to prove is the short-range nature of the direct correlation function. For a one-dimensional interface in a two-dimensional system, $c_n(r_\parallel)$ can be calculated analytically and one finds that c_{cap} is indeed short ranged.[30] For a two-dimensional interface in a three-dimensional system, this cannot be done analytically, but we expect c_{cap} to be short-ranged here as well. Using Eqs. (5.4.7), (5.7.8), (5.9.8), and (5.9.9), this may easily be verified numerically.

As a final note, we emphasize that the behavior of the direct correlation function in the neighborhood of the interface is very different from its behavior in the bulk phase away from the interface. This can be seen most clearly from the scaling behavior.[36,37] Using the density-functional method, as was done by Evans,[32] implicitly assumes the incorrect scaling behavior which is the origin of his claim that the width of the interface W does not diverge in the zero-gravity limit. The proper scaling behavior shows quite clearly why this claim is incorrect.[36,37]

VI.   TIME-DEPENDENT FLUCTUATIONS OF A LIQUID–VAPOR INTERFACE

A.   Introduction

To calculate the unequal-time correlation functions, one must set up equations of motion for the fluctuating quantities. This can be done by the addition of the appropriate random thermodynamic fluxes to the equations of motion discussed in the previous sections. This procedure is analogous to the addition of a random force to the equation of motion of a Brownian particle. The extension of this procedure to a one-component one-phase fluid was given by Landau and Lifshitz.[17] Extensions to multicomponent systems are straightforward. As we will discuss in this section, one may give an extension to a two-phase system along the same lines.

For simplicity's sake we shall restrict ourselves to the liquid–vapor interface in a one-component fluid. Extension to a system with more components is again straightforward. Landau and Lifshitz write the viscous pressure tensor and the heat flow in a one-component one-phase fluid as sums of a "systematic" and a "random" contribution:

$$\mathbf{\Pi}_{tot} = \mathbf{\Pi} + \mathbf{\Pi}_R \quad \text{and} \quad \mathbf{J}_{q,tot} = \mathbf{J}_q + \mathbf{J}_{q,R} \tag{6.1.1}$$

The systematic contributions are given by the usual linear phenomenological laws:

$$\Pi_{ij} = -\eta \left(\frac{\delta v_i}{\delta x_j} + \frac{\delta v_j}{\delta x_i} - \tfrac{2}{3}\delta_{ij} \, \text{div } \mathbf{v} \right) - \eta_V \delta_{ij} \, \text{div } \mathbf{v} \tag{6.1.2}$$

$$\mathbf{J}_q = -\lambda \, \text{grad } T \tag{6.1.3}$$

where η is the viscosity, η_V the bulk viscosity, λ the heat conductivity, and $i, j = x, y, z$. It should be stressed that $\mathbf{v}$ and T are now fluctuating quantities, so as a consequence the systematic contributions to $\mathbf{\Pi}_{\text{tot}}$ and $\mathbf{J}_{q,\text{tot}}$ also fluctuate. The sources of these fluctuations in the equations of motion are $\mathbf{\Pi}_R$ and $\mathbf{J}_{q,R}$. The average of the random fluxes is zero:

$$\langle \mathbf{\Pi}_R \rangle = 0 \quad \text{and} \quad \langle \mathbf{J}_{q,R} \rangle = 0 \tag{6.1.4}$$

The random fluxes are furthermore assumed to be Gaussian and white. It should be realized that such an assumption is correct only if it can be shown that all equal-time correlations generated by $\mathbf{\Pi}_R$ and $\mathbf{J}_{q,R}$ will approach their equilibrium values for large times. For fully linearized equations of motion, this may easily be verified. In the general nonlinear case, one finds that the so-called Onsager coefficients ηT, $\eta_V T$, and λT^2 for the one-component fluid must be constant if one is to prove that the equal-time correlations approach their equilibrium values.[41] In realistic systems the Onsager coefficients are not constant; as a consequence, the use of Gaussian white noise is consistent only if one uses the equilibrium values of the Onsager coefficients. Since the equations of motion are still nonlinear due to convective fluxes and the nonlinear nature of the equation of state the use of the equilibrium values of the Onsager coefficients is not such a severe restriction. To go beyond this, a description using the master equation becomes necessary. In the linear constitutive relations at the interface, we have systematically used the Onsager coefficients for the vectorial and scalar force–flux pairs. For the tensorial force–flux pairs the Onsager coefficient is equal to $\eta^s T^s$.

The fluctuation–dissipation theorem for the random fluxes is[17]

$$\langle \Pi_{R,ij}(\mathbf{r}, t)\Pi_{R,kl}(\mathbf{r}', t')\rangle = 2k_B T_0 [\eta(\delta_{ik}\delta_{jl} + \delta_{il}\delta_{jk} - \tfrac{2}{3}\delta_{ij}\delta_{kl})$$
$$+ \eta_V \delta_{ij}\delta_{kl}]\delta(\mathbf{r}-\mathbf{r}')\delta(t-t')$$
$$\langle J_{q,R,i}(\mathbf{r}, t)J_{q,R,j}(\mathbf{r}', t')\rangle = 2k_B T_0^2 \lambda \delta_{ij}\delta(\mathbf{r}-\mathbf{r}')\,\delta(t-t') \tag{6.1.5}$$

where the equilibrium values of η, η_V, and λ should be used. Random fluxes of a different tensorial nature are not correlated with each other. Notice that the prefactor of the δ function and the Kronecker deltas is always equal to $2k_B$ times the equilibrium value of the appropriate Onsager coefficient.

In the two-phase situation the bulk viscous pressure tensors $\mathbf{\Pi}_{\text{tot}}^\pm$ and the heat current $\mathbf{J}_{q,\text{tot}}^\pm$ may be written in a similar way as sums of a systematic and a random contribution. The systematic contributions are again given by Eqs. (6.1.2) and (6.1.3) with $+$ or $-$ superscripts to indicate the phase. The average of the random fluxes is again zero and the fluctuation–dissipation theorem is given by (6.1.5) with a $+$ or $-$ superscript to indicate the phase.

B. Fluctuation–Dissipation Theorems for Excess Random Fluxes at the Interface

At the interface one should write the total excess fluxes also as sums of a systematic and a random contribution. Thus, one has for the tensorial excess flux

$$\mathbf{\Pi}^s_{tot} = \mathbf{\Pi}^s + \mathbf{F}_{v,R} \tag{6.2.1}$$

where $\mathbf{\Pi}^s$ is given in Section IV.D. Similarly, one has for the vectorial excess fluxes

$$\mathbf{J}^s_{q,tot} = \mathbf{J}^s_q + \mathbf{F}_{q,R}$$
$$[\Pi_{n,\parallel} + (v_n - v^s_n)\,\rho \mathbf{v}_\parallel]_{+,tot} = [\Pi_{n,\parallel} + (v_n - v^s_n)\,\rho \mathbf{v}_\parallel]_+ + \mathbf{F}_{v+,R} \tag{6.2.2}$$
$$[\Pi_{n,\parallel} + (v_n - v^s_n)\,\rho \mathbf{v}_\parallel]_{-,tot} = [\Pi_{n,\parallel} + (v_n - v^s_n)\,\rho \mathbf{v}_\parallel]_- + \mathbf{F}_{v,-R}$$

where the systematic contributions are given by the linear constitutive equations in Section IV.E specialized to the one-component case. For the scalar excess fluxes, one has

$$\Pi^s_{tot} = \Pi^s + F_{v,R}$$
$$[J_{q,n} + (v_n - v^s_n)\,T\rho s]_{+,tot} = [J_{q,n} + (v_n - v^s_n)\,T\rho s]_+ + F_{q+,R}$$
$$[J_{q,n} + (v_n - v^s_n)\,T\rho s]_{-,tot} = [J_{q,n} + (v_n - v^s_n)\,T\rho s]_- + F_{q,-R}$$
$$[\Pi_{nn} + (v_n - v^s_n)\,\rho v_n - \rho(\mu - \mu^s - \tfrac{1}{2}|\mathbf{v}|^2 + \tfrac{1}{2}|\mathbf{v}^s|^2)]_{+,tot} \tag{6.2.3}$$
$$= [\Pi_{nn} + (v_n - v^s_n)\,\rho v_n + \rho(\mu - \mu^s - \tfrac{1}{2}|\mathbf{v}|^2 + \tfrac{1}{2}|\mathbf{v}^s|^2)]_+ + F_{v+,R}$$
$$[\Pi_{nn} + (v_n - v^s_n)\,\rho v_n + \rho(\mu - \mu^s - \tfrac{1}{2}|\mathbf{v}|^2 + \tfrac{1}{2}|\mathbf{v}^s|^2)]_{-,tot}$$
$$= [\Pi_{nn} + (v_n - v^s_n)\rho v_n + \rho(\mu - \mu^s - \tfrac{1}{2}|\mathbf{v}|^2 + \tfrac{1}{2}|\mathbf{v}^s|^2)]_- + F_{v-,R}$$

where the systematic contributions are given by the linear constitutive equations in Section IV.F specialized to the one-component case.

The averages of the random contributions to the excess fluxes are again zero:

$$\langle F_v \rangle = 0, \qquad \langle \mathbf{F}_{q,R} \rangle = \langle \mathbf{F}_{v+,R} \rangle = \langle \mathbf{F}_{v-,R} \rangle = 0$$
$$\langle F_v \rangle = \langle F_{q+,R} \rangle = \langle F_{q-,R} \rangle = \langle F_{v+,R} \rangle = \langle F_{v,-R} \rangle = 0 \tag{6.2.4}$$

The fluctuation–dissipation theorems for the excess random fluxes are given for the special case that the equilibrium dividing surface is the xy plane by

$$\langle F_{v,R,ij}(x, y, t)F_{v,R,kl}(x', y', t')\rangle$$
$$= 2k_B T_0 \eta^s[\delta_{ik}\delta_{jl} + \delta_{il}\delta_{jk} - \delta_{ij}\delta_{kl}]\, \delta(x - x')\, \delta(y - y')\, \delta(t - t') \quad (6.2.5)$$

where $i, j, k, l = x$ or y, for the tensorial random fluxes;

$$\langle F_{\alpha,R,i}(x, y, t)\, F_{\beta,R,j}(x', y', t')\rangle = k_B(L^s_{\alpha,\beta} + L^s_{\beta,\alpha})\delta_{ij}\delta(x - x')\, \delta(y - y')\, \delta(t - t')$$
$$(6.2.6)$$

where $\alpha, \beta = q$, $(v+)$, or $(v-)$ and $i, j = x$ or y, for the vectorial random fluxes; and

$$\langle F_{\alpha,R}(x, y, t)F_{\beta,R}(x', y', t')\rangle = k_B(L^s_{\alpha,\beta} + L^s_{\beta,\alpha})\delta(x - x')\, \delta(y - y')\, \delta(t - t')$$
$$(6.2.7)$$

where $\alpha, \beta = v$, $(q+)$, $(q-)$, $(v+)$, or $(v-)$, for the scalar random fluxes. The equilibrium values of the Onsager coefficients should be used in the above fluctuation–dissipation theorems. Random fluxes of a different tensorial nature are not correlated. For the more general case that the equilibrium interface is curved, one must replace x, y, x', y' by $\xi_{2,eq}, \xi_{3,eq}, \xi'_{2,eq}, \xi'_{3,eq}$ and divide the right-hand sides by $h_{2,eq}h_{3,eq}$ [cf. Eq. (5.3.3) for the interfacial equilibrium correlation functions].

The dynamic equations describing the fluctuations are now found by substitution of the total fluxes containing the systematic and the random contribution into the various balance equations. Some of these total fluxes, such as $[\Pi_{n,\|} + (v_n - v_n^s)\rho v_\|]_{\pm,tot}$ in Eq. (6.2.2), are used in the description of the system as boundary conditions for the equation of motion in the bulk regions. It is clear that these boundary conditions, for example, the slip condition or the temperature-jump condition, now contain random terms. That such a random term in these boundary conditions is necessary is shown explicitly by Bedeaux et al. in a paper on the derivation of the Langevin equation for the Brownian motion of a spherical particle with a finite slip coefficient in a fluid.[42]

The resulting equations of motion for the fluctuations in the bulk regions and at the interface are very complicated, due to the rather large number of phenomena that may take place at the interface. It is clear that to calculate time-dependent correlation functions for the excess densities and the normal on the dividing surface one must simplify the equations of motion. One standard procedure is to linearize them. The resulting fully linear equations are still difficult to analyze. As further simplifications, one may neglect certain phenomena. Thus one often neglects the compressibility of the bulk phases when calculating the velocity autocorrelation function. Similarly, one may neglect certain interfacial phenomena. An example is the use of either the no-slip or the perfect-slip boundary condition, obtained by choosing the appropriate Onsager coefficients to be either infinite or zero. Also, one may neglect excess currents along the interface if there is reason to believe that the excess is small.[43] There is one aspect, however, about which one should be careful. If some random fluxes are neglected, the resulting time-dependent correlation functions generated by these random fluxes will no longer necessarily have their correct equilibrium values for equal times. This problem originates in the replacement of some relaxation times by zero due to the approximations. A well-known example is the velocity autocorrelation function of a Brownian particle in an incompressible fluid. Because of the incompressibility of the fluid, the equal-time autocorrelation function for the velocity of the Brownian particle contains in the denominator the mass of the particle plus one-half the mass of the displaced fluid, rather than just the mass of the particle. Even though the reason for this is perfectly clear, it has led to some confusion.[44] A similar situation will clearly arise for the interface if one uses the no-slip condition, because it forces velocities to be equal instantaneously rather than after a short relaxation time.

References

1. J. Meixner, *Ann. Physik* **39**, 333 (1941); **41**, 409 (1942); **43**, 244 (1943); *Z. Phys. Chem.* **B53**, 235 (1943).

2. I. Prigogine, *Etude Thermodynamique des Phénomènes Irréversibles*, Dunod, Paris, and Desoer, Liège, 1947.

3. S. R. de Groot and P. Mazur, *Non-equilibrium Thermodynamics*, North-Holland, Amsterdam, 1962 (reprinted by Dover, New York, 1984).

4. A. Katchalsky and P. F. Curran, *Nonequilibrium Thermodynamics in Biophysics*, Harvard Univ. Press, Cambridge, Massachusetts, 1965; R. Defay and I. Prigogine, *Surface Tension and Adsorption*, Longmans and Green, London, 1966.

5. J. W. Gibbs, *Collected Works*, 2 vols., Dover, New York, 1961.

6. D. Bedeaux, A. M. Albano, and P. Mazur, *Physica* **82A**, 438 (1976).

7. D. C. Mikulecky and S. R. Caplan, *J. Chem. Phys.* **70**, 3049 (1966).

8. J. Kovac, *Physica* **A86**, 1 (1977); **A107**, 280 (1981).

9. P. A. Wolff and A. M. Albano, *Physica* **98A**, 491 (1979).

10. F. Vodak, *Physica* **A93**, 244 (1978); **A112**, 256 (1982); J. M. Rubi and J. Casas-Vazques, *Physica* **111A** 351 (1982); G. Bertrand, *J. Non-eq. Thermodynamics* **6**, 165 (1981); A. G. Bashkirov, *Theor. and Math. Physics* **43**, 542 (1980); G. Bertrand and R. Prudhomme, *J. Non-eq. Thermodynamica* **4**, 1 (1979); *Int. J. Quantum Chem.*, **12**, 159 (1977).

11. B. J. A. Zielinska and D. Bedeaux, *Physica* **112A,** 265 (1982).

12. F. P. Buff, R. A. Lovett, and F. H. Stillinger, *Phys. Rev. Lett.* **15,** 621 (1965).

13. D. Bedeaux and J. D. Weeks, *J. Chem. Phys.* **82,** 972 (1985).

14. For this subject and further references, see: J. S. Rowlinson and B. Widon, *Molecular Theory of Capillarity*, Clarendon, Oxford, 1982; S. Ono and S. Kondo, in *Encyclopedia of Physics*, Vol. 10, S. Flügge, Ed., Springer-Verlag, Berlin, 1960, p. 134; M. S. John, J. S. Dahler, and R. C. Desai, *Adv. Chem. Phys.* **46,** 279 (1981); R. Evans, *Adv. Phys.* **28,** 143 (1979); J. K. Percus, in *Studies on Statistical Mechanics*, Vol. VIII, E. W. Montroll and J. L. Lebowitz, Eds., North-Holland, Amsterdam, 1982, p. 31.

15. P. M. Morse and H. Feshbach, *Methods of Theoretical Physics*, Vol. I, McGraw-Hill, New York, 1953.

16. C. Truesdell and R. Toupin, *Encyclopedia of Physics*, Vol. III/3, S. Flügge, Ed., Springer-Verlag, Berlin, 1960.

17. L. D. Landau and E. M. Lifshitz, *Fluid Mechanics*, Pergamon, Oxford, 1959.

18. A. M. Albano, D. Bedeaux, and J. Vlieger, *Physica* **99A,** 293 (1979).

19. A. M. Albano, D. Bedeaux, and J. Vlieger, *Physica* **102A,** 105 (1980).

20. E. Kröger and E. Kretschmann, *Z. Phys.* **237,** 1 (1970).

21. L. Waldmann, *Z. Naturforsch.* **22a,** 1269 (1967).

22. V. G. Levich, *Physicochemical Hydrodynamics*, Prentice-Hall, Englewood Cliffs, New Jersey, 1962.

23. D. Bedeaux and I. Oppenheim, *Physica* **90A,** 39 (1978).

24. L. A. Turski and J. S. Langer, *Phys. Rev.* **A22,** 2189 (1980).

25. J. D. Gunton, M. San Miguel, and P. S. Sahni, "The Dynamics of First-Order Phase Transitions," in *Phase Transitions and Critical Phenomena*, Vol. 8, C. Domb and J. L. Lebowitz, Eds., Academic, New York, 1983; J. S. Langer, *Rev. Mod. Phys.* **52,** 1 (1980).

26. W. W. Mullins and R. F. Sekerka, *J. Appl. Phys.* **34,** 323 (1963); **35,** 444 (1964); R. F. Sekerka, "Morphological Stability," in *Crystal Growth: An Introduction*, P. Hartman, Ed., North-Holland, Amsterdam, 1973.

27. Ya. B. Zeldovic and Yu. P. Rajzev, *Physics of Shockwaves and High Temperature Hydrodynamic Phenomena*, Academic, New York, 1966.

28. W. van Saarloos and J. D. Weeks, *Physica* **D12** (*Proc. Conf. on Fronts, Interfaces and Patterns*), 279 (1984), and references therein.

29. D. Bedeaux and J. D. Weeks, *J. Chem. Phys.* **82,** 972 (1985).

30. D. Bedeaux, J. D. Weeks, and B. J. A. Zielinska, *Physica* **130A,** 88 (1985).

31. J. D. Weeks, *J. Chem. Phys.* **67,** 3106 (1977).

32. R. Evans, *Mol. Phys.* **42,** 1169 (1981).

33. D. B. Abraham, *Phys. Rev. Lett.* **47,** 545 (1981).

34. D. Sullivan, *Phys. Rev.* **A25,** 1669 (1982).

35. J. D. Weeks, D. Bedeaux, and B. J. A. Zielinska, *J. Chem. Phys.* **80,** 3790 (1984).

36. J. D. Weeks, *Phys. Rev. Lett.* **52,** 2160 (1984).

37. D. Bedeaux, "The Fluid–Fluid Interface," in *Fundamental Problems in Statistical Mechanics VI*, E. G. D. Cohen, Ed., North-Holland, Amsterdam, 1985, p. 125.

38. M. C. Wang and G. E. Uhlenbeck, *Rev. Mod. Phys.* **17,** 323 (1945).

39. M. S. Wertheim, *J. Chem. Phys.* **65**, 2377 (1976).

40. D. G. Triezenberg and R. Zwanzig, *Phys. Rev. Lett.* **28**, 1183 (1972).

41. W. van Saarloos, D. Bedeaux, and P. Maxur, *Physica* **A110**, 147 (1982); D. Bedeaux, "Hydrodynamic Fluctuation Theory for a One-Component Fluid in Equilibrium: The Non-Linear Case," in *Fundamental Problems in Statistical Mechanics V*, E. G. D. Cohen, Ed., North-Holland, Amsterdam, 1980, p. 313.

42. D. Bedeaux, A. M. Albano, and P. Maxur, *Physica* **88A**, 564 (1977).

43. M. Grant and R. C. Desai, *Phys. Rev.* **A27**, 2577 (1983).

44. See, e.g., D. Bedeaux and P. Mazur, *Physica* **78**, 505 (1974), and references therein.

THE DISSOCIATION DYNAMICS
OF ENERGY-SELECTED IONS

TOMAS BAER

Department of Chemistry
University of North Carolina
Chapel Hill, North Carolina, U.S.A.

CONTENTS

I. INTRODUCTION

State-to-state reaction dynamics! This has been the holy grail of kineticists for the past quarter century. Over the years the quest has lost some of its luster, in part because of the formidable experimental difficulties in achieving this goal, and in part because the results, at least for bimolecular reactions, often cannot be understood in terms of simply dynamical models. However, recent advances in the use of lasers and coincidence techniques have provided a wealth of experimental data, giving new insights into the dissociation dynamics of both small and large molecules and ions. Theoretical advances have kept pace, so that we are now at the threshold of exciting new developments in the field of unimolecular reactions.

A true state-to-state reaction study would involve measuring the reaction rate, lifetime, or cross section of state-selected reactants to the various final product states. Such studies have been carried out in only a few cases.[1–5] The more common studies are ones in which either the reactants or the products are state analyzed, but not both. Virtually all of the results described in this report will be ones in which only the reactants are state selected or, in the case of large ions, energy selected. The difference between state and energy selection is that in the former, the vibrational or possibly rotational states are specified, whereas in the latter only the energy is specified. State selection probably becomes impossible at high energies in large molecules or ions, because the many states within a small energy bandwidth are closely coupled.

This chapter will deal primarily with the dissociation dynamics of ions. There is a somewhat less extensive list of state-selected neutral-dissociation studies, which are obviously related to this field.[6] The separation of ions and neutral species is a very unfortunate one, the origin of which is related primarily to the nature of the experimental approaches rather than to fundamental differences in the theoretical frameworks. The different experimental approaches to neutral and ion dynamics lead to important differences in the information extracted from these studies. Energy selection of ions by the coincidence technique, described in Section II, typically yields a resolution of $\sim 30 \, meV$ ($\cong 250 \, cm^{-1}$). This resolution is sufficient for state selecting the vibrational levels of diatomic ions and some larger ions at low energy. However, no mode-selective preparation of large ions is possible. This is in contrast to recent investigations of neutral-dissociation reactions by laser excitation of the higher harmonics of a single mode, which give resolutions of a fraction of a wavenumber.[6-8] The other major difference is in the time scale accessible to the two experiments. In the ion studies, dissociation rates between 10^3 and $5 \times 10^6 \, s^{-1}$ are possible, whereas the neutral-dissociation reactions are studied typically in the 10^7–$10^9 \, s^{-1}$ range.

Although the experimental conditions for neutral and ionic studies are quite different, the questions addressed by both studies are the same. The fundamental question, in the case of dissociation on a single potential-energy surface, is whether the dissociation rate or the products produced depend on the manner in which the molecule or ion is prepared. If the former does not depend on the mode of preparation, but only on the total energy, the dissociation is statistical, according to a theory that will be discussed in a subsequent section. By far the majority of the slow dissociation reactions ($k < 10^9 \, s^{-1}$) studied to date have been consistent with the statistical theory.[6-9] In fact, in the absence of contrary evidence, it can generally be assumed that slow polyatomic dissociation reactions will be statistical. That the reaction is slow is very important in this statement, because fast dissociations from excited states are often very nonstatistical.

Interest in unimolecular reactions goes well beyond the question of statistical versus nonstatistical dissociation. A reaction involving a simple bond fission ought to be different from one involving some rearrangement of the atoms. In principle, the dissociation rate and the distribution of the energy in the product states should allow us to deduce the nature of the reaction mechanism. Considerable progress has been made in this direction for ionic reactions.

II. ION STATE-SELECTION TECHNIQUES

A. Photoelectron–Photoion Coincidence (PEPICO)

An extremely powerful and generally applicable approach to state selecting ions is that of photoelectron–photoion coincidence (PEPICO).

The ionization process

$$AB + h\nu \rightarrow AB^+ + e^- \qquad (2.1)$$

is a bound to continuum transition, so the ion AB^+ can be formed in a range of internal energy states from its ground state up to an energy $h\nu - IP$ (where IP is the ionization potential). The excess energy is released as the kinetic energy of the two particles, which, because of the small electron mass, is given primarily to the electron. Figure 1 shows the relation between the electron kinetic energy and the ion internal energy.

Polyatomic molecules not cooled in a supersonic beam often have significant thermal energy $\langle E_{th} \rangle$. Thus the ion internal energy is expressed by

$$E_{ion} = h\nu - IP + \langle E_{th} \rangle - KE(e^-). \qquad (2.2)$$

(where KE is kinetic energy). It is evident that if ions are detected in coincidence with electrons of a particular kinetic energy, the coincidence condition selects ions of a particular internal energy. However, because of the ion thermal energy $\langle E_{th} \rangle$, the uncertainty is often of the order of

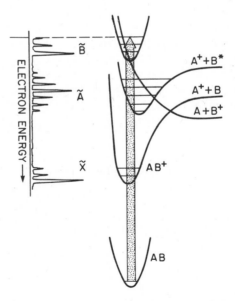

Figure 1. The distribution of the ion states produced by photoionization with light reaching the $v = 2$ level of the $\tilde{B}$ state. The photoelectron spectrum is shown on the left side of the figure.

60–100 meV. When the PEPICO technique is used with supersonically cooled molecules, this thermal-energy spread is greatly reduced.

Two types of coincidence experiments are being pursued. In one, the light-source energy is fixed, usually at 21.2 eV, which is the He(I) resonance line.[10–15] In this arrangement, the ions are detected in coincidence with electrons of selected energies. The electrons are generally selected by electrostatic energy analyzers. In the other type of coincidence experiment, the light source is a continuum that is dispersed by a vacuum UV monochromator.[16–19] In this approach, initially zero-energy electrons are detected on the basis of their large collection efficiency. Threshold electrons are easily accelerated toward the electron detector by a small electric field, while energetic electrons are scattered in all directions. The differences between and relative merits of the two approaches have been discussed in a review on coincidence techniques.[20]

A diagram of the coincidence apparatus is shown in Fig. 2. In the PEPICO experiment, the ion time of flight (TOF) is determined by measuring the time difference between the arrivals of the energy-analyzed electron and ion at their respective detectors. The electron and ion flight times are of the order of 0.1 and 10 μs, respectively. For low total ionization rates, say 10,000 counts/s, the average time between ionization events is 100 μs. Therefore a single coincidence event is generally over

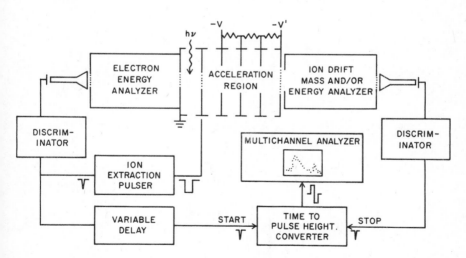

Figure 2. Experimental setup of a photoelectron–photoion coincidence (PEPICO) apparatus. The electron-energy analyzer may be of the dispersive type[10–15] or a zero-energy-electron analyzer.[16–19] The ion-extraction pulser is not used in all experiments. Taken with permission from Baer.[20]

before the next ionization event begins. However, for larger ion fluxes, say 10^6 counts/s, the average time between ionization events is only 1 μs and problems with false coincidences arise. This is particularly a problem with the use of a time-to-pulse-height converter that accepts only one stop pulse. These problems and some of their solutions have been discussed in more detail in other publications.[20–24]

All the dynamical information is contained in the ion TOF distribution. If an ion dissociates with the release of kinetic energy, the fragment ion signals will be symmetrically broadened. For slowly dissociating ions, the parent ion travels some distance in the acceleration region before dissociation takes place. This results in an asymmetrically broadened TOF distribution. Examples will be shown in the following sections.

B. State Selection by Multiphoton Ionization

In recent years, the technique of multiphoton ionization (MPI) has been used to state select ions.[25,26] The advantage of this approach is that coincidence detection is no longer necessary (or even possible.) The major drawback is that unlike PEPICO, which is completely general, MPI state selection must be very carefully applied to each molecule. The principle of MPI state selection is illustrated in Fig. 3. The first two or more photons ionize the molecule. The resulting ion can then be used

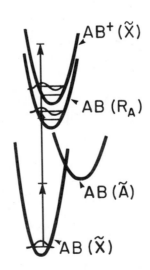

Figure 3. Schematic potential-energy curves illustrating the use of multiphoton ionization in state selection. R_A is a Rydberg state converging to the ground ionic state.

directly to study ion–molecule reactions, or it can absorb another photon, after which it will dissociate.

For MPI state selection to work, the ions must be generated with a narrow band of internal energies. Can this condition be assured? The answer appears to be affirmative if the last photon absorbed before ionization reaches a pure Rydberg state. The reason for this is that the Franck–Condon factors between such a Rydberg state and the ground ionic state to which it converges are such as to preserve the vibrational energy. Thus, if a Rydberg state in $v = 3$ is excited, the ion will end up exclusively in $v = 3$. Considerable success has been achieved with this approach for such molecules as NO,[27–29] N_2,[30] and H_2.[31–33] The appropriateness of a particular intermediate state for state selection by MPI can be assessed from the photoelectron spectrum (PES). Two MPI PES for NO are illustrated in Fig. 4 for different intermediate states. Ionization is achieved by a four-photon process, with the resonant $\tilde{A}$ state at the two-photon level. With the excitation via the $\tilde{A}(v = 1)$ state, mostly $v = 1$ ions are formed. However, a significant number of ions in $v = 8$ and other vibrational levels are formed as a result of interactions between the $\tilde{A}$ state and nearby valence states as well as the influence of accidental resonances at the three-photon level. By careful choice of the intermediate state it should be possible to obtain considerably better state selection than is illustrated in Fig. 4.

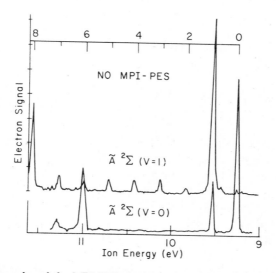

Figure 4. Examples of the MPI PES obtained when NO is ionized via a two-photon excitation of the $\tilde{A}$-state vibrational levels followed by a two-photon ionization. Adapted from Kimman, Kruit, and Van der Wiel.[28]

The results illustrated in Fig. 4 are encouraging. However, even the availability of a pure Rydberg state, with the resulting ejection of monoenergetic electrons, does not assure state selection. Suppose that the Rydberg state can be accessed only via a nonresonant two-photon transition. To have reasonable ion intensity, the laser must be tightly focused. At such high photon densities, the ion, once produced, may absorb many more photons and end up in a very high-energy state from which it will dissociate much more rapidly than will the ions of low energy. That is, it may be difficult to control the number of additional photons absorbed. In other cases, it may be necessary to use two or even three different laser colors to reach the desired Rydberg state. In such cases it will be difficult to control which of the two or three photons will ionize this state, and pure state selection will probably not be possible.

The problems associated with multicolor MPI state selection can, to some extent, be avoided by the use of vacuum UV lasers. In addition, higher-energy ion states can be reached by the use of higher-energy photons. However, it is evident that the benefits of MPI state selection do not come without considerable effort being invested in tailoring the laser excitation scheme to each molecule of interest.

The major advantage of MPI state selection over PEPICO is that ions can be formed in a much narrower range of energies. Even rotational-state selection is probably within our reach. Second, it may be possible to study dynamical processes on a much faster time scale than is now done with PEPICO. One approach is to use a second laser, delayed relative to the ionizing laser by a variable time, to probe the appearance of a fragment ion.[34] This can be done over the whole range from picoseconds to microseconds.

III. THE DISSOCIATION RATES OF POLYATOMIC IONS—PRELIMINARY CONSIDERATIONS

The dissociation rates of a large number of ions have now been measured. Because of the many vibrational degrees of freedom, the theoretical understanding of polyatomic dissociations has centered on the one theory that is capable of providing some physical insight at a reasonable cost in complexity. This is the statistical theory,[35] known either as the RRKM (after Rice, Ramsperger, Kassel, and Marcus)[36,37] or the quasi-equilibrium theory (QET).[38] The goal of the initial experimental investigations was the testing of the statistical theory, and the problem of statistical versus nonstatistical dissociation continues to be a dominant theme. In one sense, a reaction that proceeds statistically is uninteresting, because its rate and product energy distributions depend on only the molecular frequencies and the total energy. As a result, the search for nonstatistical reactions is a search for "interesting" reactions. These are reactions whose rates might depend on how the energy is initially deposited in the molecule, that is, mode-selective reactions.

It is by now evident that truly nonstatistical reactions in large molecules or ions are extremely rare. This scarcity results, in part, from the impossibility of state selecting a molecule at high internal energy, due to the strongly coupled vibrational modes at these energies. In addition, large molecules and ions often dissociate via complex sequences of reactions. As a result, the overall dissociation rate or product energy distribution is not easily interpreted.

A. What Excited States Are Produced in State-Selected Experiments?

Suppose that a molecule with a number of vibrational modes is excited from its ground electronic, vibrational, and rotational states to a very precise energy within the ground electronic state. Assume further that all the molecules are at rest, so that there is no Doppler broadening. At low energy, the density of vibrational states is quite small, so that a relatively pure mode is excited and the molecule remains in this state until it is depleted with a rate of k_f reciprocal seconds by fluorescence, or with a rate of k_c reciprocal seconds by collisions. Although there are many accidental degeneracies and perturbations caused by Fermi and Renner–Teller interactions that may complicate the assignment of the IR spectrum, the spacing of the vibrational levels is large compared with their natural line width, which is given by the uncertainty principle as $(k_f + k_c)/c$, where c is the speed of light. Assuming a sum of the collisional and radiative rates of $3 \times 10^3 \, \text{s}^{-1}$, the natural line width would be of the order of $10^{-7} \, \text{cm}^{-1}$, which is over three orders of magnitude narrower than the Doppler width at 300 K. It is, therefore, clear that at low energies, the molecule is prepared in a single state.

When the energy of excitation is increased, the definition of the state becomes less clear. Consider the density of vibrational states calculated for the bromobenzene molecule by the use of the Hase–Bunker program[39] with the Whitten–Rabinovitch semiclassical state-counting procedure[40] (Fig. 5). At an energy just below 2 eV, the density of vibrational states reaches 10^7 states/cm^{-1}. Thus, the vibrational-state spacing becomes equal to the energy uncertainty. If the coupling among the vibrational modes is strong enough, the internal energy can flow freely among the internal degrees of freedom. An effect similar to this has been observed in the laser-induced fluorescence of electronically excited naphthalene molecules.[41] There are a number of fundamental questions

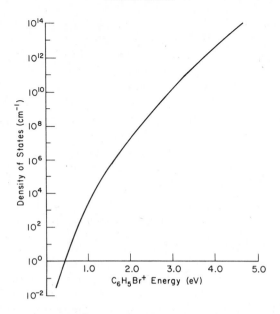

Figure 5.   Calculated density of vibrational states of bromobenzene as a function of the internal energy.

related to this problem. These concern the distinction between internal vibrational-energy redistribution (IVR) and the immediate excitation of a superposition of many states. The first process (IVR) evolves in time from the initial excitation of a single state to that of many states, while the second process involves no time evolution. This distinction is not of great concern to us here, because the time scale for the IVR is in the picosecond domain, whereas the dissociation rates are in the nanosecond-to-microsecond domain. The conclusion we draw from the density of states and state-width estimation is that at higher energy, if the coupling among internal modes is sufficiently large, it is not possible to state select the molecule even if our excitation source is infinitely narrow and the molecules are at 0 K.

In most experiments the molecules are prepared initially in a distribution of states that encompasses a large number of states. This is because the excitation source has an energy width of a few wave numbers for the case of laser excitation and as high as $200 \, cm^{-1}$ for coincidence studies. Finally, the molecules are not usually in their ground state. For these reasons, the excited state actually produced in experiments is generally a superposition of many initial states.

B. The Statistical Theory of Dissociation Rates

The dissociation, on a single potential-energy surface, of a polyatomic molecule with many vibrational degrees of freedom is a complicated event that involves many or all of the vibrational modes. This process, which has been termed vibrational predissociation, can be treated in a detailed manner only for small model compounds.[42-45] Although such studies have given us a great deal of insight into the nature of unimolecular decay, the bulk of dissociation reactions must be treated with a less detailed theory. Such a theory is the statistical model first advanced by Rice, Ramsperger, and Kassel (RRK) in the late 1920s.[36] The original RRK theory treated all vibrational modes identically and assumed that they were fully activated (the classical limit). The resulting rate $k(E)$ for a reaction with total internal energy E, an activation energy E_0, and a total of s oscillators was found to be

$$k(E) = \frac{kT}{h} \left[\frac{E - E_0}{E} \right]^{s-1} \tag{3.1}$$

Although Eq. (3.1) is a purely classical result, it is more easily derived assuming quantized harmonic oscillators. Using this language, Eq. (3.1) is based on the assumption that all arrangements having j quanta in s oscillators are equally probable. Reaction occurs whenever a minimum of m quanta (which correspond to an energy of E_0) reside in the critical reaction coordinate.

The point of describing the old RRK theory, which is virtually useless in predicting dissociation rates quantitatively, is to point out that many of the essential features of the statistical theory were contained in this 1929 version of the theory. These features include the assumption of free energy exchange among the vibrational modes, and the random nature of this energy exchange.

A simple improvement in the RRK theory that removes essentially all of its problems with prediction is to describe the dissociating molecule as a microcanonical ensemble in terms of the density of internal energy states. This improvement was introduced by Marcus and Rice as the RRKM theory,[37] as well as by Rosenstock et al.[38] as the QET. Although some authors attempt to differentiate between these two theories, they are identical. In this theory the introduction of different vibrational frequencies allows a distinction to be made between the structure of the transition state and the molecule. The final expression for the dissociation rate of a molecule at an energy E with an activation energy E_0 is

$$k(E) = \sigma \sum_0^{E-E_0} g^{\ddagger}(\varepsilon) \Big/ hN(E) = \sigma \int_0^{E-E_0} N^{\ddagger}(\varepsilon) \, d\varepsilon \Big/ hN(E)$$
$$= \sigma G^{\ddagger}(E - E_0)/hN(E) \quad (3.2)$$

The terms $g^{\ddagger}$, $N^{\ddagger}$, and $G^{\ddagger}$ in Eq. (3.2) are, respectively, the degeneracy of the internal energy states, the density of internal energy states, and the sum of the states from 0 to $E - E_0$ of the transition state. The $N(E)$ term in the denominator is the density of states for the molecule, while σ represents the number of equivalent ways the molecule can dissociate. For H loss in benzene, for instance, its value is 6.

The expressions in Eq. (3.2) have a very simple interpretation. The numerator, being the sum over the possible energy levels of the transition state, represents the number of ways the excited molecule can dissociate. Each chanel decays via a rate $1/hN(E)$, and the sum of all of these paths through the transition state is the total decay rate. At the dissociation limit, the sum in the numerator reduces to 1, so that the minimum rate of dissociation is $1/hN(E_0)$. Figure 6 shows this graphically for three different energies above the dissociation limit. Each square in the phase space, represented by the horizontal areas, corresponds to a particular way of arranging the internal energy in the excited molecule. A statistical dissoci-

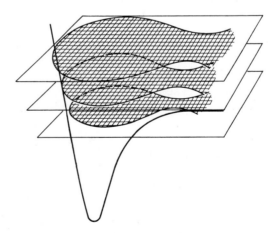

Figure 6. Diagram of the internal energy states during a dissociation as a function of the total energy. Each square corresponds to a particular arrangement of the internal energy states. The bottleneck is the position of the transition state at which the density of states is a minimum.

ation is one in which there is an equal probability of finding the system in each square in the phase space. Whenever the arrangement of internal energy corresponds to the molecule being in the bottleneck, a reaction can occur. These types of pictures have been used by Hase[46] not only to describe RRKM versus non-RRKM dissociation, but also to indicate the sorts of initial energy distributions that are formed under a variety of excitation conditions. The picture is somewhat misleading, in that in the region of the transition state, one of the oscillators has been converted into a translation. Thus, the density of internal energy states is converted into a sum over these states because the energy conservation can be maintained by means of the translational degree of freedom. This region of the phase space can be pictured either as a density of all the states, including those associated with the translational degrees of freedom, or as a sum over the internal degrees of freedom, as shown in Eq. (3.2).

C. Improvements in the Theory

1. Effect of Angular Momentum

The theory as expressed in Eq. (3.2) does not take into account the effect of angular momentum on the density of states, nor is the theory very sophisticated in terms of identifying the position of the transition state. The elimination of these shortcomings has been the goal of numerous studies during the past two decades,[44-62] and is becoming increasingly important as the experimental data become more precise.

The effect of angular momentum in the rotating molecule (to be distinguished from internal free rotations) has been investigated from several points of view. Figure 7 shows what happens to the rotational energy as a function of the reaction coordinate for a diatomic molecule with rotational constant B. The rotational energy in the molecule decreases from $J(J+1)B$, where $B = \hbar^2/2\mu r^2$ and $r = r_e$, to zero as the two fragments separate and r increases to infinity. The symbols J, r, r_e, and μ have their usual meaning of the rotational quantum number, the internuclear distance, the equilibrium internuclear distance, and the reduced mass of the separating fragments, respectively. The angular momentum, $L = \sqrt{J(J+1)}\hbar$, is conserved as the $\mu v b$ value of the dissociating atoms, where v is the relative velocity and b is the impact parameter. The barrier height near the transition state is a function of the attractive potential between the two fragments, and will therefore be very different for neutral and ionic dissociations. It is instructive to compare the centrifugal barriers for different interaction potentials of the form

$$V_{\text{eff}} = A\left[\left(\frac{\sigma}{r}\right)^{12} - \left(\frac{\sigma}{r}\right)^{s}\right] + \frac{L^2}{2\mu r^2} \tag{3.3}$$

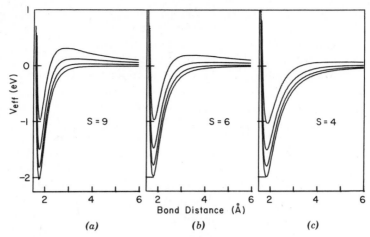

Figure 7. Calculated effective-potential-energy (V_{eff}) curves [Eq. (3.3)] for a diatomic dissociation. The total rotational energy in the molecule was assumed to be 0, 0.2, 0.5, and 1.0 eV for each indicated attractive interaction potential r^{-s}.

Figure 7 shows this function for three values of s and for four values of the rotational energy in the bottom of the potential well (0, 0.20, 0.50, and 1.0 eV). The ion-induced dipole potential characterized by the $s = 4$ value is frequently used as an appropriate potential for the dissociations of ions, whereas the $s = 6$ potential is more appropriate for neutral dissociations. As is evident from Fig. 7, both of these have rather small centrifugal potential barriers, at least in comparison with the rotational energy at the bottom of the well. The barrier for a typical average rotational energy of 0.026 eV is only 1 meV for the $s = 6$ case and is much lower still for the $s = 4$ case. However, as the potential is changed to a less attractive one at $s = 9$, the centrifugal barrier becomes significantly larger, rising to 2.5 meV for an initial rotational energy of 26 meV. The potentials of Fig. 7 are characterized as "tight" or "loose": The $s = 4$ potential is very loose, whereas the $s = 9$ potential might be termed moderately tight.

The potential curves of Fig. 7 apply to the dissociation of a rotor into two atoms, For molecules, the problem is more complex, because the angular momentum can also be accommodated by the rotations of one or both of the frgaments. However, angular momentum stored in this fashion is more energy rich than the orbital angular momentum and so it does not reduce the centrifugal barrier significantly. On the other hand, there are certain dissociations of polyatomics in which the centrifugal barrier is higher than is shown in Fig. 7. Examples are the H-atom loss from such molecules as CH_4 and NH_3. The reason for this is that the reduced mass in Eq. (3.3) varies as the molecule dissociates. Rotation of the whole molecule is converted into rotation of the $CH_3 \cdots H$ complex which has a lower moment of inertia, and therefore greater rotational energy for the same angular momentum.

Very little experimental information on centrifugal barriers is available. Rotationally resolved state selection is difficult to achieve when preparing the molecule or ion at an energy near the dissociation limit. However, it is possible to vary the average rotational energy by varying the gas temperature. In two such studies, McCulloh and Dibeler and Chupka and Berkowitz[63] measured the photoionization onset for H loss from CH_4^+ at 300 and 86 K. They found that the CH_3^+ onset shifts to higher energy precisely by the rotational energy difference, $\frac{3}{2}R\Delta T$, between the two temperatures. This demonstrates that essentially all of the rotational energy is available for overcoming the activation barrier, which indicates that there is no measurable centrifugal barrier. It suggests that s in the attractive part of the potential, r^{-s}, is equal to or less than 6.

These results on CH_4^+ dissociation, however, are in apparent conflict with other data for this reaction. Hydrogen-atom loss from methane ions has been measured to be very slow near the dissociation onset.[64-67] This process, which is considerably slower than can be predicted by the standard statistical theory, was suggested by Rosenstock[68] and by Klots[69] to occur via tunneling through the centrifugal barrier. That the CH_4^+ dissociation is considerably slower than the CD_4^+ dissociation supported this interpretation.[64,70] Careful measurements of the CH_3^+-fragment kinetic energy indicate that there is some structure in the kinetic-energy-release distribution and that the maximum kinetic energy is ~ 15 meV.[71,72] The intensity of the slow dissociation and the maximum kinetic-energy release both increase with the ion source temperature. These facts also support the model of tunneling through the centrifugal barrier, in that higher temperatures give rise to higher J values. We have then, on the one hand, the calculations of Fig. 7 and the variations of the photoionization onsets with temperature, which suggest that there is no barrier in the CH_4^+-ion dissociation. On the other hand, the slow rate is interpreted in terms of a barrier. This paradox is resolved when the intensities of the slow dissociation via tunneling are compared with those of the fast dissociation reactions. The former signal are only $1/10^5$ as great as the latter. Thus, even at room temperature, the intensity calculations of Illies et al.[70] show that there will be sufficient CH_4^+ ions with rotational energies greater than 200 meV to account for the observed slow dissociations. This reaction is the only well-documented ionic dissociation in which tunneling has an important effect on the observed dynamics.

The dissociation of CH_4^+ involves a rather loose transition state in which the critical bond is well stretched in the transition-state configuration. Other reactions, with tighter transition states, ought to have larger centrifugal barriers. The measurement of the appearance energy as a function of the temperature can be used to provide information about the potential surface, or at least about the position of the transition state on this surface.

The CH_4^+ dissociation demonstrates the importance of angular momentum in the reaction dynamics. Another example is that of H loss in NH_3^+, which is discussed in more detail in Section V.A. To take these angular-momentum effects into account, expressions for the density and sums of states in Eq. (3.2) have been derived that conserve angular momentum. The case of very loose transition states was first treated by Light and co-workers[51-54] as well as by Nikitin.[55] It was applied primarily to bimolecular reactions in which the transition state was treated statistically. Klots adapted this approach to unimolecular reactions.[48-50,69] More recently, Chesnavich and Bowers[56] derived expressions for the density and sums of vibrational and rotational states. These results can be applied to the cases of both small angular momentum, which is characteristic of unimolecular reactions, and large angular momenta, which are important in bimolecular reactions.

Particularly interesting demonstrations of the role of angular momentum are given by comparisons of the dissociation of some molecule or ion with the breakup of a collision complex of the same molecularity. Such studies have included the dissociation of $C_4H_8^+$ produced by photoionization of C_4H_8 and through the ion–molecule reaction of $C_2H_4^+$ and C_2H_4,[73] as well as similar reactions in the $C_6H_6^+$ system.[74]

2. Detailed Treatments of the Transition State

One of the major problems in the use of the statistical theory lies in characterizing the transition state. Figure 7 shows very clearly that for the case of low angular momentum and a moderately attractive potential, there is no obvious internuclear distance that can be associated with the structure of the transition state. In addition, we often have only a vague idea about the structure of the transition state, which makes choosing the vibrational frequencies difficult. Finally, even when the structure of the dissociating species is known, which is more or less the case for the fragmentation of such ions as CH_3X^+ and $C_6H_5X^+$ (X = Cl, Br, or I), it is not possible to identify a unique structure of the transition state, because this varies with the amount of angular momentum and the total energy.

According to Wigner,[75] the search for the transition-state structure is the search for the configuration through which the reaction trajectory passes once and only once. This structure is generally assumed to be the one with the minimum density of states, or more precisely, the minimum flux. Miller[57] and, more recently, Chesnavich et al.[76] have suggested that there are at least two such structures. The first one, associated with the "tight" transition state, is a result of the conversion of vibrational energy into potential energy as the critical bond is stretched. If the attractive part of the potential is steep, as is the one illustrated in Fig. 7a, this transition state occurs at short internuclear distances. As the bond is stretched, the vibrational density of states decreases. But at the same time, two rotational and a translational degree of freedom are beginning to form, which tends to increase the density of states. Thus the first minimum in the flux (the tight transition state) occurs where these two opposing tendencies balance. The second transition state (the orbiting transition state) is at the centrifugal barrier (see Fig. 7), where the minimum in the flux is a result of the conversion of rotational energy to translational energy. As pointed out by Chesnavich et al.,[76] the minimum flux at low total energy occurs at the orbiting transition state. At higher total energies, the minimum shifts to the tight transition state. Although this model has been used to calculate the dissociation rate of several ionic fragmentations, the data have not been of sufficient quality to detect or verify the presence of the two transition states. The switch from orbiting to tight transition state occurs very close to the reaction threshold. At these energies ionic reactions are often too slow ($\leq 10^3 \, s^{-1}$) to be measurable. Also, the energy range over which the switch of transition states occurs is quite narrow, so that excellent energy resolution would be required to observe it. An important consequence, which will be discussed in more detail in a later section, is that the tight-transition-state energy barrier is below the thermochemical dissociation limit. Hence, the common use of the dissociation limit as the transition-state energy may be in error. This difference between the dynamic and thermochemical dissociation onsets has not yet been verified experimentally.

A more detailed treatment of the transition state, the adiabatic channel theory, has been developed by Troe and Quack.[59-62] This approach considers the "interesting" aspects of the reaction in detail while using the statistical approach for the rest of the degrees of freedom. The example of the halobenzenes illustrates the approach. The benzene ion has 30 vibrational normal modes and no internal rotations, and the dissociated products, $C_6H_5^+$ and X, have 27 vibrational normal modes. Three degrees of freedom have been converted into translations and orbital angular momenta. The usual RRKM–QET formulation treats only the one translation of the critical coordinate uniquely. In the adiabatic channel theory, all three degrees of freedom, the C–X stretch and the two

C–X bends, are analyzed in detail from the molecular structure to the dissociated states. In more complicated reactions, where normal modes change considerably from the reactant to the products, the vibrational frequencies are linearly interpolated through the region of the transition state.

Considerable progress has been made in molecular-orbital calculations of reacting molecular structures.[77-82] Waite et al.[82] have analyzed the fate of the vibrational normal modes of formaldehyde during its dissociation to H_2 and CO. They found that some frequencies remain approximately constant, whereas others vary considerably so that at the transition state, their values differ from those of the molecule by as much as 60%. The problem now is to find a reacting system, that is simple enough to analyze by these more refined treatments and whose dissociation rates and kinetic-energy releases can be studied in sufficient detail to provide us with a good test of the theory. The simple bond-cleavage reactions of ions such as the halobenzenes are ideal candidates, because their rates can be measured over a very wide internal-energy range and because the dissociations proceed via a loose transition state in which the switching of vibrations to rotations and translations occurs over a wide range of internuclear distances.

One problem in the application of these detailed theories is that the structure and vibrational frequencies of the reactants and products must be known. These have been determined only for neutral molecules, usually in their ground electronic states. The frequencies of very few ions are known, in part because the major spectroscopic tool for ions, photoelectron spectroscopy, does not have sufficient resolution to resolve the congested spectra often observed. Also, selection rules, which allow only symmetric vibrations to be excited, limit the number of frequencies that can be observed.[83] Fluorescence from polyatomic ions is largely limited to the relatively rigid acetylenes, diacetylenes, and fluorinated benzenes.[84] This is because radiationless transitions to the ground electronic state rapidly depopulate the excited states. A new development in photoelectron spectroscopy by multiphoton ionization, MPI-PES, is beginning to provide us with some of the vibrational frequencies.[85-90] The phenol ion has been investigated by MPI-PES and as many as 14 vibrational frequencies have been measured.[90] The key to this experimental approach is based on the same principle as that of state selection by MPI, described previously. If a molecule is excited to a particular vibrational level of an excited state (not a Rydberg state), the ionizing transition will populate the ions in a range of vibrational levels determined by the Franck–Condon factors connecting the excited intermediate and the final ion states. By changing the intermediate vibrational level, a new series of final vibrational states can be reached. Thus, selected series of vibrational levels can be excited and analyzed. The analysis is greatly aided if the intermediate neutral-state spectroscopy is well known.

Complications in the identification of a unique transition state when the reaction is a complex one have been discussed by Lorquet.[77-81] The potential-energy surface of even a simple reaction, such as the loss of H or H_2 from CH_2CH_2 (ethylene), includes several minima and maxima and often involves more than one electronic surface.[91] Bent configurations can mix electronic states, resulting in conical intersections rather than simple curve crossings.

D. The Statistical Kinetic-Energy-Release Distribution

There are two major experimental observables in unimolecular reactions. The first is the dissociation rate, as discussed in the previous sections. In cases where there are several products, branching ratios are often measured; however, these are just ratios of rates. The second experimental observable is the kinetic-energy-release distribution (KERD) and its average value.

The product kinetic energies can be best understood in terms of the product phase space. Consider a sphere with an angular-momentum quantum number of J_0 breaking apart into an atom and a sphere of angular-momentum quantum number J and an orbital angular momentum of L. The probability $P(\varepsilon_t, E)$ of the products recoiling with translational energy ε_t at a total energy E above the dissociation limit will be related to the number of combinations of final product vibrational and rotational states. Klots[50] found analytical expressions for these distributions and showed that if the restrictions imposed by the centrifugal barrier are neglected, the unnormalized probability is given by

$$P(\varepsilon_t, T) = \int_0^{E-\varepsilon_t} \rho_v(x)[\rho_r(E-x-\varepsilon_t)]\, dx = \int_0^{E-\varepsilon_t} \rho_v(x)\, dx \qquad (3.4)$$

where ρ_v and ρ_r are the vibrational and rotational densities of states. The rotational degeneracy for a sphere is $(2J+1)^2 \cong 4J^2$. Because the energy also increases as J^2, the rotational density of states ρ_r is just a constant. When the ion is dissociating into products with r final rotational degrees of freedom, the expression in terms of $s = (r-1)/2$ is

$$P(\varepsilon_t, E) = \int_0^{E-\varepsilon_t} \rho(x)[E-\varepsilon_t-x]^{s-1}\, dx \qquad (3.5)$$

Equations (3.4) and (3.5) are approximate, not only because the rotational density of states is treated approximately, but more importantly because the effect of the centrifugal barrier is neglected. The restriction imposed by the centrifugal barrier is important at low translational energies and when the long-range potential is only weakly attractive. In the limit of zero translational energy, the probability $P(\varepsilon_t)$ must go to zero, because the dissociating complex cannot pass over the centrifugal barrier. As is seen in Fig. 7, the less attractive potentials have higher centrifugal barriers. When Eq. (3.5) is corrected for this effect, the result, which is valid for the case of attractive potentials $(s < 6)$, becomes

$$P(E, \varepsilon_t) = \int_{E-\varepsilon_t-\sqrt{s\beta'\varepsilon_t}}^{E-\varepsilon_t} \rho_v(x)[E - \varepsilon_t - x]^{s-1} \, dx \qquad (3.6)$$

The parameter β' has been defined in terms of the reduced mass of the separating fragments, the polarizability of the neutral fragment, and the moments of inertia. It is evident that when $\varepsilon_t = 0$, the integral in Eq. (3.6) goes to zero, because the upper and lower limits are identical. At higher values of ε_t, the integral approaches the result of Eq. (3.5), because $\varepsilon_t \ll \sqrt{s\beta'\varepsilon_t}$.

The formulation of Klots has the advantage of describing the KERD in terms of analytical expressions. However, the approximations made must be adjusted for various conditions such as low- and high-angular-momentum regimes. The more recent work of Chesnavich and Bowers[56] treats the problem exactly, but at the expense of simple analytical expressions.

The average translational energy for the dissociation can be obtained either from Eqs. (3.4)–(3.6) or, more simply, from the following considerations, again due to Klots.[49] Suppose that a dissociating system at a total energy of E forms products with r rotational and v vibrational degrees of freedom. Although this is a microcanonical system, we can define an approximate temperature T^* and express the total energy as the sum of the average translational, rotational, and vibrational energies by

$$E = RT^* + \frac{r-1}{2} RT^* + \sum_{i=1}^{v} \frac{h\nu_i}{\exp(h\nu_i/RT^*) - 1} \qquad (3.7)$$

in which the vibrational frequencies of the products are the ν_i's. Conservation of momentum has reduced the three translational and rotational degrees of freedom to two each. The translational energy of the products is just the first term in Eq. (3.7), which is RT^*. Although Eq. (3.7) cannot be solved for RT^* in terms of E, it is easily evaluated for a number of values of RT^*.

E. Under What Circumstances Might Dissociations Be Statistical?

Figure 8 shows some typical situations encountered in the decay of ionic species. In Fig. 8a, the ion is initially prepared in an excited electronic state. Often, a rapid, radiationless transition (internal conversion) changes the electronic energy into vibrational energy of the ground electronic states, from which a slow dissociation takes place. The statistical theory treats only the problem of vibrational predissociation on a single potential-energy surface. Specifically, the assumption of a rapid, radiationless transition from the excited to the ground electronic state is not inherent in the statistical theory, as is often assumed. The initially excited state can decay by fluorescence, by a direct dissociation, by electronic predissociation, or by radiationless transition. Furthermore, all of these processes are in competition. Some examples of such decays are known, but in the majority of polyatomic-ion dissociations, internal conversion to the ground electronic state is the dominant process. Once the ion has converted its electronic energy to vibrational energy, the statistical process of dissociation can take place.

The initially prepared state in Fig. 8a is bound. In Fig. 8b the excited state of the ion is dissociative. This type of fragmentation is not a vibrational predissociation and thus cannot be treated by the statistical theory. That is, there are no combinations of vibrational quantum numbers that correspond to a stable molecular structure. Such a dissociation will be extremely fast, of the order of a vibrational frequency.

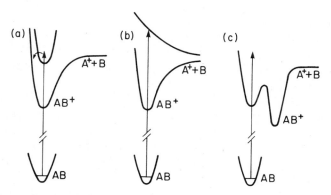

Figure 8. Three different possible potential-energy curves (in reality they are surfaces) commonly encountered in the dissociation of molecules and ions. The dissociations for (a) and (c) can be treated by the statistical theory, whereas the direct dissociation of (b) is not a statistical process.

Figure 8c shows a situation frequently encountered in ionic dissociations, namely that isomerization barriers, which lie below the dissociation barriers, cause the ion to rearrange prior to dissociation. In such cases, the dissociation rate will be characteristic of the rate from the lowest-energy isomer. Both parts of the reaction, the isomerization and the dissociation steps, can be treated with the statistical theory. However, it is evident that if isomerization to the more stable state is not recognized, the dissociation might be judged to be nonstatistical. In this regard, we note that a reaction that is slower than statistically predicted may still be statistical, due to isomerization to lower-energy states. On the other hand, a reaction rate faster than statistically predicted would have to be considered nonstatistical. A particularly interesting case of isomerization is one in which there is competition between direct dissociation and isomerization followed by a slow dissociation. As will be shown in Section IV.C.3 for the pentene-ion dissociations, this situation leads to nonexponential decay. However, all of it is consistent with the statistical theory.

IV. THE EXPERIMENTALLY MEASURED DISSOCIATION RATES

Dissociation rates of ions that are on the time scale of typical mass spectrometers, 10^{-3}–10^{-7} s, can be measured by both PEPICO and MPI techniques. One method of measuring such a rate is based on the fact that product ions that are formed in an acceleration field leave that field with kinetic energies that depend on the position of dissociation. Suppose that a parent ion of mass m_p starts from rest in an acceleration field of V volts per centimeter with a length of d centimeters. At the end of the d centimeters, the parent ion will have a kinetic energy of Vd electron volts. Similarly, a daughter ion of mass m_d formed at rest and accelerated over d centimeters will have kinetic energy of Vd electron volts. But a daughter ion formed during the acceleration process, say after traveling s centimeters as a parent ion, will have a kinetic energy of $(m_d/m_p)sV + (d-s)V$ electron volts. This is because at the time of dissociation, the parent ion breaks up into a neutral fragment and an ion, so that the share of the energy going to the ion is just the ratio of the masses, m_d/m_p. This decreased kinetic energy can be measured either by an electrostatic energy analyzer or by TOF. In either case, the position of dissociation can be deduced from the fragment-ion kinetic energy. This position is in turn related to the time of dissociation. In a TOF measurement, the daughter-ion peak shape is distorted toward longer times and has the general shape of an exponential. It is not a true exponential, because the time of dissociation and the shift in the fragment-ion TOF are not linearly related. An analysis of this asymmetric TOF distribution yields the ion-dissociation rate. In addition, it is possible to determine whether the reaction is best described by a single-exponential decay or a two- (or more) exponential decay.

Another approach to measuring the rates is to prepare the ions in a small and nearly field-free region for a prescribed time after the arrival of the energy-analyzed electron. The ion is then extracted by applying a large-voltage pulse, after which it is mass analyzed and detected. If the parent ion has dissociated before the application of the extraction pulse, it will be collected as a daughter ion. The probability of detecting a daughter ion for an extraction-pulse delay of τ s is given by

$$\text{probability} = \int_0^\tau e^{-kt}\, dt \Big/ \int_0^\infty e^{-kt}\, dt \tag{4.1}$$

By varying the time delay τ between electron detection and the ion-extraction pulse, the integral of the dissociation rate can be determined. In practice, only two or three values of τ are explored; these are in the range 1–50 μs.

A. Test of the Statistical Theory—Dissociation Rates

The statistical theory has now been applied to numerous ion-dissociation-rate studies and it has been found to account rather successfully for the data both qualitatively and quantitatively. Because the time scale is of the order of microseconds, these studies have all been carried out on large molecular ions that dissociate on this time scale. Two examples of the good agreement between theory and experiment are discussed below.

1. $C_6H_5Br^+$ Dissociation Rates

One of the assumptions of the statistical theory is that all vibrational degrees of freedom are involved in the flow of internal energy during the dissociation of the molecule. The loss of the Br atom from the ions $C_6H_5Br^+$ and $C_6D_5Br^+$ is one of the simplest unimolecular reactions. It proceeds without any reverse activation energy. Because the structure of the $C_6H_5^+$ ion is very similar to that of benzene itself, the reaction is not unlike that of the dissociation of a diatomic molecule. The big difference is that the phenyl ring acts as an energy sink for storing and redistributing the internal energy. The reaction energetics for $C_6H_5Br^+$ and $C_6D_5Br^+$ are virtually identical, because the CH vibrational frequencies do not change significantly as the system goes from the ion to the transition state. Thus, the only difference between the two reactions is that the density of vibrational states is considerably higher for the deuterated molecule, because all the CH vibrational frequencies are reduced approximately by $\sqrt{2}$. Therefore, if all of these CH vibrations participate in the distribution of the internal energy, these frequency differences should be reflected in the rates.

An example of the experimental data at a photon energy of 12.66 eV is shown in Fig. 9.[92] The asymmetric TOF distributions are shifted with respect to each other because of the different masses of the two fragments. However, the important difference is in the two dissociation rates, which were determined from the data by fitting calculated TOF distributions to the experimental points. The only adjustable parameter was the indicated dissociation rate. Similar data at several ion internal energies led to the plot in Fig. 10, which shows the dissociation rate as a function of the ion internal energy.

The solid lines in Fig. 10 are the rates calculated by use of the statistical theory. The parameters needed for doing these calculations are the vibrational frequencies and the moments of intertia of the ion and of the transition state, as well as the activation energy. Most ion frequencies are not known, and the transition-state frequencies are known to an even lesser extent. Nevertheless, reasonable guesses can be made. The activation energy for this reaction was not well known until recently, because the $C_6H_5^+$ heat of formation was not established. However, this problem is

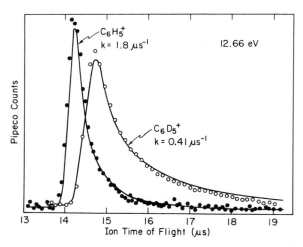

Figure 9. PEPICO fragment-ion TOF distributions from bromobenzene and deuterated bromobenzene at a photon energy of 12.66 eV. The points are the experimental data, while the solid lines were calculated for the indicated ion-dissociation rates. Taken with permission from Baer and Kury.[92]

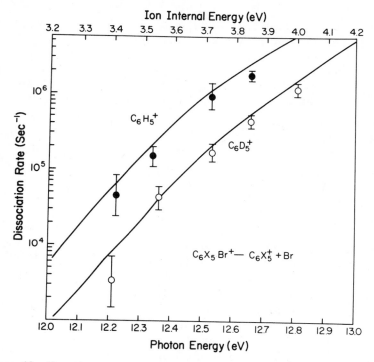

Figure 10. Bromobenzene and deuterated bromobenzene ion-dissociation rates as a function of ion internal energy. The points are the experimental results, and the solid lines are the statistical-theory results. The vibrational frequencies and activation energy used in the statistical-theory calculation for the $C_6H_5Br^+$ reaction were identical to those used by Rosenstock et al.[93] Taken with permission from Baer and Kury.[92]

now nearly resolved. Assuming a value of 1133 ± 5 kJ/mol for $\Delta H_{f0}{}^0$, the 0 K value for E_0 is 2.76 eV.[93] With reasonable frequencies and this value for E_0, the RRKM–QET result is the line through the solid points.

When the CH frequencies are reduced by $\sqrt{2}$, the calculated line falls perfectly on top of the data for the dissociation of deuterated bromobenzene. This perfect fit indicates that all of the CH modes do take part in the energy flow, even though these modes are unrelated to the reaction coordinate. It should be pointed out that this type of isotope effect is neither a primary nor a secondary isotope effect. Those isotope effects show up only if there is a change in vibrational frequency between the molecule (or ion) and the transition state. In bromobenzene, there is probably no significant change in these CH frequencies. In fact, according to the statistical theory, the isotope effect illustrated in Fig. 10 is observable only in the rates of energy-selected molecules or ions, and not in the canonical rates $k(T)$. The reason for this is that in the canonical transition-state theory, in which the rate constant is expressed in terms of canonical partition functions, the partition functions are evaluated at a common temperature for the molecule and the transition state.[94] As a result, all the partition functions of vibrational frequencies that are identical in the molecule and the transition state will cancel. This is not true for the microcanonical rates of Eq. (3.2), in which not only are the numerator and the denominator evaluated at different energies, but the former is a sum over states while the latter is the density of states.

Besides the choice of the vibrational frequencies for the ion and the transition state, one additional assumption was made in carrying out the statistical-theory calculations, namely that a rapid, radiationless transition converted the initially excited electronic state into vibrational energy of the ground electronic state prior to dissociation. This assumption was introduced by Rosenstock et al.[38] when the theory was first applied to the problem of metastables in mass spectrometry. Quantitative rate studies have shown that all slow reactions (i.e., those in the microsecond time range) are preceded by such radiationless transitions. The assumption appears to fail only when the reaction is very rapid.

Although the data and the calculations of the dissociation rates of energy-selected ions are done in terms of microcanonical ensembles, it is instructive to calculate their canonical counterparts in transition-state theory in order to compare the entropy of activation for the ionic reaction with that for neutral reaction. The entropy of activation ΔS can be calculated as follows from the partition functions $Q^{\ddagger}$ and Q for the transition state and precursor ion respectively:

$$\Delta S = k \ln \frac{Q^{\ddagger}}{Q} = k \ln(\prod q_i^{\ddagger} / \prod q_i) \qquad (4.2)$$

The total partition function Q is just the product of the individual vibrational partition functions, $q_i = [1 - \exp(-h\nu_i/kT)]^{-1}$. The entropy of activation is a convenient parameter for defining the terms "loose" and "tight." If the vibrational frequencies in the transition state and ion are identical, $\Delta S = 0$. A loose transition state will give rise to a $\Delta S > 0$, while a tight transition state has a $\Delta S < 0$. The best statistical-theory fit to the bromobenzene data of Fig. 10 gives $\Delta S(1000 \text{ K}) = 7.38 \text{ cal/mol}$. This translates into a preexponential factor $[(kT/h)\exp(\Delta S/R)]$ of $9.5 \times 10^{14} \text{ s}^{-1}$. As pointed out by Rosenstock et al.,[93] this is very close to the value of $2 \times 10^{15} \text{ s}^{-1}$ for the dissociation of the neutral bromobenzene molecule.[95] It is interesting that the ionic transition state is, if anything, "tighter" than its neutral counterpart.

The $C_6H_5Br^+$-dissociation rates in Fig. 10 are known quite accurately, since they have been measured by two groups with very different approaches. Rosenstock et al.[93] used the method illustrated by Eq. (4.1), while Kury and Baer[92] used analysis of the asymmetric peak shapes (Fig. 9). The two sets of data agree perfectly. This would therefore seem to be an appropriate reaction to analyze in terms of the refined statistical theories discussed earlier. However, the major problem is that we do not know any of the vibrational frequencies of the bromobenzene ion, and the assumption that the ion and neutral frequencies are the same is surely suspect. Until these are available, use of the more sophisticated calculations does not seem warranted.

2. $C_6H_5I^+$ Dissociation Rates—A Benchmark Study

The most thoroughly studied ionic dissociation is that of iodobenzene. Not only have several groups measured the rates of these processes, but they have been investigated over a very wide internal-energy range. It is particularly significant that the rates have been measured by the appearance energies of metastables in photoionization[96] and by threshold PEPICO[97,98] and He(I) PEPICO.[98] The latter two techniques differ in one very important respect, which is illustrated in Fig. 11. The solid line is a He(I) photoelectron spectrum (PES) of iodobenzene, for which the abscissa (energy scale) corresponds to $21.2\,eV - E_{el}$ where E_{el} is the photoelectron energy. The dotted line is a threshold PES, for which the abscissa is just the photon energy. In addition to the normal iodobenzene

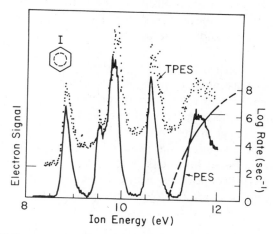

Figure 11. He(I) PES (solid line) and threshold PES (dotted data) of iodobenzene. The measured dissociation rate as a function of the ion energy is shown on the right. Spectra and rate data adapted from Dannacher et al.[98]

ion electronic states, threshold ionization, represented by the TPES, leads to ion states that are inaccessible by PES. This is clearly shown by the rising base line of the dotted curve, which represents about 50% of the electron signal at 11.5 eV. This failure of the threshold electron signal to go to zero between electronic states can be attributed to autoionization.

$$AB + h\nu \rightarrow AB^+ + e^- \qquad \text{direct ionization} \qquad (4.3)$$

$$AB + h\nu \longrightarrow AB^* - \begin{cases} \rightarrow AB^+ + e^- & \text{autoionization} \qquad (4.4) \\ \\ \rightarrow A + B & \text{predissociation} \qquad (4.5) \end{cases}$$

In the direct ionization of Eq. (4.3), those ion states that have good Franck–Condon factors connecting the ground neutral and ionic states are produced. On the other hand, in autoionization, the Franck–Condon factors between the intermediate AB^* state and the ion state are important. In addition, Guyon et al.[99,100] have clearly shown that predissociation and autoionization are strongly coupled, so that ions with up to several electron volts of vibrational energy are initially formed. Thus, although the ions produced in direct ionization and autoionization have the same internal energy, they are not initially in the same states.

The data of Fig. 12 for the iodobenzene ion dissociation were collected by measuring the relative abundance of $C_6H_5^+$ daughter ions as a function of the ion internal energy. These curves were obtained for several time delays between ionization (the arrival of the energy-selected electron) and the ion-extraction pulse. By fitting the dissociation rate $k(E)$ to these curves and by the use of Eq. (4.3), the rates shown in Fig. 11 were obtained.

The rates derived from the He(I) and threshold PEPICO data agree extremely well. This indicates that the dissociation rate is a function only of the total energy, and not of the particular microstates initially populated. In addition, the experimental data and the statistical theory agree well over a rate range from 5×10^2 to $1.3 \times 10^6 \, \text{s}^{-1}$. Finally, it is evident from Fig. 11 that the initial state excited is primarily the fifth electronic state of $C_6H_5I^+$ (the second peak is comprised of two partially resolved electronic states). Yet the dissociation rate is entirely consistent with dissociation from the ground electronic state. This demonstrates once more that the radiationless transition rate is more rapid than the dissociation rate.

The entropy of activation for this reaction was calculated to be 6.4 cal/mol K, which is very similar to the value found for the bromobenzene ion dissociation. However, no neutral iodobenzene dissociation data

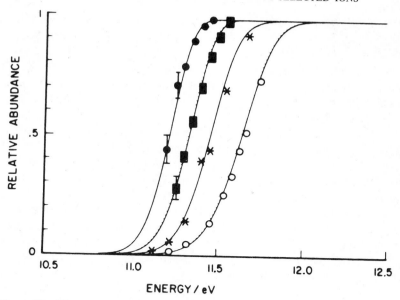

Figure 12.   The relative abundance of $C_6H_5^+$ ions from energy-selected iodobenzene as measured by PEPICO at four sampling times: 1 μs (open circles), 5.9 μs (asterisks) 21 μs (filled squares) and 57 μs (filled circles). Taken with permission from Dannacher et al.[98]

are available for comparison. As with bromobenzene, the iodobenzene-ion vibrational frequencies have not been measured, so that the best we can do is to use the simple statistical theory.

B.   The Kinetic-Shift Problem

One of the standard methods for determining the heat of formation ΔH_f^0 of a fragment ion A^+ is to measure the fragment-ion signal from some molecule AB as a function of the ionizing-photon energy. The onset, or appearance potential (AP), is related to the fragment ΔH_f^0 value by the thermochemical cycle:

$$\Delta H_f^0(A^+) = AP + \Delta H_f^0(AB) - \Delta H_f^0(B) \qquad (4.6)$$

When appropriate care is exercised in taking into account the thermal energy of the parent molecule AB, heats of formation at both 0 and 298 K can be obtained. However, the whole cycle is based on a good knowledge of the AP, and on the appropriateness of associating this AP with the thermochemical dissociation limit. Clearly, if there is a reverse activation energy, the AP will be too high. However, such activation barriers can often be detected by measuring the fragment-ion kinetic energy (see Section V). A more insidious problem arises when the dissociation rate at the thermochemical dissociation limit is so slow that an insufficient number of parent ions dissociate prior to detection. The result is that the observed onset is shifted to higher ion internal energies. This shift in the onset was recognized a number of years ago by Chupka, who called it the *kinetic shift*.[101]

This shift has now been experimentally verified in a number of different fashions. A direct approach is to utilize a pulsed ionization source, trap the ions for a specific time to let them dissociate, and then to pulse them out of the source for analysis. This has been done in coincidence experiments (see, e.g., Fig. 12) with delay times as long as 57 μs, as well as in ion traps that store the ions for as long as milliseconds.[102] Lifshitz and others have used the latter technique to measure the decrease in the AP as a function of trapping time for such reactions as H loss from $C_6H_6^+$,[103] HCN loss from $C_5H_5N^+$,[104] CO loss from $C_6H_5OH^+$,[105] and HNC loss from $C_6H_5NH_2^+$.[106] The data for the aniline dissociation are shown in Fig. 13. In these experiments, the ionizing source was electron impact with an energy resolution of ~0.2 eV. Because of the rather modest energy resolution, these experiments gave only approximate values of the true appearance energies. Coincidence experiments, with their continuously working light sources, are not very useful

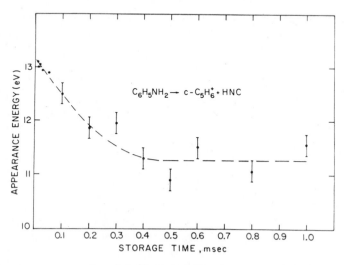

Figure 13.   Appearance energy of $C_5H_6^+$ from aniline ions formed by electron impact as a function of the ion storage time. Taken with permission from Lifshitz et al.[106]

when long trapping times are desired, because ions are continuously being made; thus, when the extraction pulse is applied, there will always be an ion (or more likely several ions) in the trap. The use of a conventional pulsed vacuum-UV photon source has improved the resolution considerably,[107] but the technique suffers from a low signal level. A better approach would be to use a pulsed vacuum-UV laser, because the timing would then be automatically built in. However, such lasers are not yet able to reach the photon energies necessary to dissociate ions at energies between 12 and 14 eV.

1.   The Dissociation of Aniline Ions

The dissociation limit for the aniline-ion fragmentation has also been determined from a PEPICO experiment.[108] Figures 14 and 15 show the

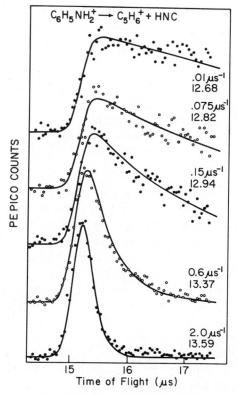

Figure 14.   Coincidence TOF distributions of $C_5H_6^+$ fragment ions from aniline. The solid lines are calculated distributions in which the indicated rates were assumed. Taken with permission from Baer and Carney.[108]

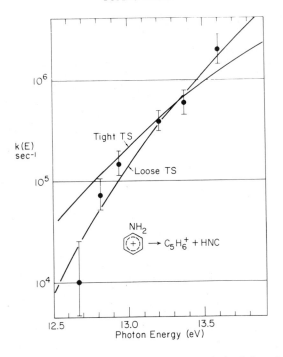

Figure 15.   The dissociation rates obtained from the analysis of the coincidence data of Fig. 14. The solid lines are statistical-theory rates calculated using a tight and a loose transition state. Taken with permission from Baer and Carney.[108]

data and the deduced dissociation rates. Two parameters were varied in comparing the experimental and statistical-theory dissociation rates: the looseness of the transition state and the dissociation limit. As long as the rate is known at only a single energy, the two parameters cannot be distinguished. However, when the data are known over a large energy range, the difference is clearly discernible (Fig. 15). In the case of aniline, the loose transition state that consisted of frequencies characteristic of an open-chain $C_6H_7N^+$ ion gave a much better fit than a tight transition state that had the same frequencies as the aniline molecule itself. There are more ways to pass through a loose transition state than a tight one; thus the former's predicted rate should be higher. On the other hand, this higher rate can be compensated by raising the activation energy. For this reason the tight transition state was consistent with a 10.43-eV dissociation limit, whereas the loose transition state fit the data best when a limit of 11.2 eV was assumed.

Figure 16 shows the energy diagram appropriate for the aniline ion. It

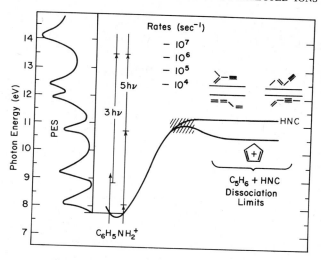

Figure 16.   Potential-energy curve for the aniline-ion dissociation. The actual dissociation limit is to the HNC structure via the loose transition state at 11.1 eV (see text). The
hatched region near the transition state is the range of dissociation limits determined from
the theoretical fit to the measured rates (Fig. 15). Taken with permission from Baer and
Carney.[108]

is evident that neither the tight nor the loose dissociation limit is consistent with any of the noncyclic $C_5H_6^+$ product ions. This is important,
because previous studies of this reaction found the dissociation onset to
be between 12.0 and 12.6 eV, very near the energies of some of these
straight-chain isomers. As a result, it was assumed that the $C_5H_6^+$ products
have the open-chain structure. However, this energy range also corresponds to a fragmentation rate of about 10^3–$10^4 s^{-1}$, which means that at
energies below this, the rate becomes too slow to give an observable
product yield. Thus the kinetic shift in aniline is ~1 eV.

Both $C_5H_6^+$ and HCN have well-known heats of formation. Their sum
minus the heat of formation of the aniline molecule is 10.43 eV, which is
considerably below the 11.20 eV obtained from the loose-transition-state
fit to the rate data. On the other hand, the result of Lifshitz in Fig. 13,
11.26 ± 0.2 eV, is consistent with the PEPICO result. Lifshitz et al.[106]
resolved this paradox by suggesting that the neutral product is HNC
rather than HCN. The former has been calculated to be 0.63 eV less
stable than HCN.[109] Thus the dissociation limit would be 11.06, which is
very close the 11.2 eV predicted by the loose-transition-state calculation.

Other recent investigations of the HCN/HNC structure produced in the fragmentation of the aniline ion have shown without doubt that the structure is HNC. The identification in emission of NH and CN from products of the aniline- as well as the pyridine-ion dissociation suggested that the neutral fragment from the aniline reaction is HNC whereas that from the pyridine ion is HCN.[110] These results have been confirmed in a very clever experiment by Holmes et al.[111,112] in which the metastable aniline or pyridine ions are accelerated to several kilovolts. When they dissociate in the drift region of the mass spectrometer, both the ionic and the neutral fragment are traveling at high velocities. After deflection of the ions, the neutral fragments are collided with He. Enough energy is imparted that they ionize and fragment further. The observation of a large CH^+ peak in the study of pyridine and no CH^+ peak in the study of aniline clearly indicated that the neutral fragment in the aniline dissociation is HNC. This structure is also the most reasonable one from a mechanistic point of view, since the H atom is already attached to the nitrogen atom in the aniline molecule.

The best fit for the transition-state frequencies shows it to be moderately loose, with a $\Delta S(1000\,K)$ value of 4.27 cal/mol. This corresponds to a preexponential factor at 1000 K of $2\times10^{14}\,s^{-1}$. The fact that the ΔS value is less than it is for the halogen loss from the halobenzene ions is also consistent with the more complicated transition state involved in the conversion of a six-membered ring into a five-membered ring.

2. The Dissociation of Benzonitrile Ions

In contrast to the aniline-ion fragmentation, which produces the higher-energy HNC molecule, the benzonitrile ion ($C_6H_5CN^+$) dissociates into the products $C_6H_4^+$ and HCN. These rates have been determined by Rosenstock et al.[113] using the PEPICO delayed-ion-extraction technique. As shown in Fig. 17, these results agree perfectly with the earlier He(I) PEPICO data of Eland and Schulte[114] and the statistical-theory calculations of Chesnavich and Bowers.[115] The dissociation limit derived from the statistical-theory fit to the rate data is 12.72 eV. This is just slightly lower than the thermochemical limit of dissociation to the cyclic benzyne ion ($C_6H_4^+$) and the HCN molecule of 12.76 eV. Could this difference be a result of the transition state lying 40 meV below the dissociation limit (recall section III.C.2)? The 1000 K entropy of activation was found to be 2.9 eV, which is rather tight, considering that all the products are formed in their ground states. It is not known from which ring position the H atom leaves to attach itself to the carbon atom. However, these data certainly will provide guidelines for any future molecular-orbital calculations on this ionic dissociation. The measured onset for HCN loss from

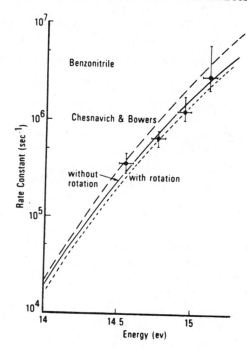

Figure 17. Calculated and measured dissociation rates for the decay of benzonitrile ions to $C_6H_4^+ + HCN$. Points from Eland and Schulte,[114] solid and short-dashed lines from Rosenstock et al.,[113] and long-dashed line from Chesnavich and Bowers.[115] Taken with permission from Rosenstock et al.[113]

the benzonitrile ion is ~14.3 eV, whereas the true dissociation limit is 12.76. The difference of 1.5 eV is a result of the kinetic shift.

Dissociation reactions that have large activation energies and large densities of vibrational states have extremely slow rates near the reaction threshold. In the case of the aniline ion, the rate is calculated to be $6 \times 10^{-3} s^{-1}$. That is, the lifetimes exceed 100 s! There is an interesting problem associated with measuring ion lifetimes down to milliseconds, or even seconds. In principle, long-lifetime measurements are possible in an ion cyclotron resonance (ICR) spectrometer.[116] However, ions can decay by IR fluorescence with rates of the order of $10^2 s^{-1}$.[108,117–120] Therefore, this decay process becomes more rapid than dissociation. Infrared fluorescence of highly excited polyatomic ions is a most interesting experimental and theoretical problem. The actual observation of IR photons appears to be very difficult, because ion densities must be kept very low to avoid ion–molecule collisions. Thus, the occurrence of IR emission has been inferred rather than directly measured.[117–120]

3. *The Dissociation of Chlorobenzene Ions*

Chlorobenzene is the third of the halobenzene ions for which dissociation rates have been investigated using energy-selected ions. The C–X bond energies for the three halogens are 3.19 (Cl), 2.76 (Br), and 2.34 eV (I). Thus, at a given energy above the dissociation limit, the chlorobenzene ion will dissociate with the slowest rate, and will therefore exhibit the largest kinetic shift. This is illustrated in Fig. 18, in which the onset of 13.0 eV measured by photoionization mass spectrometry is 0.74 eV above the dissociation limit of 12.26 eV.

The dissociation rates of energy-selected chlorobenzene ions have been measured by several techniques. In the first determination, by Baer et al.,[97] the rates were extracted from the asymmetric $C_6H_5^+$ TOF peak shapes. Because of the low electron-energy resolution and a calculational error in analyzing these early results, the derived rates were about a factor of 2 higher than they should have been. Rosenstock et al.[122] repeated these measurements using the delayed-pulsing technique and obtained data similar to those of Fig. 12. Pratt and Chupka[96] used the derivative of the metastable peak intensity in a photoionization mass spectrometer to estimate the ion-dissociation rate. These measurements allowed them to obtain the rate only over a small energy range. However, because this range is at a low energy, the data complement those from the

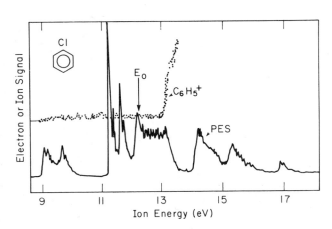

Figure 18.   Chlorobenzene PES from Kimura et al.[121] and photoionization coincidence curve for $C_6H_5^+$ ions from Baer et al.[97]

other techniques in a most valuable manner. Finally, Durant et al.[26] energy selected the ions by MPI and measured the rates from the TOF peak asymmetry. All these data are illustrated in Fig. 19. Aside from the early PEPICO data (open squares), agreement among the data is quite good.

The dissociation of chlorobenzene may represent our best hope of applying some of the more refined versions of the statistical theory. We now know the dissociation thermochemistry extremely well. In addition, chlorobenzene is a relatively small molecule, so it may be possible to carry out ab initio calculations of the potential-energy surface during the dissociation. Finally, we have some good data on the ion vibrational frequencies from the work of Anderson et al.[89] using MPI-PES. A total

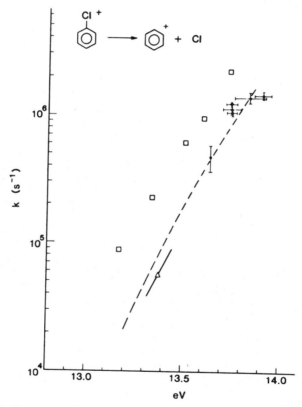

Figure 19.   Rate of chlorobenzene ion dissociation to $C_6H_5^+$ as a function of ion internal energy. Open triangle from Pratt and Chupka,[96] open squares from Baer et al.,[97] dashed line from Rosenstock et al.,[122] and solid points from Durant et al.[26] Taken with permission from Durant et al.[26]

of five vibrational frequencies were determined and compared with those of the neutral ground-state molecule and the excited $^1B_2$-state molecule. As for phenol, the ion frequencies were closer to those of the excited neutral molecule than to those of the ground state. Some of the changes were quite significant. For instance, the 18b and 6b modes change from 294.7 and 614.9 cm^{-1} in the molecule to 226 and 451 cm^{-1} in the ion. Such changes in the lower frequencies tend to have a large effect on the calculated rates.

C. Competition between Isomerization and Dissociation

As indicated in Fig. 8c, excited ions may isomerize prior to dissociation. This situation has been recognized in mass spectrometry for a number of years.[123] In fact, it is this rapid isomerization that makes the analysis of isomers very difficult, because they have similar mass-spectral patterns. Often the isomerization involves the making and breaking of several bonds, as, for instance, in the rearrangement of $CH_3CCCCCH_3^+$ (2,4-hexadiyne) to an ion having the benzene structure.[124] In principle, the 2,4-hexadiyne ion could fragment by losing a methyl group, which would distinguish it from the benzene ion. However, it is evident that isomerization to the benzene ion is preferred energetically over direct fragmentation. We have then the classic competition between the lowest-energy path and the lowest-entropy path. At low ion internal energies, the energetic considerations are dominant. However, as the ion internal energy is raised, the entropically favored direct dissociation ought to dominate. Thus, whether an ion isomerizes or dissociates directly is a function of the ion internal energy and the height of the isomerization barrier.

1. *Isomerization Rate ≫ Dissociation Rate*

The condition of isomerization rates exceeding the direct-dissociation rates is experimentally exhibited in two ways. First, the dissociation rate of two isomers prepared at the same total energy will be identical and should be consistent with the statistically predicted rate for the lower-energy isomer. Second, the dissociation products, and the fraction of each one produced, ought to be identical.

a. *The Hydrocarbons*: $C_3H_4^+$, $C_4H_6^+$, $C_6H_6^+$, $C_8H_8^+$

These highly unsaturated hydrocarbons have very low isomerization barriers.[124–131] The smallest, the $C_3H_4^+$ ions, which include allene, propyne, and cyclopropene, have relatively fast dissociation rates even at threshold.[126] The lowest-energy dissociation path is H loss to form the cyclopropenium ion, $C_3H_3^+$. The shift in the onset as a function of the delay time in ion extraction has been measured for allene.[126] The dissociation rates derived from this shift have not been published as such; however, they can be calculated from the reported entropy of activation of -2.37 cal/mol K. This negative activation entropy indicates a tight transition state, which is consistent with the formation of the cyclic ionic product.

Other, higher-energy dissociation products are $C_3H_2^+$ and C_3H^+. Figure 20 shows the breakdown diagram for the three ions. The fractional abundances of the products are plotted as a function of the parent-ion internal energy. The energy scale is based on the total energy $[\Delta H_f^0$ (molecule) $+ h\nu]$ to allow all three molecules to be plotted on the same scale. Between total energies of 15 and 18.5 eV, the $C_3H_2^+$ ion is formed by the direct loss of the H_2 molecule from the parent ion. At ~ 19 eV the $C_3H_2^+$ ion can also be formed by H loss from the $C_3H_3^+$ ion. The change in slope is a result of this reaction. At about the same energy, the C_3H^+ ion is formed as a secondary product either by H loss from $C_3H_2^+$ or by H_2 loss from $C_3H_3^+$. All three $C_3H_4^+$ isomers have the same products, in the same intensities, with identical onsets. The perfect correspondence in the breakdown diagrams of the three isomers is a strong indication that the ions isomerize to a common structure prior to dissociation. Among the neutral isomers, propyne is the most stable. However, among the ions, the lowest-energy (by ~ 0.6 eV) isomer has the allene structure. Thus, the other two isomers will isomerize to it to maximize their entropy.

It is significant that the breakdown diagrams in Fig. 20 are identical up to 8 eV above the dissociation limit, especially considering that the products are formed via a tight transition state. We know this because the

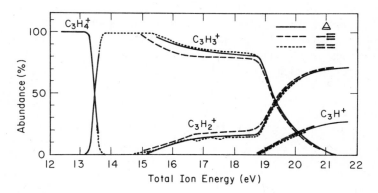

Figure 20.   Breakdown diagram for three isomers of $C_3H_4^+$ on a common energy scale of $h\nu + \Delta H_f^0(C_3H_4)$. Adapted from figures in Parr et al.[125,127,128]

$C_3H_3^+$ ion has the cyclopropenium structure.[132] For this product to be formed, the transition state must also have this structure. In fact, the cyclopropene ion is probably a good candidate for the transition-state structure. Even though the direct fragmentation of propyne to form CH_3 with $\Delta H_f^0 = 34.8$ kcal/mol (reference 133) and C_2H^+ with $\Delta H_f^0 = 389.9$ kcal/mol (reference 134) is energetically possible at a total energy of 18.4 eV, it does not occur. The isomerization to allene is too rapid for the direct bond cleavage to compete.

The break in the $C_3H_3^+$ curve occurs at about 18.8 eV. This is 5.3 eV above the onset of the formation of $C_3H_3^+$ at 13.5 eV. A typical C–H bond energy is ~ 100 kcal/mol, or 4.34 eV. Thus, the observed onset of H loss is 1 eV above the calculated one. However, if we take into account that the linear form of $C_3H_3^+$ is 1 eV less stable than the cyclic form and that the $C_3H_2^+$ ion is probably linear, the expected energy is indeed 5.3 eV, as experimentally found.

Six isomers of $C_4H_6^+$ have been investigated by PEPICO.[129,135] These are, in decreasing order of stability, 1,3-butadiene (the lowest-energy isomer), 1,2-butadiene, cyclobutene, 2-butyne, 1-butyne, and methyl cyclopropene. They all dissociate to the products, $C_3H_3^+ + CH_3$, with the same rate, even though their activation energies vary from 2.3 eV (1,3-butadiene) to 0.5 eV (methyl cyclopropene). This is clear evidence that when the parent ions are prepared with moderate excess energies, these various structures isomerize prior to dissociation. It would be interesting to measure the breakdown diagrams of the $C_4H_6^+$ isomers to determine whether they continue to rearrange at higher ion internal energies. A recent study of Bombach et al.[130] suggests that the appearance of $C_2H_4 + C_2H_2$ from 1,3-butadiene at an energy 1.3 eV above the $C_3H_3^+$ onset is inconsistent with the statistical theory. It would be surprising if all the isomers behaved similarly in this respect.

The benzene ion and its numerous isomers (in principal, there are over 100 of these) have been investigated more often than any other ion. At low energies, the benzene ion dissociates to the products shown in Eqs. (4.7)–(4.10), in which the 0 K dissociation limits[136] are given in parentheses:

$$C_6H_6^+ \rightarrow \begin{cases} C_6H_5^+ + H & (12.654\ eV) & (4.7) \\ C_6H_4^+ + H_2 & (12.934\ eV) & (4.8) \\ C_4H_4^+ + C_2H_2 & (13.722\ eV) & (4.9) \\ C_3H_3^+ + C_3H_3 & (13.608\ eV) & (4.10) \end{cases}$$

If this reaction is consistent with the statistical theory, the four reaction channels should be in competition with one another and the relative rates should be determined by the activation energy and the entropy of reaction.

To a large extent the great interest in this reaction was the result of an early investigation by Andlauer and Ottinger[137,138] in which the ions were energy selected by electron transfer, a method based on the assumption that all the energy of a reaction such as $Xe^+ + C_6H_6 \rightarrow Xe + C_6H_6^+$ remains as internal energy. By using a variety of ions with different ionization potentials, it is possible to prepare benzene ions with a variety of energies. Andlauer and Ottinger's study found that the dissociation rates measured with respect to the H-loss and C_3H_3-loss channels were different. Just as fluorescence lifetimes probe the excited-state lifetime with respect to all decay paths, so too are the dissociation rates a measure of the sum of all decay processes. For this reason, the decay rates measured by the appearance of $C_6H_5^+$ or $C_3H_3^+$ should be identical. The contrary observation by Andlauer and Ottinger led them to conclude that there were two types of parent ions, perhaps two different electronic states, that decay via independent paths, that is, that the various dissociation paths were not in competition.

Even though there is a considerable body of evidence that indicates that the original data for the H-loss rate are in error, the controversy is still not completely resolved.[136] The difficulty in measuring the H-atom loss is part of the problem. Because of the small mass difference between $C_6H_6^+$ and its product $C_6H_5^+$, parent and daughter ions have nearly the same TOF and the asymmetry of the daughter-ion TOF peak is not measurable. The major argument against the existence of two $C_6H_6^+$ precursors is an indirect one. At least two other $C_6H_6^+$ isomers are known to rearrange to the benzene structure prior to dissociation. This is deduced on the basis of the identical dissociation rates for all three isomeric ions as measured with respect to the loss of both C_3H_3 and C_2H_2. It seems most unlikely that the two higher-energy isomers, which have structures very different from benzene, isomerize to an excited state of benzene from which they dissociate to $C_4H_4^+$ and $C_3H_3^+$. In fact, Baer et al.[124] were able to fit all of the dissociation rates to the statistical theory by assuming that the dissociation proceeded from the ground state of benzene.

The isomerization of the benzene isomers takes place at relatively low ion internal energies. At higher energies there is evidence (Fig. 21) that 1,5-hexadiyne dissociates directly to the product $C_3H_3^+$ without prior isomerization to the benzene structure. Most likely, this higher-energy channel produces the linear $C_3H_3^+$ ion rather than the cyclopropenium ion.

Only a few of the benzene-ion isomers have been investigated by

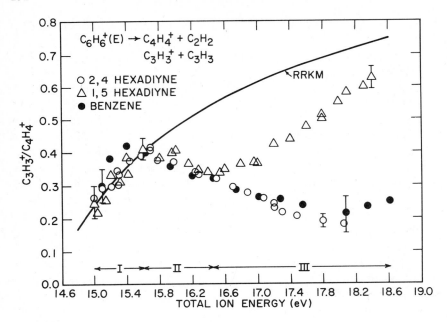

Figure 21.   Ratio of the $C_3H_3^+$ and $C_4H_4^+$ signals as a function of the total $C_6H_6^+$ ion internal energy. The RRKM line is the result of a calculation that fits the dissociation data near threshold. Additional dissociation channels at higher energies (e.g., at 15.8 eV) are not taken into account. Taken with permission from Baer et al.[124]

PEPICO. These are 2,4-hexadiyne,[124,139,140] 1,3-hexadiyne,[141] 1,5-hexadiyne,[124] and benzene itself.[114,124] All of these are known to isomerize to benzene prior to dissociation. There are several other isomers whose ionic heats of formation are known from PES studies and neutral heats of formation (see Fig. 22). Some of these surely take part in the reaction of the $C_6H_6^+$ system. In fact, the isomerization among these structures may account for the lack of fluorescence from the excited states of benzene.[84,142] In any case, the benzene-ion dissociation offers a bewildering richness of dissociation mechanisms.

b. $C_4H_5N^+$ Isomers

Another group of ions whose isomerization barrier is lower than their dissociation limits is the series of isomers of pyrrole:

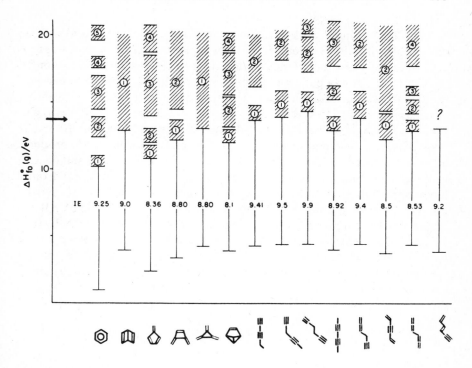

Figure 22. Energies and electronic states of a number of benzene-ion isomers on a common energy scale. Taken with permission from Rosenstock et al.[136]

It is difficult to imagine four molecules more structurally different than these, yet their dissociation rates were found to be identical and were consistent with those calculated by assuming the lowest-energy pyrrole structure.[143] Under these circumstances it is to be expected that the dissociation products are also in their lowest-energy structures. The structure of the major product, $C_2H_3N^+$, illustrates this point. In the neutral form, the most stable molecule with this formula is methyl cyanide, CH_3CN. However, the onset for the $C_2H_3N^+$ product from pyrrole and its isomers is nearly 2 eV lower than expected on the basis of the methyl cyanide ion. The structure was determined by ab initio molecular-orbital calculations to be CH_2CNH^+.[143] Because of the loose transition state (a linear structure gave a good fit), it is nearly certain that this is the lowest-energy structure of this ion.

c.   Dimethyl disulfide ($CH_3SSCH_3^+$)

The PES of dimethyl disulfide has a broad first electronic band with no discernible vibrational structure.[121] This suggests that the ion structure is very different from that of the neutral molecule. As a result it is difficult to determine the adiabatic ionization potential (IP), which is the energy of the 0,0 transition. The approximate photoionization onset is 8.33 eV.

This ion dissociates into two sets of products, $C_2H_5S^+ + HS$ and $CH_2S^+ + CH_3SH$, with rather well-defined onsets at 10.08 and 10.15 eV, respectively.[144] The parent-ion rates of dissociation to these two products have been measured by PEPICO (see Fig. 23). Dimethyl disulfide is a relatively small ion, so that a slow dissociation rate can result only from a large activation energy. In the previously mentioned case of the pyrrole ion, which is of a similar size, the activation energy is 3.43 eV. If we assume that the dimethyl disulfide ion activation energy is just $AP - IP = 1.73$ eV, the statistically expected rates are 1000 times greater than the experimentally determined rates. However, if we assume that the parent ion rearranges to a different, more stable structure prior to dissociation, then the effective activation energy could be much larger and the predicted rates slower.

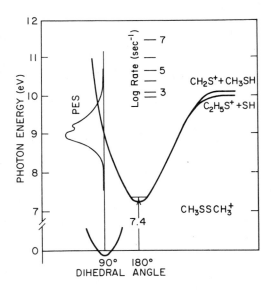

Figure 23.   Potential-energy diagram and rate constants for the dimethyl disulfide ion dissociation. The energy of the minimum was determined from the statistical-theory fit to the rate data. Taken with permission from Butler et al.[144]

The ground-state energy of the isomerized $CH_3SSCH_3^+$ ion was determined by fitting the statistical-theory rates to the experimental rates using the activation energy as a variable parameter. This results in a ground-state ion energy of 7.4 eV, nearly 1 eV below the adiabatic IP. The most likely structure of this isomerized ion is one in which the dihedral angle has opened up from 90° to 180°, as shown in the structures

Me   Me
| | +
Ө—Me   Ф ion
neutral |
               Me

The structure of dimethyl disulfide is similar to that of hydrogen peroxide, HOOH, which also is known to change its dihedral angle from about 90 to 180° on ionization.[145,146] The calculated adiabatic IP of 7.4 eV has not been verified by either molecular-orbital calculations or other thermochemical measurements.

2. Isomerization Rate ≪ Dissociation Rate—Examples of Tight Transition States

A number of isomeric systems have rearrangement barriers that exceed their dissociation limits. The consequence of this is that the various isomers dissociate to different products, and that their rates of dissociation differ as well. This means that the lowest-energy products are not always formed, and that the fragments may not even be in their lowest-energy configurations. Finally, one can expect that the transition states will be tight.

a. $C_4H_4O^+$ Isomers: Furan and 3-Butyn-2-one

These molecules are related to the previously mentioned pyrrole ($C_4H_4NH^+$) system. However, whereas the $C_4H_4NH^+$ ions isomerize to pyrrole prior to dissociating via a loose transition state, the ions of furan and 3-butyn-2-one dissociate in completely different ways.[147] Table I shows the products and the experimental appearance energies for the two isomers. The only two products that the two isomers have in common are $C_3H_3^+$ and $C_3H_4^+$. The major product from 3-butyn-2-one ($HCCC(O)CH_3$) is formed by loss of the methyl group. This appears to be a direct bond cleavage, and from the onset at 11.00 eV, a heat of formation of 970 kJ/mol was determined for the $HCCCO^+$ ion. All of the other appearance energies are considerably higher than the thermochemical dissociation limits, which suggests that this ion dissociates by a tight transition state.

The dissociation of furan has an activation energy of 2.64 eV. This is

TABLE I
Fragment Appearance Energies (AE) from $C_4H_4O^+$ Ions

	AE (measured)[a] (eV)	AE (thermochemical)[b] (eV)
Furan (IP = 8.88 eV)		
$C_2H_2O^+ + C_2H_2$	11.80 ± 0.1	11.70
$C_3H_4^+ + CO$	11.60 ± 0.1	10.88(allene)
$C_3H_3^+ + CHO$	12.10 ± 0.1	11.85
$CHO^+ + C_3H_3$	13.2 ± 0.1	12.32
3-Butyn-2-one (IP = 10.19 eV)		
$CHCCO^+ + CH_3$	11.00 ± 0.1 eV	—[c]
$CH_3CO^+ + C_2H$	12.10 ± 0.1	10.97
$C_3H_4^+ + CO$	10.68 ± 0.05	10.00
$C_3H_3^+ + CHO$	11.55 ± 0.1	10.97

[a] Reference 147.
[b] See reference 147 for heats of formation used in these calculations.
[c] Derived $\Delta H_f^0(CHCCO^+) = 970$ kJ/mol.

sufficiently large to cause this ion to dissociate on a microsecond time scale. These rates were measured and compared with the statistical-theory results.[147] A transition state with vibrational frequencies identical to those of the molecular ion of furan gave the best fit. Thus the reaction has a moderately tight transition state with an entropy of activation of about zero.

b. $C_4H_8O_2^+$ Isomers

The $C_4H_8O_2$ molecule has numerous isomers. Only three of these have been investigated by PEPICO:[148–150]

    5 $CH_3CH_2CH_2C{-}OH$ $CH_3{-}C{-}O{-}C_2H_5$
                             6   **7**

These are dioxane (**5**) butanoic acid (**6**), and ethyl acetate (**7**). Although these ions are very different from each other, they are no more so than are the pyrrole isomers or those of benzene. Yet the dioxane isomers behave completely independently. Table II shows the products formed from each of these isomers and their appearance energies. In each case, the major fragmentation path is unique. The dominant fragment in the dioxane dissociation is $C_3H_6O^+$, which appears only from this ion. Similarly, the major fragment from butanoic acid is $C_2H_4O_2^+$, which appears

TABLE II

Fragment Appearance Energies (AE) from $C_4H_8O_2^+$ Ions

	AE (measured) (eV)	AE (thermochemical) (eV)
1,4-Dioxane (IP $= 9.19 \pm 0.01$ eV)[a]		
$C_2H_4O^+ + C_2H_4O$	10.39 ± 0.05	10.27^b
$C_2H_5O^+ + C_2H_3O$	10.46 ± 0.05	—[c]
$C_3H_6O^+ + CH_2O$	10.56 ± 0.10	—[d]
$C_3H_5O^+ + CH_3O$	11.20 ± 0.10	—[c]
Ethyl acetate (IP $= 10.01 \pm 0.05$ eV)[e]		
$C_4H_6O^+ + H_2O$	10.31 ± 0.1	—[c]
$C_3H_5O_2^+ + CH_3$	10.60 ± 0.1	—[c]
$C_2H_5O_2^+ + C_2H_3$	10.67 ± 0.08	10.5^f
$C_2H_5O^+ + C_2H_3O$	10.70 ± 0.1	—[c]
n-Butanoic acid (IP $= 10.17$ eV)[g]		
$C_2H_4O_2^+ + C_2H_4$	10.42 ± 0.05	10.38^h
$C_3H_5O_2^+ + CH_3$	10.48 ± 0.1	—[c]
$C_3H_7^+ + COOH$	10.96 ± 0.05	10.85
$C_3H_6^+ + HCOOH$	11.35 ± 0.1	10.77

[a] Reference 148.

[b] Based on the calculated energy of $CH_2OCH_2^+$ (see reference 148).

[c] Not enough is known about the product structures to determine a thermochemical onset.

[d] See text.

[e] Reference 149.

[f] Assumes protonated acetic acid structure.

[g] Reference 150.

[h] Assumes the enol of acetic acid.

only from this ion. Finally, the loss of H_2O to give the ion $C_4H_6O^+$ is unique to the ethyl acetate ion dissociation.

The interesting aspect in these dissociations is that these ions do not isomerize to one another even though they dissociate very slowly. Furthermore, the reactions are not just simple bond cleavages, but in some cases must involve rather tortuous paths through the transition state. The loss of H_2O from ethyl acetate is a complicated reaction the path for which is not yet fully understood. In fact, not even the structure of the product ion, $C_4H_6O^+$, is known with any degree of certainty.

Although the butanoic acid ion is the most stable and might therefore be expected to dissociate with the slowest rate, the activation energy for the formation of the acetic acid product ($C_2H_4O_2^+ + C_2H_4$) is only 0.27 eV.

As a result, the dissociation is too rapid to be measurable on the microsecond time scale.[150] However, there is convincing evidence that when this ion is formed by electron impact rather than by photoionization, a slowly dissociating (metastable) ion is produced.[151] This problem remains a paradox to be resolved.

The ethyl acetate ion dissociation to $C_4H_6O^+ + H_2O$ also has a very small activation energy of only 0.30 eV. Thus this ion should dissociate rapidly. Yet it is a well-known metastable ion in mass spectrometry and its dissociation rate was measured by PEPICO. Fraser-Monteiro et al.[149] concluded that the ethyl acetate ion must rearrange to a more stable structure prior to dissociation, and on the basis of their measured rates, they obtained the energy of this isomerized structure. However, they were unable to identify this more stable form with any of the known isomers of $C_4H_8O_2^+$.

Perhaps the most interesting of the three isomers investigated by PEPICO is dioxane. With a heat of formation of 602 kJ/mol, it is the most energetic isomer of the three. The heats of formation of the butanoic acid and ethyl acetate ions are 506 and 523 kJ/mol, respectively. Figure 24 shows the energy diagram for the dissociation of dioxane to the

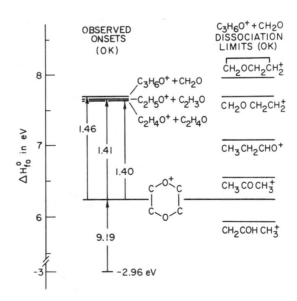

Figure 24. Potential-energy diagram of the dioxane ion dissociation, and some of the $C_3H_6O^+ + CH_2O$ products. The product ion formed in the dissociation has the $CH_2OCH_2CH_2^+$ structure. Taken with permission from Fraser-Monteiro et al.[148]

possible $C_3H_6O^+ + CH_2O$ products. There is little doubt that the neutral fragment is the formaldehyde molecule. However, the identity of the ionic species, which has the chemical formula of acetone, is more difficult to establish. As illustrated in the figure, the heat of formation of the acetone ion and formaldehyde is considerably below the observed onset for this ion. An even lower energy is obtained if the enol form of acetone is assumed.

If the $C_3H_6O^+$ product has one of the more stable structures, some of the available energy will be partitioned into translational energy. Thus, to decide which of the energetically allowed $C_3H_6O^+$ ions was being formed, Fraser-Monteiro et al. measured the kinetic energy released on dissociation as a function of excess energy.[148] Their data clearly indicate that the kinetic energy goes to zero at the dissociation limit of 7.69 eV (using the energy scale of Fig. 24). This is strong evidence that the ions being formed have the $CH_2OCH_2CH_2^+$ structure. Of course, this is not a stable ion and its heat of formation has never been measured experimentally. But its energy has been calculated by ab initio methods and is placed on the appropriate scale in Fig. 24.[152] It coincides exactly with the dissociation onset.

The dissociation rate of the dioxane ion with respect to the three products shown in Fig. 24 has also been measured and modeled with the statistical theory.[148] Good agreement is possible if the transition state has nearly the same vibrational frequencies as those of the dioxane ion itself. That is, the transition state is a tight one with a small ΔS value. The transition states to the other two product ions, $C_2H_5O^+$ and $C_2H_4O^+$, are even tighter. These results are entirely consistent with the evidence for the formation of the excited state of $C_2H_6O^+$. In fact, this fragment ion can be formed directly from dioxane with the breaking of only two ring bonds and no additional rearrangement. The unusual aspect of this dissociation is that the dioxane ion dissociates very slowly (minimum rate $\cong 10^2 \, s^{-1}$) to the very unstable $CH_2OCH_2CH_2^+$ product.

3. Isomerization Rate $\cong$ Dissociation Rate

If the barrier for isomerization is of the same magnitude as the barrier for dissociation, the overall reaction becomes more complicated, in that the rates for four reactions, as shown in Fig. 25, are involved in the final rate expression. The various possible rate combinations have been discussed in some detail in a previous publication.[153] We will only consider the case in which isomer A is initially prepared at some energy above the dissociation and isomerization limits. In addition, we will assume that the rates k_1 and k_2 are fast and approximately equal, whereas $k_3 = k_4 \ll k_1$.

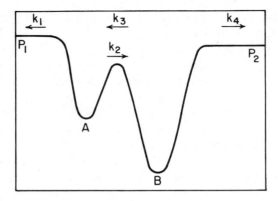

Figure 25. Potential-energy curve for a dissociation reaction in competition with isomeri-zation. Taken with permission from Baer et al.[153]

The products P_1 and P_2 are formed with the rates

$$\frac{d[P_1]}{dt} = k_1[A] \quad \text{and} \quad \frac{d[P_2]}{dt} = k_4[B] \tag{4.11}$$

To determine the rates, we need to know the time dependences of $[A]$ and $[B]$. These are given by the coupled, linear, homogeneous differential equations

$$\frac{d[A]}{dt} = (-k_1 - k_2)[A] + k_3[B]$$

$$\frac{d[B]}{dt} = k_2[A] + (-k_3 - k_4)[B] \tag{4.12}$$

The solutions of Eqs. (4.12) for $[A](t)$ and $[B](t)$ is given as the sum of two exponentials. When these are substituted in Eqs. (4.11) and the previously mentioned conditions are placed on the rate constants, the rates of production of P_1 and P_2 are given by

$$\frac{d[P_1]}{dt} = k_1[A] = k_1\left(\frac{k_3}{k_1 + k_2}e^{-\alpha t} + \frac{k_1 + k_2}{k_2}e^{-\beta t}\right)$$

$$\frac{d[P_2]}{dt} = k_4[B] = k_4(e^{-\alpha t} - e^{-\beta t}) \tag{4.13}$$

where $\alpha = k_3(2k_1 + k_2)/(k_1 + k_2)$ and $\beta = k_1 + k_2 + k_2k_3/(k_1 + k_2)$. Since k_1

and k_2 are much larger than k_3 and k_4 and about equal to each other, the phenomenological rate will be the sum of two terms: a fast rate β, and a slow rate α. Product P_1 will be formed with both slow and fast rates, while product P_2 will be formed only at the slow rate. The reason for this is that the fast component with rate β is of negligible importance except at very short times. Often, the two products P_1 and P_2 are the same, so that the measured dissociation rate will be the sum of the two rates in Eqs. (4.13).

A similar analysis of the situation in which isomer B is initially formed shows that the dissociation rate will be characterized by a single exponential and will be equal to the slow component for the dissociation of isomer A. An example in which these conditions appear to apply is the system of pentene isomers whose thermochemistry is illustrated in Fig. 26. Five of the six isomers investigated by PEPICO dissociate with two-component decay rates.[153,154] Figure 27 shows the data at selected energies for these

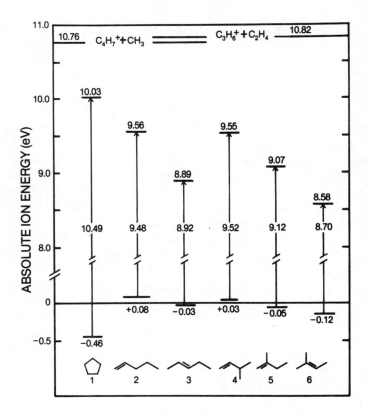

Figure 26.   Energies of some $C_5H_{10}^+$ isomers and their dissociation limits. Taken with permission from Brand and Baer.[154]

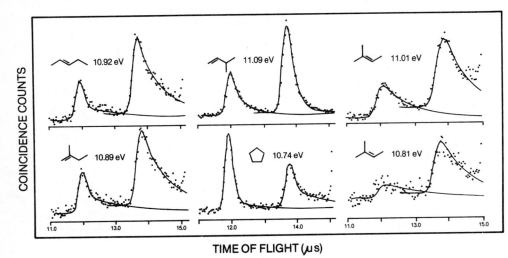

Figure 27. PEPICO data for the $C_3H_6^+$ and $C_4H_7^+$ products from the pentene isomers. The solid lines are TOF distributions calculated assuming two-component dissociation rates for all isomers except 2-methyl-2-butene. This latter, most stable ion dissociates with a one-component decay rate. Taken with permission from Brand and Baer.[154]

isomers. The sixth, the most stable isomer, to which all the others rearrange, is characterized by a single exponential, as shown in Fig. 27.

Two products are formed in the low-energy dissociation reaction of the pentenes. The mechanism for this dissociation is certainly complicated and not yet understood. Nonetheless, the model illustrated in Fig. 25 and Eqs. (4.13) is capable of providing us with an overall understanding of these reactions. One consequence of the model is that at a given ion internal energy, the slow rate components of all of the reactions must be identical, and they must be equal to the rate constant of the lowest energy (2-methyl-2-butene) isomer. This is demonstrated in Fig. 28, in which the rates are plotted as a function of the ion internal energy.

The data of Figs. 26–28 provide an overall framework for understanding this complex dissociation–isomerization reaction. A more detailed understanding may come from additional studies with isotopically labeled molecules, and perhaps from ion-photodissociation studies, some of which have already been performed.[155] In addition, PEPICO studies with supersonically cooled beams will result in narrower TOF distributions, which will allow a more precise analysis of the two rate components.

Another set of isomers with two-component decay rates are the butenes. As with the pentenes, the slower of the two rates is the same for all the isomers.[156] In addition, the breakdown diagram indicates that the

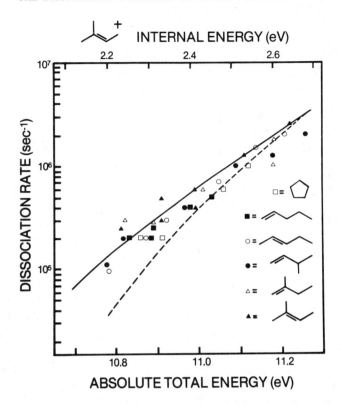

Figure 28. The rate for the slow components of the dissociations of six pentene isomers as a function of the total parent-ion internal energy. Taken with permission from Brand and Baer.[154]

isomerization is incomplete, in that two of the isomers have different product-ion branching ratios from those of the other three.[157]

Incomplete isomerization has been observed in smaller ions as well. The nitromethane ($CH_3NO_2^+$) and methyl nitrite (CH_3ONO^+) system is particularly interesting and complex.[158–160] The potential-energy surface for the several reactions is shown in Fig. 29. The lower-energy methyl nitrite ion dissociates only to the products on the right-hand side of the figure, whereas the higher-energy nitromethane ion dissociates only to those products on the left. The only exception is that nitromethane ions also fragment to the NO^+ product. The KERDs for the formation of NO^+ both from nitromethane and from methyl nitrite have been measured at several energies above the dissociation limit.[161] They were found to be identical and very close to the statistically expected distributions. For

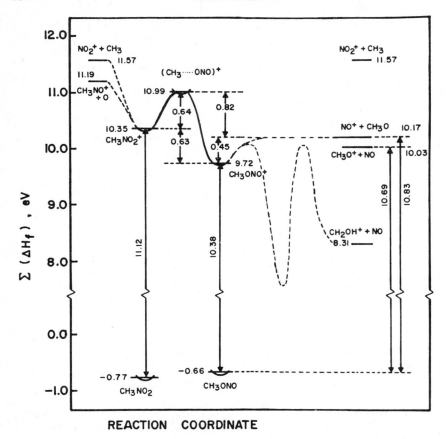

Figure 29. Potential-energy diagram for the $CH_3NO_2^+$ dissociation. This figure was adapted from Gilman et al.[159] by adding the deep well at 8 eV. The existence of this well is supported by the metastability of CH_3ONO^+ ions near the dissociation threshold and by ab initio calculations. See text for discussion.

these reasons it has been proposed that nitromethane ions can dissociate directly to the products CH_3NO^+ and NO_2^+ or isomerize to methyl nitrite and dissociate via those paths open to the isomerized ion.[159] Because at an energy of 11 eV the rates are already too fast to be measurable by PEPICO, no rate data are available to corroborate these ideas. However, the dissociation rates have been measured for the CH_3ONO^+ reactions near the dissociation limit.[158,160] These results are extremely interesting and will be discussed in the part of Section IV.E.2 dealing with nonstatistical decay.

D. The Complex Dissociations of Phenol and Nitrobenzene Ions

Many of the slow dissociation reactions involve benzene derivatives. The requirements for a large activation energy (i.e., a stable parent ion) and a large number of vibrational degrees of freedom make these molecules excellent candidates for PEPICO investigations. Most of the previously mentioned dissociations are ones with no reverse activation energy and with loose to moderately loose transition states. The structure of the products is therefore reasonably well established on the basis of the dissociation energy. Two reactions that do not fall into this category are those of phenol and nitrobenzene. In both cases, the products and reactant ion are separated by a substantial energy barrier. It is tempting to attribute this barrier to the complex nature of the reaction, for indeed a number of bonds must be broken and re-formed in the course of the reaction. However, it must be pointed out that these reactions are no more complicated than those of HNC loss from aniline, C_2H_2 loss from benzene, and CH_3 loss from 1,3-butadiene. Why these latter reactions have no reverse activation energy, whereas the phenol and nitrobenzene ion reactions do, is not readily understood, and no explanation will be offered here.

The phenol ion potential-energy curve for the dissociation is shown in Fig. 30, and the measured dissociation rates are illustrated in Fig. 31.[162] From the statistical-theory fit to the measured rates, the activation energy

$$C_6H_5OH^+ \longrightarrow C_5H_6^+ + CO$$
$$\longrightarrow C_5H_5^+ + H \qquad (4.14)$$

was determined to be 3.12 eV. The considerable kinetic shift had prevented an accurate measurement of the onset from the $C_5H_5^+$ appearance energy. However, Lifshitz et ai.[163] were able to obtain an onset with their delayed-pulsing technique, much as was done for the aniline reaction (Fig. 13). Their result yields an E_0 value of 3.1 ± 0.15 eV, in excellent agreement with the PEPICO-derived value.

A number of the phenol ion vibrational frequencies have been determined by Anderson et al.[90] using the MPI-PES method. As in chlorobenzene, the ion frequencies are considerably different from the neutral ground-state molecular frequencies, but are quite close to those of the excited $^1B_2$ state. Some of the transition-state frequencies were lowered to fit the steep $k(E)$ versus E curve. Use of Eq. (4.2) yields a $\Delta S(1000 \text{ K})$ value of 3.5 cal/mol K, making this a moderately tight transition state.

In view of the moderately tight transition state, in which the frequencies are basically those of the phenol ion, it is expected that the final

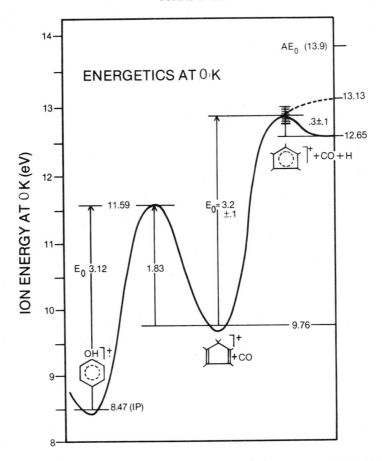

Figure 30.   Phenol ion potential-energy diagram. The dissociation barrier of 3.12 eV was determined from the statistical-theory fit to the measured dissociation rates (see Fig. 31). Taken with permission from Fraser-Monteiro et al.[162]

product ion will have a structure similar to that of phenol, namely the cyclopentadiene structure shown in Fig. 30. However, some linear isomers (see Fig. 16) have energies that fall just about at the phenol ion dissociation onset of 11.6 eV. Whether these higher-energy products are being formed can be determined by measuring the kinetic energy released in the dissociation. The result from numerous determinations[162-166] is that this reaction releases 0.5–0.8 eV of translational energy. In view of these data, it is clear that the cyclopentadiene ion is formed, at least at lower energies. That a single transition-state configuration fits the rate data so

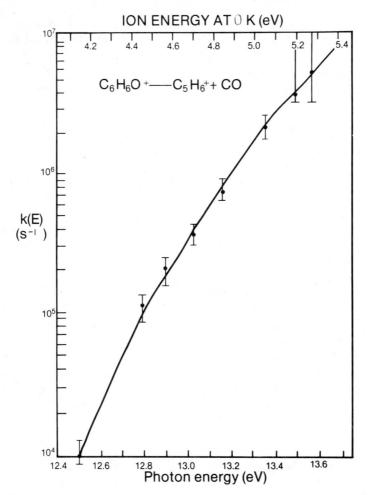

Figure 31. Phenol ion dissociation rate as a function of the phenol ion internal energy. The points are experimental results, and the line is the statistical-theory fit to the data. Taken with permission from Fraser-Monteiro et al.[162]

well over a large energy range suggests that the cyclopentadiene product is formed even up to 2 eV above the dissociation onset.

In an energy-selected study of the phenol dissociation, Fraser-Monteiro et al.[162] measured the KERD at a single energy of 13.6 eV and found it to be uncharacteristic of a statistical distribution. This is not unexpected. Most ionic dissociation reactions with large reverse activation barriers appear to have "early" transition states that channel much of the available energy into the translational degrees of freedom. Although the PEPICO investigation is the only study involving energy-selected phenol ions, the average release energy has been measured as a function of the ion lifetime in several mass-spectrometric studies.[163-166] In an electron-impact ionization source, ions of all internal energies are produced. Three classes of ions are then extracted from the ion source. First, there are the fragment ions produced from the dissociation of the parent ions of high internal energy. Second, there are the phenol ions that are formed at an energy below the dissociation limit and therefore remain intact. The third class of ions are the metastable phenol ions, which have enough internal energy to dissociate, but not enough energy to do so rapidly. They are transported through the various elements of the mass spectrometer and dissociate randomly during their flight. The lower the internal energy, the longer the ions will live, on the average. Thus by varying the time at which the kinetic-energy release is measured, it is possible to obtain the average release energy as a function of the ion internal energy by use of the data in Fig. 31.

Some of the mass-spectrometric measurements had suggested that the kinetic energy remains constant, or even decreases, as the ion internal energy is raised.[165] These results have recently been confirmed by Lifshitz et al.[164] using the delayed-pulsing technique. It appears, therefore, that there is an inverse relation between translational energy released and the phenol ion internal energy. Two explanations are possible. First, the product ion structure may change as the ion energy increases. Thus if more of the linear products were formed at higher phenol ion energies, the measured kinetic energy would decrease. However, as pointed out previously, the rate data suggest that the product-ion structure remains that of cyclopentadiene. The explanation proposed by Lifshitz et al.[164] is that the nature of the transition state changes with increasing energy such that it shifts from an "early" toward a "late" transition state. In view of the complicated dissociation paths calculated for the far simpler reactions of $C_2H_4^+$, CH_2O^+, and others,[77-81] this explanation has considerable merit.

The nitrobenzene ion dissociation is more complex than that of the phenol ion. The following reactions (with 0 K appearance energies in parentheses) have been observed by photoionization at low energies:[167]

$$C_6H_5NO_2^+ \rightarrow \begin{cases} C_6H_5O^+ + NO & (11.12 \text{ eV}) & (4.15) \\ NO^+ + C_6H_5O & (11.18 \text{ eV}) & (4.16) \\ C_6H_5^+ + NO_2 & (11.28 \text{ eV}) & (4.17) \\ C_4H_3^+ + NO + C_2H_2O & (11.54 \text{ eV}) & (4.18) \\ C_5H_5^+ + CO + NO & (11.44 \text{ eV}) & (4.19) \end{cases}$$

Although the ground-state products of the various reactions have very different energies, their onsets are quite close together, because of the large activation barriers shown in Fig. 32. Two of the products have energies below that of the nitrobenzene parent ion. In addition, there is a thermodynamically even more stable set of products, $C_5H_5N^+$ (pyridine) + CO_2, with a thermochemical dissociation limit of 6.63 eV. However, this reaction is evidently dynamically unfavorable. It would involve exchange of the nitrogen and carbon atoms.

Reactions (4.15) and (4.16) are closely related, differing only in the position of the positive charge. In modeling the experimental decay rates of these two reactions with the statistical theory, it was necessary to increase the transition-state vibrational frequencies relative to those of the parent ion.[167] The resulting tight transition state is equivalent to a $\Delta S(1000\text{ K})$ value of -4.2 cal/mol K. The structure of the $C_6H_5^+$ ion is mose likely that of the phenoxy radical. Thus the structures of the products for reactions (4.15) and (4.16) are almost certainly alike. As a result, the transition states are probably also very similar. That the best statistical-theory fit to the rate data uses identical transition-state fre-

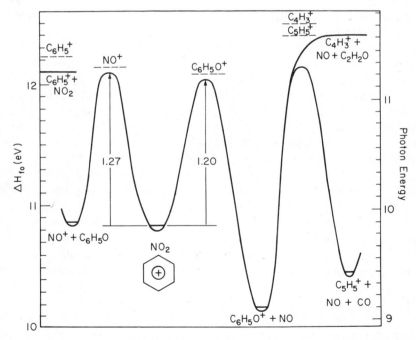

Figure 32.   Nitrobenzene ion potential-energy diagram. Taken with permission from Panczel and Baer.[167]

quencies for the two reactions is further evidence that the same, or nearly the same, transition state is involved in this reaction.

There is one complicating factor in the interpretation of the NO^+ formation. This is that the decay rate is clearly two component. The slow part is identical to the decay rate as measured on the basis of the $C_6H_5O^+$ product. The fast component is faster than $10^7\,s^{-1}$. These findings, in addition to the observation of a fast NO_2 loss from the nitrobenzene ion, will be discussed further in the following section on nonstatistical dissociations.

The large activation energy for the formation of $C_6H_5O^+$ allows this ion to be formed with sufficient internal energy to react further by the loss of CO. It is only by virtue of the fact that a considerable amount of the available energy is released as translational energy that a significant fraction of the products remains as $C_6H_5O^+$. The measured translational-energy distribution shows that the $C_6H_5O^+$ ions are formed with a wide range of internal energies.

E. Evidence for "Nonstatistical" Decay and Other Anomalies

The question of statistical versus nonstatistical decay is often one of semantics. Consider the direct dissociation of an ion or molecule on a repulsive potential-energy surface. Such a reaction cannot be described as a vibrational predissociation, and therefore the statistical theory of dissociation is not applicable. Certainly, the rate of such a reaction cannot be calculated by the statistical theory. On the other hand, the product energy distribution can be measured and the results can be compared with those of the phase-space theory described by Eqs. (3.4)–(3.7). In such a situation, the statistical theory is useful for calculating the extent of randomness in the energy distribution, even though the reaction cannot be expected to be statistical. With this strict interpretation, the term "non statistical behavior" is reserved for those reactions to which the theory ought to apply, but in fact does not. No well-documented ionic reaction has been shown to be nonstatistical in this sense. The examples described in this section are all nonstatistical in the purely phenomenological sense. All of them involve direct dissociation from excited electronic, or repulsive, states. For these reasons, the statistical theory can be used as a diagnostic tool for determining if a reaction is proceeding in some interesting fashion, such as via isomerization or direct dissociation from an excited state.

There are many examples of rapid and direct dissociation of ions from excited electronic states. Such behavior is evident from the distribution of product kinetic energies, and some examples will be discussed in Section V.B. In addition, some of these excited states decay exclusively to specific products that are often not the energetically most stable ones. In general, such reactions can easily be understood in terms of the character of the excited state. In this section we will discuss only those anomalous reactions that involve metastable ions, that is, ions that decay on the microsecond time scale.

1. NO_2 Loss in Nitrobenzene Ions

One of the major reaction products in the dissociation of nitrobenzene ions is formed by the simple C–N bond cleavage that yields $C_6H_5^+$ and NO_2. As shown in Fig. 32, the activation energy for this reaction is slightly higher than that for the loss of NO. It was pointed out previously that if a parent ion can decay via several channels, the total decay rate is simply the sum of all the individual decay rates, and its lifetime is the reciprocal of that sum. The parent ion lifetime can be measured from the appearance rate of any one of the fragments. All of them should give the same result. However, in the case of nitrobenzene, the measured rates are not the same.[167] The rate as measured on the basis of the NO_2 loss is so rapid that only a lower limit of $10^7 s^{-1}$ can be determined. At the same parent ion internal energy, the rate as measured with respect to NO loss is found to be $6 \times 10^5 s^{-1}$. This clearly indicates that these two reactions have different immediate precursor nitrobenzene ions. Furthermore, the parent ion that produces $C_6H_5^+$ so rapidly cannot be the ground state of the nitrobenzene ion, because the rate is orders of magnitudes faster than that expected on the basis of the statistical theory.

The explanation of these rates is very simple, and not unlike that given for the two-component decay rates (Figs. 25–28). The difference is that we are dealing here with different electronic states, and not different isomers. Initially, the nitrobenzene is produced in an excited electronic state. It then dissociates rapidly at a rate k_d with the loss of NO_2, or it undergoes a radiationless transition at a rate k_{rt} to the ground electronic state as shown in Eq. (4.20):

$$C_6H_5^+ + NO_2$$

$$\Big\uparrow k_d$$

$$C_6H_5NO_2 + h\nu \longrightarrow C_6H_5NO_2^{+*} + e^- \qquad (4.20)$$

$$\Big\downarrow k_{rt}$$

$$C_6H_5NO_2^{+*}(\tilde{X}) \longrightarrow \begin{array}{l} C_6H_5O^+ + NO \\ NO^+ + C_6H_5O \end{array}$$

The ground-state ions then dissociate slowly to form $C_6H_5O^+$ and NO^+. The rates of radiationless transition of many ions have been measured.[84]

Typical rates for substituted diacetylenes and halogen-substituted benzene ions are in the 10^6–$10^8\,s^{-1}$ range. However, there is an even larger list of ions that do not fluoresce because the radiationless transitions are too rapid. Typical fluorescence rates are $10^7\,s^{-1}$ and typical sensitivities in detecting this fluorescence allow quantum yields as low as 10^{-5} to be measured.[168] Thus the absence of fluorescence implies that the radiationless transition rates are greater than $10^{12}\,s^{-1}$. From the breakdown diagram and from the absence of a reported fluorescence spectrum, we conclude that nitrobenzene ions do not fluoresce, so that the internal conversion rate is at least $\sim 10^{12}\,s^{-1}$. Hence, the direct-dissociation rate must also be of this magnitude to account for the large yield of $C_6H_5^+$ ions.

2. The Methyl Nitrite Ion Dissociation

Figure 29 shows the potential-energy surface for the dissociation of methyl nitrite. There are a number of experimental observations concerning the production of NO^+ and CH_3O^+ for which we have no adequate explanation. These observations are summarized as follows:

1. Between the total energies of 10.17 and 10.43 eV the dissociation rate of CH_3ONO^+ is fast $(k > 10\,\mu s^{-1})$ when measured with respect to NO^+ appearance, whereas it is slow $(1.0 < k < 2.5\,\mu s^{-1})$ when measured with respect to CH_3O^+ formation.[158-160]

2. The potential-energy diagram of Fig. 29 indicates that the E_0 value for CH_3O^+ formation is only 0.31 eV. This is insufficient for a statistical-theory rate in the 1-μs^{-1} range.

3. There is evidence that the structure of the CH_3O^+ ion is H_2COH^+, which would place the dissociation limit to $H_2COH^+ + NO$ well below the energy of the methyl nitrite ion.[169]

4. There is an extremely strong deuterium isotope effect (see Fig. 33) in this reaction that causes the CD_3O^+ intensity to be very small.[160]

We begin the discussion of this reaction with the hypothesis that although it may be phenomenologically nonstatistical, the reaction is inherently statistical. The fact that NO^+ and CH_3O^+ are produced at different rates indicates that their immediate CH_3ONO^+ precursors are different. The small activation energy is consistent with a fast reaction. The fact that at higher CH_3ONO^+ internal energies the reaction goes exclusively to NO^+ suggests that this reaction is a simple bond-cleavage reaction producing the neutral CH_3O radical. Thus the rapid production of NO^+ is a totally expected process.

The reaction that produces CH_3O^+ is more complicated. First, it should not be slow, because of its small activation energy. Second, the

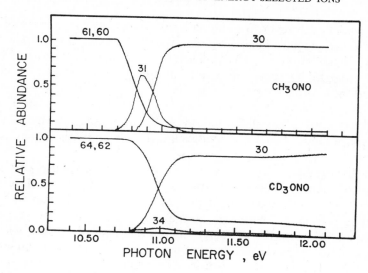

Figure 33. Breakdown diagrams of methyl nitrite and deuterated methyl nitrite ions. The parent ion masses 61 and 64 could not be resolved from ions formed by H (or D) loss resulting in ions of mass 60 and 62. Mass 30 is NO^+, mass 31 is CH_3O^+, and mass 34 is CD_3O^+. Taken with permission from Gilman, Hsieh, and Meisels.[160]

structure of this ion is certainly H_2COH^+. This was established by a collisional activation experiment in which the CH_3O^+ ions were reacted at ~8 keV with He neutrals.[169] The major products formed were H_2CO^+ and CO^+ ions, whereas no CH_3^+ fragment ions were observed. These products are characteristic of the H_2COH^+ structure.[170] To verify this conclusion, the collisional activation experiment should be repeated with an ion known to have the CH_3O^+ structure. Such ions are difficult to prepare, because they rapidly rearrange to the more stable H_2COH^+ structure. However, as a negative ion, the CH_3O^- structure is the more stable one. When such negative ions were used in a charge-reversal collisional activation experiment in which two electrons were stripped off and the resulting positive ion was fragmented, one of the major products was CH_3^+.[170] This establishes then that the m/z 31 product from the methyl nitrite ion has the H_2COH^+ structure.

The slow reaction producing the H_2COH^+ ion can come about by only two mechanisms. Either it is a case of electronic predissociation, as suggested by Gilman et al.,[158] that is so slow as to be on the microsecond time scale, or the ion isomerizes to a more stable structure prior to dissociation. There could be a barrier to isomerization close to the NO^+ onset such that there is competition between direct dissociation and isomerization. However, for such a small ion to have a lifetime of 1 μs with respect to vibrational predissociation would require an activation energy of the order to 2–3 eV. Furthermore, the structure of this ion would have to be very different from that of the product H_2COH^+, because there would have to be a large reverse activation barrier for the formation of this product. A possible candidate for such a structure is $HNC(OH)_2^+$ with a calculated energy 3 eV lower than the energy of the methylnitrite ion.[169] This potential well is illustrated by the dotted line in Fig. 29.

The hypothesis of an isomerized structure of the methyl nitrite ion explains the slow dissociation. However, it does not necessarily account for all of the detailed aspects of the rate data. These are that the decay rate rises very slowly with ion internal energy, and that the deuterium isotope effect increases with increasing internal energy. A slow electronic predissociation might have rates that do not vary rapidly with ion internal energy, and it might have an isotope effect such as the one observed. On the other hand, it is difficult to reconcile the formation of a structurally different H_2COH^+ product ion with such a slow electronic predissociation. Slow curve crossings are expected when the molecules are rigid and symmetries are strictly observed. Typical curve crossing reactions in small ions have rates in the 10^8–10^{11} s^{-1} range.

The final unusual aspect of this reaction is the dramatic isotope effect, which virtually stops CD_3O^+ production. If we assume the mechanism in which there is a competition between direct dissociation to NO^+ and isomerization to a deep well prior to dissociation to H_2COH^+, it may well be that the isomerization reaction involves a critical H or D shift in the transition state. Either tunneling through a barrier or the necessity of a large-amplitude CH bend might explain the isotope effect.

In summary, the loss of NO from the methyl nitrite ion and its deuterated analog is a fascinating and complex reaction. The observed slow rate may be the result either of an electronic predissociation or of an isomerization preceding the fragmentation step. Neither of these explanations violates the fundamental hypotheses of the statistical theory, except in the strictly phenomenological sense. Molecular-orbital calculations, which have been so successful in aiding our understanding of the reactions of $C_2H_4^+$, H_2O^+, CO_2^+, and CH_3OH^+,[77–81] may help shed more light on methyl nitrite dissociation as well.

3. The Dissociation of Acetone and Its Isomers

The dissociation of acetone, $CH_3C(O)CH_3$, to $CH_3CO^+ + CH_3$ has all the appearances of a very simple fragmentation reaction. Trott et al.[171] determined both the ionization potential of acetone (9.694 eV) and the

CH_3-loss onset (10.54 eV) at 0 K using a cooled beam of acetone. Because the dissociation onset is identical to the thermochemical dissociation limit, the potential-energy surface has no barrier for this reaction. This is not unexpected, in that the lowest-energy form of CH_3 is certainly the methyl radical, and the lowest-energy isomer of the m/z 43 ion has the CH_3CO^+ structure[172] (the CH_2COH^+ ion is ~1.6 eV less stable[173]) However, Bombach et al.[174] have recently pointed out several remarkable features in this reaction, some of which are reminiscent of those of the methyl nitrite dissociation.

The activation energy for CH_3 loss from the acetone ion is 0.85 eV and the minimum RRKM–QET calculated dissociation rate is considerably larger than $10^7 s^{-1}$. This is consistent with the experimental findings of threshold rates that are too fast to be measured by PEPICO.[175] It is, however, most surprising to find some undissociated acetone ions with 2.2 eV of internal energy in the breakdown curve. These metastable acetone ions are produced in only a small energy region ($\cong$9.6–9.9 eV) in the vicinity of the $\tilde{A}$ state. Above and below these energies, the rates are fast.

A related dissociation is that of the isomer 1,2-epoxypropane (9), in which the oxygen atom is bonded in a three-membered ring to two carbon atoms. It has an ion heat of formation that places it more than 1 eV above the dissociation limit for CH_3 loss. Yet this ion has been observed to dissociate into CH_3CO^+ and a number of other fragments with rates as low as $10^4 s^{-1}$.[174] Furthermore, these slow reactions take place with epoxypropane ions having no more than 0.4 eV of internal energy (see Fig. 34). Finally, the epoxypropane ions dissociate with the same rate as do the slowly dissociating acetone ions when both are prepared at the same total energy. On the basis of these findings, Bombach et al.[174] concluded that both of these ions rearrange to the same structure prior to dissociation. Two candidates for this common isomer that have very stable ionic structures are the enol of acetone (11) and 1-propenol (10). However, to account for the slow reactions, not only must they be more stable than the acetone ion, they must also have a considerable barrier to dissociation.

The energetics of the several isomers and the acetone ion $\tilde{A}$ state are illustrated on a common 0 K energy scale in Fig. 34. All of the energies

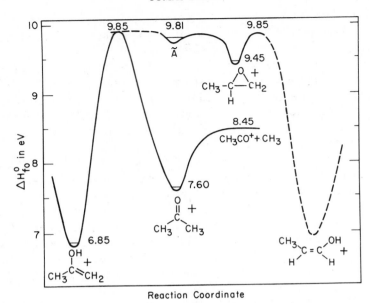

Figure 34. Potential-energy diagram of the acetone ion dissociation showing the enol of acetone at 6.85 eV, the excited Ã state of acetone at 9.81 eV, and two other isomers. Some lines are dashed to indicate that they are less certain than the solid lines.

except the energy of isomer **10** are reasonably well known. In particular, the barrier between the acetone enol and acetone is now quite well established.[176] The slow dissociation rates can be readily explained if 1,2-epoxypropane as well as some of the acetone Ã-state ions rearrange to the enol structure prior to dissociation. Although such a mechanism is not unreasonable, it needs further verification, especially in the form of ab initio calculations. Figure 34 is only diagrammatic. The real surface is multidimensional and probably involves a complex sequence of conical intersections. Furthermore, this mechanism does not account for all of the observations, one of which is that ion **9** dissociates slowly to a variety of products that are not observed when acetone ions dissociate. Bombach et al.[174] suggested that some of the acetone Ã-state ions isomerize to the 1-propenol structure **10** via 1,2-epoxypropane. This is drawn in Fig. 34 as a dashed line to indicate our relative ignorance about it.

The acetone enol (**11**) has its own peculiar way of dissociating. As a neutral molecule, the enol structure is not very stable. However, it is extremely stable as an ion. It can be formed as a dissociation product in the source of a mass spectrometer from a larger parent ion such as 2-pentanone. Both deuterium-labeled and ^{13}C-labeled enol ions of acetone have been prepared and used to determine the dissociation products of the metastable ions.[177] These are enol ions that are formed in the ion source just above the barrier for isomerization to acetone and subsequently dissociate. The overall reaction sequence is presumed to be

$$[CH_3—\overset{\overset{\displaystyle OH}{|}}{C}={}^*CH_2]^+ \longrightarrow [CH_3—CO—{}^*CH_3]^+ \longrightarrow CH_3CO^+ + CH_3 \quad (4.21)$$

The experimental findings are that the loss of the $^*CH_3$ group is preferred over that of the unlabeled methyl group by a factor of $3:2$.[177] This indicates that the energy is not randomly distributed prior to reaction. That is, once the enol ions pass through the barrier (Fig. 34) to the acetone structure, they dissociate so fast that the internal energy is not redistributed randomly. Lifshitz and Tzidony[172] estimate the lifetime of the intermediate acetone ion to be only 5×10^{-13} s. A consequence of this may be that the KERD is two component and not statistical.

We can summarize the acetone dissociation as follows. The slow dissociation of ions at an energy of 1.7 eV above the dissociation limit results from an isomerization reaction that takes place only when the acetone ion is formed in the lowest vibrational levels of the first excited electronic state. Above and below these energies, the ion rapidly converts to, or is already in, the ground electronic state from which dissociation takes place. Similarly, the 1,2-epoxypropane ion isomerizes to a more stable structure, possibly to the same one as do the acetone Ã-state ions, from which it dissociates slowly. By comparing the KERDs of the products from acetone and 1,2-epoxypropane ions with that of the enol ion, it may be possible to establish whether the isomerized structure is the enol ion.

V. THE EXPERIMENTALLY MEASURED KINETIC-ENERGY-RELEASE DISTRIBUTIONS

The KERDs have been measured for a number of ionic dissociation reactions. Because these measurements are not limited to slow reactions in the microsecond time domain, the KERDs can be measured over a much larger range of parent-ion internal energies than can the dissociation rates. The principle of measuring the KERD in coincidence experiments is based on deconvoluting the more or less symmetrically broadened TOF distribution. The distribution is close to symmetric as long as the kinetic-energy release is small compared with the energy gained by the ion in the extraction field. Measurement of the KERD by this method is limited to ions that decay rapidly. If ions dissociate slowly, the TOF distributions are asymmetrically broadened. This interferes with the analysis of the symmetric broadening caused by the kinetic-energy release.

The total translational energy E_t is partitioned between the two fragments according to their masses such that linear momentum is conserved. This means that when two fragments of masses m_1 and m_2 are formed, the energy of the particle with mass m_1 is given by $E_1 = m_2 E_t/(m_1 + m_2)$. Because the precision with which the KERDs can be measured is relatively low, most studies have dealt with reactions in which a massive neutral fragment is lost, thereby imparting a large fraction of the total energy to the fragment ion. As a result, many of the KERD studies have involved reactions in which a halogen atom such as I, Br, or Cl is lost.

A. Dissociations from Ground Electronic States

Most of the dissociation reactions that proceed from the ground electronic state with no reverse activation energy are well described by statistical KERDs. This is true also of excited states that decay by internal conversion to the ground electronic state. There are two criteria for establishing whether a reaction is statistical on the basis of the translational-energy release. One is the average energy release, given by Eq. (3.7), while the other is the KERD itself, given by Eq. (3.6). There are several examples in which the KERD is statistical while the average energy release is not.

1. The Dissociation of CH_3I^+

Examples of ions that dissociate from the ground electronic state with statistical KERDs are members of the series RI^+, where $R = CH_3$, C_2H_5, or C_3H_7.[178-181] The CH_3I^+ dissociation will be used as an example. This ion, whose ionization energy is 9.54 eV, dissociates to a number of products by the following reactions (among others):

$$CH_3I^+ \rightarrow \begin{cases} CH_3^+ + I(^2P_{3/2}) & (12.24\ eV) & (5.1) \\ CH_3^+ + I(^2P_{1/2}) & (13.18\ eV) & (5.2) \\ CH_2I^+(^1A') + H(^2S) & (12.74\ eV) & (5.3) \\ I^+(^3P_2) + CH_3(^2A_2'') & (12.87\ eV) & (5.4) \end{cases}$$

The electronic states of methyl iodide are, in order of increasing energy, $\tilde{X}\ ^2E_{3/2}$ (9.54 eV), $\tilde{X}\ ^2E_{1/2}$ (10.17 eV), $\tilde{A}\ ^2A_1$ (11.95 eV), and $\tilde{B}\ ^2E$ (13.90 eV). Thus the first dissociation limit at 12.24 eV lies above the onset of the $\tilde{A}$ state. However, that the CD_3I^+ fragmentation rate was found to be $\sim 10^7\, s^{-1}$ near the dissociation onset suggests that the $\tilde{A}$-state ions decay be internal conversion to the ground electronic state.[178] This rate is consistent with the rate predicted by the statistical theory using an activation energy of 2.7 eV. Because of the faster dissociation rate, the CH_3I^+ dissociation rate could not be measured, but it is reasonable to assume that the $\tilde{A}$ state of this ion also converts to the ground state prior to dissociation. This statistical behavior is reflected in the KERD[178,179] as well as in the long CD_3I^+ lifetime.

Figure 35 shows some CH_3^+ TOF distributions at selected CH_3I^+ internal energies. Because of the many final internal-energy states of the CH_3^+ ions, the kinetic-energy release is in the form of a distribution from zero to the maximum energy possible. The data were therefore deconvoluted to obtain the KERD. The solid lines in Fig. 35 underneath the data points are calculated TOF distributions for single energy releases. They serve as basis functions for fitting the experimental data. That ions with significant translational energies perpendicular to the ion-extraction field hit apertures and are lost was taken into account. This discrimination

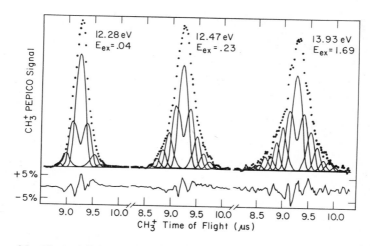

Figure 35. Time-of-flight distributions of CH_3^+ ions produced from the dissociation of CH_3I^+ at the indicated excess energies E_{ex}. The points are the experimental data, and the lines underneath the points are calculated TOF distributions for several assumed single release energies. The release energies (in millielectron volts) are given by the relation $\varepsilon = 6n^2$ ($n = 1-9$). The bimodal nature of the higher-energy TOF distributions is a result of experimental discrimination against perpendicularly ejected CH_3^+ ions. The difference between the sum of the calculated curves and the experimental points is given near the bottom of the figure.

results in the bimodal distributions for the higher release energies (forward and backward ejection of the CH_3^+ ion). The coefficients for the basis functions form the KERDs. These are shown in Fig. 36 along with the statistical-theory results. The agreement is very good at low internal energies, but becomes less good at higher energies.

The departure from the statistical theory is better seen in a plot of average energy release as a function of ion internal energy (Fig. 37). Here the data from two laboratories are combined. They clearly show that the departure from the statistical theory occurs at ~ 12.85 eV, well below the onset at 13.18 eV of reaction (5.2). Although the values of the experimental KERDs at high energies are lower than statistically expected, their shape is still rather statistical. Powis[182] introduced a very powerful method of plotting the KERDs to point out such inherent similarities or differences, namely the use of reduced KERDs. If the KERDs of Fig. 36 and additional results from references 178 and 179 are plotted as $P(E/\bar{E})$ versus $E/\bar{E}$, all of the KERDs fall on the same line (Fig. 38). In addition, the statistically expected reduced KERD agrees very well with these reduced data. This, then, is an example of a KERD that appears statistical but whose average energy is not statistical.

An energy release greater than that expected on statistical grounds can be explained by assuming that the reaction is so fast that only a subset of all the vibrational modes participates in the energy-randomization process. (This would be an inherently nonstatistical distribution of the energy.) However, the I atom loss from CH_3I^+ is just the opposite. The system acts as though it has more oscillators than its rightful $3n - 6 = 9$. An explanation proposed by Mintz and Baer[178] is that reaction (5.4), with

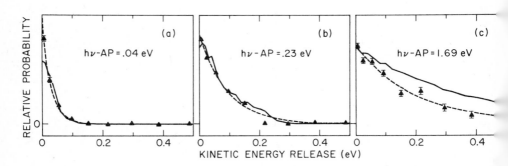

Figure 36   The KERDs derived from the data of Fig. 35. The points are the experimental results, and the solid lines are the results of the statistical-theory calculations. The dashed lines are the best-fit exponential functions through the points. Taken with permission from Mintz and Baer.[178]

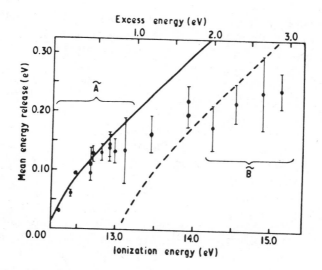

Figure 37. The average kinetic-energy release for the dissociation of CH_3I^+ to $CH_3^+ + I$. The data are a combination of results from Mintz and Baer.[178] and Powis.[179] Solid and dashed lines are the statistical-theory predictions assuming formation of $I(^2P_{3/2})$ and $I(^2P_{1/2})$, respectively. Taken with permission from Powis.[179]

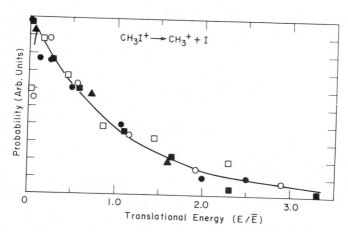

Figure 38. The reduced KERD for the formation of CH_3^+ from CH_3I^+ at several excess energies from the onset to 3 eV above the onset. The solid symbols represent values from Mintz and Baer,[178] and the open symbols, values from Powis.[179] The line through the data is the scaled statistical-theory prediction. The scaling parameter is the average release energy.

its onset at 12.87 eV, is responsible for this break in the deviation of the average energy release. If some ions initially proceed via the $I^+ + CH_3$ channel, and the electron hops to form the CH_3^+ ion when the fragments are already separated, internal energy can no longer be redistributed and the internal energy remains as vibrational energy in the CH_3^+ product. Above 13.18 eV the higher spin orbit state of the I atom can be formed, and the average energy is further depleted by this reaction channel. The statistically expected average energy, if reaction (5.2) were the only open channel, is shown as a dashed line in Fig. 37. Although Powis concluded that $I(^2P_{1/2})$ atoms are formed in the dissociation of $\tilde{B}$ state CH_3I^+ ions, the scatter in the data of Fig. 36 would appear to be too large to justify such an assertion.

The reaction channel for the loss of the H atom [reaction [5.3]] is a most interesting one. Although the dissociation limit for this reaction is 12.74 eV, the observed onset is more than 1 eV greater, and coincides with the origin of the CH_3I^+ $\tilde{B}$ state. This is similar to the situation for the other halomethane ions, in which H loss in photoionization appears only from excited states.[183,184] Powis measured an average energy release for H loss from CH_3I^+ of 0.82 eV. This release energy remains constant over the whole $\tilde{B}$ state energy range. These results indicate that there is a direct dissociation from the excited state along a repulsive potential surface.

The dissociation of $CH_3I^+(\tilde{B})$ to $I^+ + CH_3$ [reaction (5.4)] proceeds with a small and nearly statistical energy release.[179] At an excess energy of 2 eV, the release energy is only 0.2 eV. We have then the interesting situation in which the $\tilde{B}$ state ions dissociate more or less statistically to $CH_3^+ + I$, but very nonstatistically to CH_2I^+. It could be that there is competition between internal conversion to the $\tilde{A}$ and $\tilde{X}$ states from which the reaction proceeds more or less statistically and a direct dissociation from the $\tilde{B}$ state that is very nonstatistical.

The dissociation KERDs for the other alkyl iodides,[180,181] as well as those for such ions as acetaldehyde and ethylene oxide,[185] the previously discussed acetone ions,[182] butadiene,[186] CH_4^+,[187] and vinyl bromide,[186] are rather statistical. However, the average energy in a number of cases[182,186,188] is not equal to the statistically predicted one. All of these reactions appear to proceed from the ground electronic state, even though the initial excitation is to an excited electronic state.

2. Rotational Predissociation in NH_3^+ and ND_3^+

Very close to the dissociation limit, the only internal energy states available to the product fragments are rotational levels. In addition, when ions are prepared near the dissociation limit, the thermal rotational energy present in the molecule at 300 K becomes an important part of the total energy used in the fragmentation process. Powis[189,190] has investigated the dissociation of NH_3^+ and ND_3^+ with respect to the loss of H or D near the dissociation limit.

The 0 K onset for dissociative ionization to NH_2^+ from NH_3 is 15.768 eV, whereas the same onset for ND_2^+ from ND_3 is 15.89 eV.[191] These onsets lie in the middle of the second PES band (Fig. 39). There is no reverse activation energy, so the transition state is relatively loose. The effect of the thermal rotational energy is clearly evident in Fig. 40, which represents the breakdown diagram for the ND_2^+ fragment, that is, the fraction of the ions that have dissociated before reaching the detector. The ND_2^+ signal is observed well below the 0 K threshold. Because the vibrational energy spacings in NH_3 and ND_3 are rather large, the ND_2^+ signal below 15.89 eV is attributable entirely to rotational energy. The derived KERDs for this dissociation are shown in Fig. 41. The solid lines are the statistical-theory fits to the data. The Langevin ion-induced dipole potential (Fig. 7) was assumed for this calculation, with rotational constants B and C for ND_3 of 5.14 and 3.16 cm^{-1}.[190] Because the rotational constants of slightly bent ND_2^+ are 57.6 and 4.3 cm^{-1}, the angular momentum can effectively be accommodated only with rotation about the C_2 axis of ND_2^+ and the orbital angular momentum of the two separating fragments. The energy tied up in the latter is transformed into product translational energy. The excellent fit to the experimental data in Fig. 40

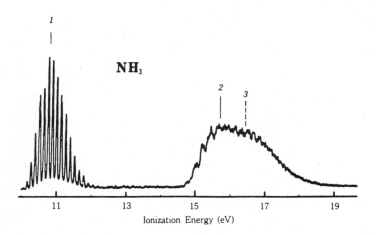

Figure 39.   Photoelectron spectrum of ammonia. Taken with permission from Kimura et al.[121]

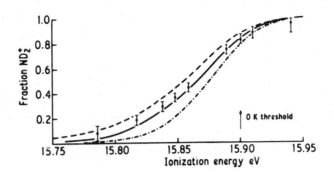

Figure 40. Fractional abundance of ND_2^+ ions from ND_3^+ in the vicinity of the dissociation threshold. Almost all of the signal below the 0 K onset is a result of rotational hot bands. Calculated curves for rotational temperatures of 200 K (–·–·), 300 K (——), and 400 K (– – –) are included for comparison. Taken with permission from Powis.[190]

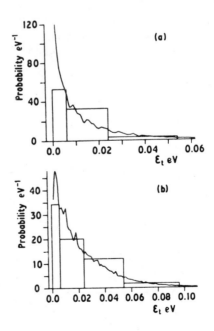

Figure 41. KERDs for the dissociation of ND_3^+ to $ND_2^+ + D$ at ion energies of 15.85 (a) and 15.91 (b) eV. Experimental points are presented as histograms and calculated distributions as continuous curves. Taken with permission from Powis.[190]

indicates that angular momentum is fully available for redistribution in the dissociating ion so long as total angular momentum is conserved.

The average translational energy as a function of the nominal ion internal energy is shown in Fig. 42. The interesting feature is that the translational energy does not go to zero. This can be understood in terms of the onset in Fig. 40 and the potential curves of Fig. 7. Let us ignore the finite resolution of the PEPICO state-selection technique and assume that the ion internal energy is exactly known. At the energy of the 0 K onset, all the ions will have sufficient energy to dissociate, even those derived from an ND_3 molecule in its ground vibrational and rotational states. Most of the ions that have a modest amount of rotational energy because they are produced in a distribution of J states determined by the temperature of the gas. On the average, this rotational energy is just $\frac{3}{2}RT$. As the photon energy is lowered below the 0 K onset, the ions that have a total energy greater than the dissociation threshold must have a greater proportion of their energy as rotational energy. Thus at 15.82 eV, which is nearly 0.1 eV below the 0 K threshold, only those ions with a rotational energy in excess of 0.1 eV can dissociate. These ions will have a significant centrifugal barrier to overcome. Most of this energy will go into translational energy of the products. As the photon energy is decreased further, the fraction of the energy in rotation will increase still more, and thus one may expect the translational energy actually to increase as the photon energy is decreased. This effect is evident in Fig. 42 from the

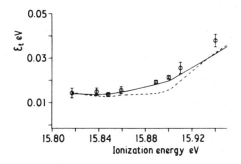

Figure 42.   The average translational energy released in the $ND_3^+ \rightarrow ND_2^+ + D$ dissociation. The points are experimental results, and the curves are results of statistical-theory calculations. The solid line includes the effect of the instrumental sampling function, whereas the dashed line does not. Note the rise in the average-energy release as the energy is lowered near 15.82 eV. Taken with permission from Powis.[190]

dashed line, which is the calculated energy release with no instrumental broadening effects included. In addition, the KERDs for the NH_3^+ dissociation clearly show a broadening as the energy is decreased. These effects of the rotational energy are extremely dependent on the form of the interaction potential. Therefore, the excellent fit of the experimental and calculated data lends strong support to the use of the Langevin potential ($s = 4$) for such ionic dissociation reactions. This is not unexpected, since the centrifugal barrier is located at long internuclear separations, at which the interaction is expected to vary as r^{-4}.

B. Dissociations from Excited States

1. *The Dissociation of* $C_2H_5Br^+$

The direct dissociation from excited electronic states of small ions is often faster than radiationless transition to the ground state. The KERDs

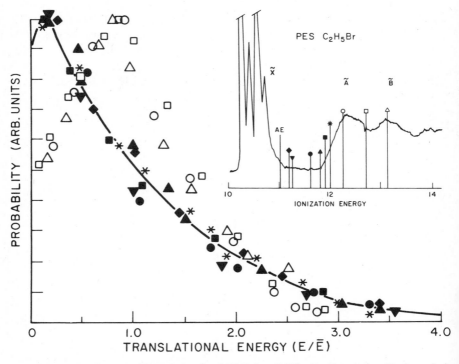

Figure 43. The scaled KERD for the $C_2H_5Br^+ \rightarrow C_2H_5^+ + Br$ dissociation. Each symbol in the KERD corresponds to an internal energy at which the data were taken. The PES is from Kimura et al.,[121] and the KERD from Miller and Baer.[188]

for these direct dissociations are usually very nonstatistical. A good example of the change from a statistical to a nonstatistical dissociation is given by the fragmentation of the ethyl bromide ion. Figure 43 shows the scaled KERDs taken at various energies within the $C_2H_5Br^+$ $\tilde{X}$, $\tilde{A}$, and $\tilde{B}$ states. It is evident that the ground-state dissociation, represented by the filled data points, proceeds statistically, whereas the excited states, represented by the open symbols, dissociate with considerably greater kinetic-energy release, and that the distribution is nonstatistical.

2. The Dissociation of CF_3Cl^+

The ions CF_3X^+, where $X = Cl$ or Br, are further examples of ions that dissociate statistically from the ground electronic state and nonstatistically from the excited states. The CF_3Cl^+ dissociation is particularly rich in its variety of fragmentation mechanisms. Its PES[192] is shown in Fig. 44. As a result of the great stability of the CF_3^+ fragment, the Cl-loss onset is 12.8 eV, just above the adiabatic ionization energy of ~ 12.7 eV. Although this reaction must be very fast, it has a statistical KERD over the whole ground-state energy range up to 14.5 eV.[193] That is, the probability for energy release is peaked at near-zero translational energy, much as in

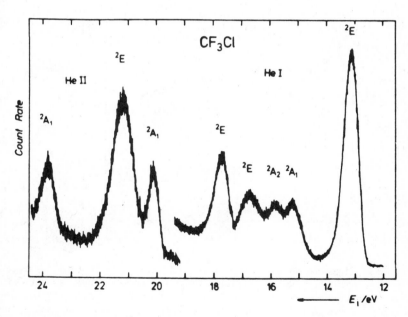

Figure 44.   The He(I) and He(II) PES of CF_3Cl. The dissociation onset for the loss of Cl is just 0.1 eV above the ionization onset. Spectra adapted from Cvitas et al.[192]

the KERDs for the methyl iodide dissociation of Figs. 36 and 38. However, in the region of the first $^2A_1$ state at 15 eV, the KERDs (Fig. 45) are very nonstatistical. In addition, when the ionizing light is an He(I) lamp with a photon energy of 21.217 eV, the KERDs are bimodal. A low-energy, statistical dissociation is superimposed on a direct nonstatistical reaction. This might indicate that some of the ions undergo an internal conversion to the ground electronic state, from which the dissociation will be statistical. However, it is most peculiar that when the ionizing radiation is a Ne resonance lamp with a photon energy of 16.848 eV, the low-energy component disappears.

The variation of the PEPICO results with photon energy for precisely the same nominal ion internal energy indicates that different ion states are being prepared. One way this can come about is through the previously discussed autoionization (Section IV.A.2). In autoionization, the first step involves the production of an excited neutral, which then ionizes to a variety of final ion states [Eqs. (4.4), (4.5)]. Often, the final ion states produced are very different from those formed by direct ionization. Because the autoionization process is very sensitive to the photon energy,

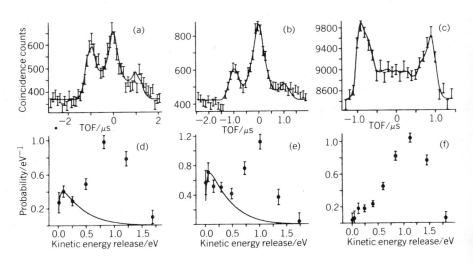

Figure 45.   Experimental TOF distributions and KERDs for the CF_3Cl^+ ($^2A_1$) dissociation to CF_3^+ under the following conditions: (a, d) 15.1 eV using He(I) radiation; (b, e) 15.4 eV using He(I) radiation; and (c, f) 15.4 eV using Ne radiation. In the KERDs, points represent experimental values, and lines, statistical-theory predictions. Taken with permission from Powis.[193]

it is possible that the use of two different photon energies will produce different ion states. Of the two major experimental approaches to PEPICO, the He(I) and threshold electron techniques, it is the latter that is most likely to involve ion preparation via autoionization (see Fig. 11). The autoionization process can prepare vibrationally excited ions in lower electronic states by a mechanism that involves the ejection of the electron while the autoionizing molecule is predissociating.[100] In this manner, highly non-Franck–Condon distributions of ion internal energies can be produced. Although many reactions have been investigated by both He(I) PEPICO and threshold PEPICO, no clear-cut examples of differences have been noted. For this reason, it is rather ironic that such a large effect of photon energy has been observed in the dissociation of the CF_3Cl^+ $^2A_1$ state when the two ionizing sources are He(I) and Ne(I), both of which involve the analysis of energetic electrons. The irony is even greater considering that the He(I) experiment appears to be the one that is subject to autoionization, because it is in this experiment that the low-kinetic-energy component appears.

Autoionizing states are usually Rydberg states that lie 0–3 eV below the ion states to which they converge. In many molecules there are no ion states between 21 and 24 eV, so that the interferences from autoionizing states are avoided by using the 21.2-eV He(I) source. Inspection of the CF_3Cl PES in the 21–24 eV region indicates that this ion has, in fact, an electronic state at 23.8 eV that could account for Rydberg states at 21.2 eV. To account for the low-kinetic-energy-release component in Fig. 45, this autoionizing state would have to produce ground-state CF_3Cl^+ ions with up to 3 eV of vibrational energy.

The $^2A_2$ state of CF_3Cl^+ dissociates to a mixture of CF_3^+ and CF_2Cl^+, both of which give large kinetic-energy releases.[193] In contrast, the ^2E states at 16.72 and 17.71 eV dissociate exclusively to CF_2Cl^+, and they do so with a broad KERD peaking at 1 eV, similarly to the high-energy component of the KERDs shown in Fig. 45. This dissociation is nonstatistical in at least two senses. First, the KERD has a nonstatistical shape. Second, the absence of the energetically more favorable CF_3^+ fragment ion indicates that energy is not randomized prior to dissociation. Rather, the excited state dissociates rapidly to a specific product.

If we take the thermochemical onset for the $CF_2Cl^+ + F$ formation to be 13.8 eV,[194] then the available energies for partitioning when the CF_3Cl^+ ion is prepared in the two ^2E states are 2.9 and 3.9 eV, respectively. Thus the fractions of the available energy released as translational energy are 38 and 28%, respectively. This fraction has been compared[193] with that predicted by the impulsive dissociation model[195] in which the energy release for the reaction $ABC \rightarrow A + BC$ is given by

$$\varepsilon_t = \frac{\mu_{A,B}}{\mu_{A,BC}} E_{avl} \qquad (5.5)$$

where E_{avl} is the total available energy, $\mu_{A,B}$ is the reduced mass of the atoms A and B, and $\mu_{A,BC}$ is the reduced mass of the fragments A and BC. The impulsive model is based on the assumption that the available energy is simply partitioned between the two atoms (A and B) adjacent to the bond that breaks, and that the bond between B and C during the dissociation process is "soft." The percentage of the energy released as translational energy is calculated by use of Eq. (5.5) as 47%.

A similar analysis of the CF_3^+ production from the $^2A_1$ state shows that the fraction of the available energy going into translation varies from 24 to 43% whereas the impulsive model predicts 38%.[193] The range of experimental percentages arises from the range of CF_3Cl^+ vibrational energies selected. The energy release was found to be independent of the vibrational level, which means that the vibrational excitation must reside totally within the CF_3 part of the ion and that this energy remains in this fragment during dissociation.

3.   The Dissociation of NF_3^+

The dissociation of the NF_3^+ ion is another reaction that proceeds statistically from the ground electronic state and nonstatistically from the higher states. However, as will be pointed out below, from at least one of the excited states, the reaction is again statistical. The PES in Fig. 46

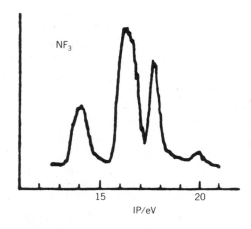

Figure 46.   Photoelectron spectrum of NF_3. Adapted from data of Bassett and Lloyd.[196]

exhibits five bands, the second and third of which overlap at 16 eV.[196] The $\tilde{X}$ state with its origin at 13.5 eV is formed by the removal of a nitrogen lone pair electron. Because of the change from pyramidal to planar geometry, the first PES band is very broad. The first dissociation onset to $NF_2^+ + F$ lies at 14.12 eV, which is precisely equal to the thermochemical dissociation limit. The KERD near the onset is illustrated in Fig. 47. Although the scatter in the data is significant, the distribution is clearly statistical.[197]

The KERDs change completely in the $\tilde{A}$ and $\tilde{B}$ states. An example is shown in Fig. 48. At an excess energy of 2 eV the dissociation releases ~ 1.1 eV of the available energy in translation. This is precisely equal to the predictions of the impulsive model [Eq. (5.5)]. It is surprising how well the impulsive model appears to predict the kinetic-energy release of some of these dissociation reactions. On the one hand, it may be a reflection of the extreme rapidity of the dissociation and of the simplicity of the potential surface. On the other hand, these large translational energies could arise also from a complicated potential-energy surface. More detailed studies are needed.

The production of NF_2^+ from the $\tilde{C}$ state proceeds with relatively little energy release. Furthermore, the KERD is again statistical in its shape.

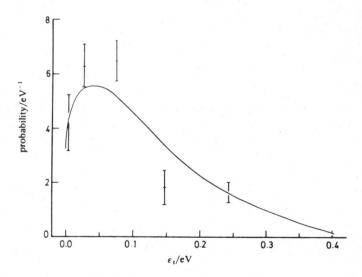

Figure 47. Normalized KERD for the dissociation of NF_3^+ ($\tilde{X}$) to $NF_2^+ + F$ at an ion energy of 14.39 eV. The smooth curve is the statistical-theory fit to the data. Taken with permission from Mansell et al.[197]

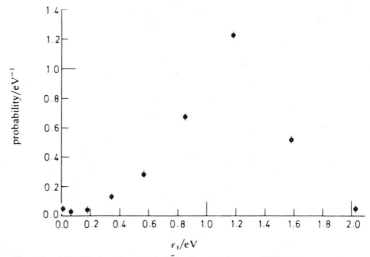

Figure 48. The KERD for the NF_3^+ ($\tilde{A}$) dissociation to $NF_2^+ + F$ at an ion energy of 16.35 eV. Taken with permission from Mansell et al.[197]

However, the products are clearly not formed in their ground state. Rather, the KERD is consistent with the formation of an excited NF_2^+ ion that is 1.5 eV less stable than the ground state $^1A_1$ NF_2^+. This implies that perhaps the parent ion dissociates from an excited state. This would be the only known example of an excited-state dissociation that is statistical. All other statistical reactions are ones in which the excited-state dissociation is preceded by an internal conversion to the ground electronic state.

4. *Dissociation to Three Fragments—Simultaneous or Sequential?*

At an energy of 19.6 eV, the products of the NF_3^+ reaction are NF^+ and 2F. These products can arise either from a simultaneous dissociation into three products [Eq. (5.6)] or from the stepwise reactions (5.7) and (5.8):

$$NF_3^+ \rightarrow NF^+ + F + F \qquad (E_0 = 4.62 \text{ eV}) \qquad (5.6)$$

$$NF_3^+ \rightarrow NF_2^+ + F \qquad (E_0 = 1.20 \text{ eV}) \qquad (5.7)$$

$$NF_2^+ \rightarrow NF^+ + F \qquad (E_0 = 3.42 \text{ eV}) \qquad (5.8)$$

How can we differentiate between the two on the basis of the translational-energy release? There are really two problems associated with this question. First, there is the partitioning of the total energy among all the degrees of freedom, including translation, rotation, and vibration. The second problem is the partitioning of the total kinetic energy among the translational degrees of freedom of three or more products. The second problem does not arise in the dissociation to two products because conservation of momentum uniquely determines the partitioning of the translational energy.

These questions have been discussed in the framework of the statistical theory by Baer et al.[198] The partitioning of the energy among the translational, rotational, and vibrational degrees of freedom for the case of two fragments is given by Eq. (3.7). Conservation of angular momentum with respect to the three rotational axes results in the loss of one degree of freedom each from the translational- and rotational-energy terms. The third component can be conserved with the impact parameter of the dissociating pair. Now when the molecule dissociates to n fragments, there will be $n-1$ impact parameters that can vary in order to conserve the angular momentum. Therefore, the translational and rotational degrees of freedom will no longer have to be decreased. For three products, the available energy is given approximately by

$$E = \frac{3}{2} RT^* + \frac{r-1}{2} RT^* + \sum_{i=1}^{v} \frac{h\nu_i}{\exp(h\nu_i/RT^*) - 1} \qquad (5.9)$$

in which only the rotational energy has had a degree of freedom reduced.

The manner in which a given total amount of translational energy is partitioned is more subtle. When there are only two products, conservation of linear momentum determines precisely how much energy goes to each fragment. However, when there are more than two particles, the problem is ambiguous. An infinite number of combinations are possible as long as the total energy remains constant. However, on the average, and in the spirit of the statistical theory, the fragment of mass m_i will receive an amount, $E_{t,i}$ given by

$$E_{t,i} = E_t \frac{1 - m_i/M}{f - 1} \qquad (5.10)$$

where E_t is the total translational energy, M is the total mass of all fragments, and f is the number of fragments formed.[198] It should be noted that when $f = 2$, the expression reduces to the well-known result for the dissociation to two products. The usual way of reporting energy release is in terms of the total energy, not the energy of the fragment that was measured. For the case of three products, the total energy cannot be determined exactly unless it is known whether the reaction is sequential or simultaneous.

The consequences of simultaneous and sequential dissociations of the

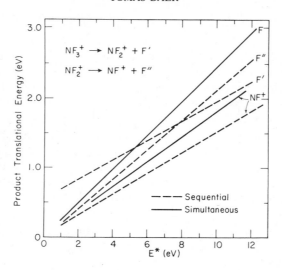

Figure 49. The predicted average fragment kinetic energies for the simultaneous and sequential dissociations of NF_3^+ to $NF^+ + F + F$. Taken with permission from Baer et al.[198]

NF_3^+ ion for the average translational-energy release are shown in Fig. 49. It is evident that the major differences arise in the energy carried off by the F fragment. In addition, the two F fragments are predicted to have the same kinetic energy in the simultaneous dissociation, but quite different energies in the sequential dissociation. The experimental data of Mansell et al.[197] were obtained at only one energy, which was 2 eV in excess of the $NF^+ + 2F$ onset. They found that the NF^+ fragment was produced with a translational energy of 0.38 eV. The calculated energy assuming a simultaneous dissociation is 0.38 eV, whereas the sequential dissociation mechanism predicts an NF^+ translational energy of 0.34 eV. The precision with which the energies were measured is probably insufficient to distinguish between the two possibilities in this particular case, although it is tempting to conclude that this reaction is simultaneous.

Most reactions near the dissociation threshold probably dissociate sequentially. However, there are special situations in which a simultaneous dissociation is probable. One such case may be the fragmentation of glyoxal, which has been predicted on the basis of ab initio calculations to dissociate simultaneously to $CH_2 + 2CO$ even at low energies.[199] At high energies, simultaneous reactions to several products may occur routinely. In addition, reactions of doubly charged ions such as $SO_2^{2+} \rightarrow S^+ + O^+ + O$, which have been observed in photoion–photoion coincidence (PIPICO) experiments may well be simultaneous.[200] To date, no reactions have been confirmed to dissociate simultaneously.

VI. FUTURE DIRECTIONS

No chapter of this magnitude would be complete without a few words concerning future directions in ion state selection. PEPICO is now about 15 years old. The number of systems investigated and the amount of information obtained about ion-dissociation reactions is impressive. The contribution of PEPICO has been particularly valuable in providing broad pictures of how large ions dissociate. In addition, it has been amply demonstrated that the statistical theory is capable of quantitative predictions of the dissociation rates as well as of the KERDs. We have learned that small is not necessarily simple, as in the cases of such ions as acetone and methyl nitrite. By contrast, the dissociation of the halobenzene ions is refreshingly simple. A great deal of additional work is necessary if we are to understand the reactions of such systems as the pentenes, in which isomerization and dissociation compete. Even such vital information as the energy barriers to isomerization is unknown. In addition, the structures of the ions are often not well understood, making the interpretation of the data ambiguous. The narrower TOF distributions expected with the use of supersonic beams will undoubtedly increase our ability to understand these complicated reactions. However, new approaches to the study of ion dynamics are clearly called for.

Some of the problems related to ion structures and isomerization barriers may be resolved by an extension of the PEPICO technique in which the structures of energy-selected ions are probed by photodissociation spectroscopy. The energy-analyzed photoelectron can be used to trigger a laser, which photodissociates the single ion that has been produced. A similar approach may also make possible the laser probing of dissociation products.

Although much has been learned about ionic dissociation reactions, our comprehension of individual reactions is still at a rudimentary stage. Only a few reactions are understood in sufficient detail to even begin to be analyzed in terms of the new theoretical ideas alluded to in this chapter. A major limitation has been the definition of the ion internal energy. The ultimate energy resolution in PEPICO will always be limited by our ability to energy analyze electrons. With present technology, this places a limit of ~ 5 meV (40 cm^{-1}) on energy resolution, and lowering of this limit is not foreseen. To improve on this, it will be necessary to use an approach in which the resolution is limited only by the photon bandwidth. The development of vacuum UV lasers will certainly open up new avenues. One can envision an experiment in which a tunable UV laser is used to produce an ion by a one-photon transition in the ground state with a very narrow range of internal energies, possibly over a few rotational states. A second laser, in the visible or near UV range, could then be used to dissociate this ion. Such an approach not only would improve the energy resolution, it would also make possible the measurement of dissociation rates up to $10^9 \, s^{-1}$.

Acknowledgments

Much of the work described in this chapter has come out of our own laboratory. For this I am deeply grateful to my co-workers, who have labored many long hours, suffered through the tribulations, and provided us with our (few) triumphs. In addition, I am grateful to colleagues elsewhere, in particular Professors Gerry Meisels, Chava Lifshitz, John Holmes, Mike Bowers, and Paul Guyon, for many fruitful discussions that have helped generate new ideas. Finally, we owe a debt to the late Henry Rosenstock, who for many years was a guiding light to all of us working in the field of ion-dissociation dynamics. On a more mundane, but equally important level, we are grateful to the National Science Foundation and the Department of Energy for generous financial support of the experiments in our laboratory.

References

1. P. R. Brooks and E. F. Hayes, Eds., *State to State Chemistry* (Am. Chem. Soc. Symp. Ser. 56), American Chemical Society, Washington, D.C., 1977.

2. R. Vasudev, R. N. Zare, and R. M. Dixon, *J. Chem. Phys.*, **80,** 4863 (1984); C. T. Rettner, E. E. Morino, and R. N. Zare, in *Int. Conf. Phys. Electron. Atom. Collisions*, J. Eichler, I. V. Hertel, and N. Stolterfoht, Eds., North Holland, Amsterdam, 1983, pp. 51–61.

3. J. C. Whitehead, *Comp. Chem. Kinet.* **24,** 357 (1982).

4. G. Hall, K. Liu, M. J. McAuliffe, C. F. Giese, and W. R. Gentry, *J. Chem. Phys.* **78,** 5260 (1983).

5. B. H. Rockney and E. R. Grant, *J. Chem. Phys.* **77,** 4257 (1982).

6. F. F. Crim, *Ann. Rev. Phys. Chem.*, **35,** 657 (1984).

7. T. R. Rizzo, C. C. Hayden, and F. F. Crim, *Faraday Discuss. Chem. Soc.* **75,** 223 (1983).

8. J. R. Beresford, G. Hancock, A. J. MacRobert, J. Catanzarite, G. Radhakrishnan, H. Reisler, and C. Wittig, *Faraday Discuss. Chem. Soc.* **75,** 211 (1983).

9. S. Ruhman, Y. Haas, J. Laukemper, M. Preuss, D. Feldman, and K. H. Welge, *J. Phys. Chem.* **88,** 5162 (1984).

10. J. Dannacher and J. P. Stadelmann, *Chem. Phys.* **48,** 79 (1980).

11. J. H. D. Eland, *Int. J. Mass Spectrom. Ion Phys.* **9,** 397 (1972).

12. C. S. T. Cant, C. J. Danby, and J. H. D. Eland, *J. Chem. Soc. Faraday Trans. II* **71,** 1015 (1975).

13. I. Powis, P. I. Mansell, and C. J. Danby, *Int. J. Mass Spectrom, Ion Phys.* **32,** 15 (1979).

14. Y. Niwa, T. Nishimura, and T. Suchiya, *Int. J. Mass Spectrom, Ion Phys.* **42,** 91 (1982).

15. A. V. Golvin, M. E. Akopyan, and Y. L. Sergeev, *Khim. Vys. Energ.* **15**, 387 (1981).

16. R. Stockbauer, *J. Chem. Phys.* **58**, 3800 (1973).

17. B. P. Tsai, A. S. Werner, and T. Baer, *J. Chem. Phys.* **63**, 4384 (1975).

18. C. Batten, J. A. Taylor, B. P. Tsai, and G. G. Meisels, *J. Chem. Phys.* **69**, 2547 (1978).

19. I. Nenner, P. M. Guyon, and T. Baer, *J. Chem. Phys.* **72**, 6587 (1980).

20. T. Baer, in *Gas Phase Ion Chemistry*, Vol. 1, M. T. Bowers, Ed., Academic, New York, 1979, Ch. 5.

21. J. H. D. Eland, *Int. J. Mass Spectrom. Ion Phys.* **8**, 143 (1972).

22. M. E. Gellender and A. D. Baker, *Int. J. Mass Spectrom. Ion Phys.* **17**, 1 (1975).

23. J. Dannacher, J. P. Stadelmann, and J. Vogt, *Int. J. Mass Spectrom. Ion Phys.* **37**, 203 (1981).

24. T. L. Bunn, W. S. Woodward, and T. Baer, *Rev. Sci. Instrum.*, **55**, 1849 (1984).

25. R. Proch, D. M. Rider, and R. N. Zare, *Chem. Phys. Lett.* **81**, 430 (1981).

26. J. L. Durant, D. M. Rider, S. L. Anderson, and R. N. Zare, *J. Chem. Phys.* **80**, 1817 (1984).

27. J. Miller and R. Compton, *J. Chem. Phys.* **75**, 22 (1981).

28. J. Kimman, P. Kruit, and M. J. Van der Wiel, *Chem. Phys. Lett.* **88**, 576 (1982).

29. M. G. White, W. A. Chupka, M. Seaver, A. Woodward, and S. Colson, *J. Chem. Phys.* **80**, 678 (1984).

30. S. T. Pratt, P. M. Dehmer, and J. L. Dehmer, *J. Chem. Phys.* **80**, 1706 (1984).

31. S. T. Pratt, P. M. Dehmer, and J. L. Dehmer, *J. Chem. Phys.* **78**, 4315 (1983).

32. S. L. Anderson, G. D. Kubiak, and R. N. Zare, *Chem. Phys. Lett.* **105**, 22 (1984).

33. S. T. Pratt, P. M. Dehmer, and J. L. Dehmer, *Chem. Phys. Lett.* **105**, 28 (1984).

34. D. A. Gobeli, J. R. Morgen, R. J. St. Pierre, and M. A. El-Sayed, *J. Phys. Chem.* **88**, 178 (1984).

35. W. Forst, *Theory of Unimolecular Reactions*, Academic, New York, 1973; P. J. Robinson and K. A. Holbrook, *Unimolecular Reactions*, Wiley-Interscience, London, 1972; K. A. Holbrook, *Chem. Soc. Rev.* **2**, 163 (1983).

36. O. K. Rice and H. C. Ramsperger, *J. Am. Chem. Soc.* **49**, 1617 (1927); L. S. Kassel, *J. Phys. Chem.* **32**, 225 (1928).

37. R. A. Marcus and O. K. Rice, *J. Phys. Colloid Chem.* **55**, 894 (1951).

38. H. M. Rosenstock, M. B. Wallenstein, A. L. Warhaftig, and H. Eyring, *Proc. Natl. Acad. Sci. USA* **38**, 667 (1952).

39. Available as program 234 through the Quantum Chemistry Program Exchange, Indiana University, Bloomington, Ind.

40. G. Z. Whitten and B. S. Rabinovitch, *J. Chem. Phys.* **38**, 2466 (1963); **41**, 1884 (1964); D. C. Tardy, B. S. Rabinovitch, and G. Z. Whitten, *J. Chem. Phys.* **48**, 1427 (1968).

41. S. M. Beck, D. L. Monts, M. G. Liverman, and R. E. Smalley, *J. Chem. Phys.* **70**, 1062 (1979).

42. K. S. Nordholm and S. A. Rice, *J. Chem. Phys.* **61**, 203, 768 (1974).

43. M. D. Morse and K. F. Freed, *J. Chem. Phys.* **74**, 4395 (1981).

44. A. J. Lorquet, J. C. Lorquet, and W. Forst, *Chem. Phys.* **51**, 241, 253, 261 (1980).

45. R. J. Wolf and W. L. Hase, *J. Chem. Phys.* **72,** 316 (1980); **73,** 3010, 3779 (1980); **75,** 3809 (1981).

46. W. L. Hase, in *Dynamics of Molecular Collisions,* Vol. 2, W. H. Miller, Ed., Plenum, New York, 1976, Ch. 3.

47. E. V. Waage and B. S. Rabinovitch, *Chem. Rev.* **70,** 377 (1970).

48. C. E. Klots, *Z. Naturforsch.* **27A,** 553 (1972).

49. C. E. Klots, *J. Chem. Phys.* **58,** 5364 (1973).

50. C. E. Klots, *J. Chem. Phys.* **64,** 4269 (1976).

51. J. C. Light, *J. Chem. Phys.* **40,** 3221 (1964).

52. P. Pechukas and J. C. Light, *J. Chem. Phys.* **42,** 3281 (1965).

53. J. C. Light and J. Lin, *J. Chem. Phys.* **43,** 3209 (1965).

54. J. C. Light, *Discuss. Faraday Soc.* **44,** 14 (1967).

55. E. Nikitin, *Theor. Exp. Chem.* **1,** 83, 90, 275 (1965).

56. W. J. Chesnavich and M. T. Bowers, *J. Chem. Phys.* **66,** 2306 (1977).

57. W. H. Miller, *J. Chem. Phys.* **65,** 2216 (1976).

58. E. Pollak and P. Pechukas, *J. Chem. Phys.* **70,** 325 (1980).

59. H. Gaedtke and J. Troe, *Ber. Bunsenges. Physik. Chem.* **77,** 24 (1973).

60. M. Quack and J. Troe, *Ber. Bunsenges. Physik. Chem.* **78,** 240 (1974).

61. M. Quack, *J. Phys. Chem.* **83,** 150 (1979).

62. J. Troe, *J. Chem. Phys.* **75,** 226 (1981); **79,** 6017 (1983).

63. K. E. McCulloh and V. H. Dibeler, *J. Chem. Phys.* **64,** 4445 (1976); W. A. Chupka and J. Berkowitz, *J. Chem. Phys.* **54,** 4256 (1971).

64. Ch. Ottinger, *Z. Naturforsch.* **20A,** 1232 (1965).

65. L. P. Hills, M. L. Vestal, and J. H. Futrell, *J. Chem. Phys.* **54,** 3834 (1971).

66. R. D. Smith and J. H. Futrell, *Int. J. Mass Spectrom. Ion Phys.* **20,** 425 (1976).

67. J. P. Flamme, J. Momigny, and H. Wankenne, *J. Am. Chem. Soc.* **98,** 1045 (1976).

68. H. M. Rosenstock, in *Advances in Mass Spectrometry,* Vol. 4, E. Kendrick, Ed., Institute of Petroleum, London, 1968, p. 523.

69. C. E. Klots, *Chem. Phys. Lett.* **10,** 422 (1971); *J. Chem. Phys.* **75,** 1562 (1971).

70. A. J. Illies, M. F. Jarrold, and M. T. Bowers, *J. Am. Chem. Soc.* **104,** 3587 (1982).

71. J. P. Flamme and J. Momigny, *Chem. Phys.* **34,** 303 (1978).

72. J. P. Flamme, H. Wankenne, R. Locht, J. Momigny, P. J. C. M. Nowak, and J. Los, *Chem. Phys.* **27,** 45 (1978).

73. G. G. Meisels, G. M. L. Verboom, M. J. Weiss, and T. C. Hsieh, *J. Am. Chem. Soc.* **101,** 7189 (1979).

74. M. F. Jarrold, W. Wagner-Redeker, A. J. Illies, N. J. Kirchner, and M. T. Bowers, *Int. J. Mass Spectrom. Ion Processes* **58,** 63 (1984).

75. E. Wigner, *Trans. Faraday Soc.* **34,** 29 (1938).

76. W. J. Chesnavich, L. Bass, T. Su, and M. T. Bowers, *J. Chem. Phys.* **74,** 2228 (1981).

77. M. Vaz Pires, C. Galloy, and J. C. Lorquet, *J. Chem. Phys.* **69,** 3242 (1978).

78. M. Desouter-Lecomte, C. Galloy, and J. C. Lorquet, *J. Chem. Phys.* **71,** 3661 (1979).

79. D. Dehareng, X. Chapuisat, J. C. Lorquet, C. Galloy, and G. Raseev, *J. Chem. Phys.* **78,** 1246 (1983).

80. C. Galloy, C. Lecomte, and J. C. Lorquet, *J. Chem. Phys.* **77,** 4522 (1982).

81. J. C. Lorquet, *Org. Mass Spectrom.* **16,** 469 (1981).

82. A. Waite, S. K. Gray, and W. H. Miller, *J. Chem. Phys.* **78,** 259 (1983).

83. J. H. D. Eland, *Photoelectron Spectroscopy,* 2nd ed., Butterworths, London, 1983.

84. J. P. Maier and F. Thommen, in *Gas Phase Ion Chemistry,* Vol. 3, M. T. Bowers, Ed., Academic, New York, 1983.

85. J. T. Meek, R. K. Jones, and J. P. Reilley, *J. Chem. Phys.* **73,** 3503 (1980).

86. J. T. Meek, S. R. Lond, and J. P. Reilley, *J. Phys. Chem.* **86,** 2809 (1982).

87. S. R. Long, J. T. Meek, and J. P. Reilley, *J. Chem. Phys.* **79,** 3206 (1983).

88. Y. Achiba, K. Sato, K. Shobatake, and K. Kimura, *J. Chem. Phys.* **77,** 2709 (1982).

89. S. L. Anderson, D. M. Rider, and R. N. Zare, *Chem. Phys. Lett.* **93,** 11 (1982).

90. S. L. Anderson, L. Goodman, K. Krogh-Jespersen, A. G. Ozkubak, R. N. Zare, and C.-F. Zheng, *J. Chem. Phys.* **82,** 5329 (1985).

91. C. Sannen, G. Raseev, C. Galloy, G. Fauville, and J. C. Lorquet, *J. Chem. Phys.* **79,** 894 (1983); J. C. Lorquet, C. Sannen, and G. Raseev, *J. Am. Chem. Soc.* **102,** 7976 (1980).

92. T. Baer and R. Kury, *Chem. Phys. Lett.* **92,** 659 (1982).

93. H. M. Rosenstock, R. L. Stockbauer, and A. C. Parr, *J. Chem. Phys.* **73,** 773 (1980).

94. R. E. Weston and H. A. Schwarz, *Chemical Kinetics,* Prentice-Hall, Englewood Cliffs, New Jersey, 1972, p. 109; P. C. Jordan, *Chemical Kinetics and Transport,* Plenum, New York, 1979, p. 213.

95. S. W. Benson and H. E. O'Neal, *Kinetic Data on Gas Phase Unimolecular Reactions* (Natl. Bur. Stand. Ref. Data Ser., NBS 21), National Bureau of Standards, Washington, D.C., 1970.

96. S. T. Pratt and W. A. Chupka, *Chem. Phys.* **62,** 153 (1981).

97. T. Baer, B. P. Tsai, D. Smith, and P. T. Murray, *J. Chem. Phys.* **64,** 2460 (1976).

98. J. Dannacher, H. M. Rosenstock, R. Buff, A. C. Parr, R. L. Stockbauer, R. Bombach, and J. P. Stadelmann, *Chem. Phys.* **75,** 23 (1983).

99. T. Baer, P. M. Guyon, I. Nenner, A. Tabche-Fouhaille, R. Botter, L. F. Ferreira, and T. Govers, *J. Chem. Phys.* **70,** 1585 (1979).

100. P. M. Guyon, T. Baer, and I. Nenner, *J. Chem. Phys.* **78,** 3665 (1983).

101. W. A. Chupka, *J. Chem. Phys.* **30,** 191 (1959).

102. C. Lifshitz, *Mass Spectrom. Rev.* **1,** 309 (1982).

103. S. M. Gordon and N. W. Reid, *Int. J. Mass Spectrom. Ion Phys.* **18,** 379 (1975); C. Lifshitz, A. M. Peers, M. Weiss, and M. J. Weiss, *Adv. Mass Spectrom.* **6,** 871 (1974).

104. C. Lifshitz, *J. Phys. Chem.* **86,** 606 (1982).

105. C. Lifshitz, S. Gefen, and R. Arakawa, *J. Phys. Chem.* **88,** 4242 (1984).

106. C. Lifshitz, P. Gotchiguian, and R. Roller, *Chem. Phys. Lett.* **95,** 106 (1983).

107. C. Lifshitz, M. Goldenberg, Y. Malinovich, and M. Peres, *Int. J. Mass Spectrom. Ion Phys.* **46,** 269 (1983); Y. Malinovitch, R. Arakawa, G. Haase, and C. Lifshitz, *J. Phys. Chem.,* **89,** 2253 (1985).

108. T. Baer and T. E. Carney, *J. Chem. Phys.* **76,** 1304 (1982).

109. P. K. Pearson, H. F. Schaefer, and V. Wahlgren, *J. Chem. Phys.* **62,** 350 (1975).

110. A. Inone, S. Yoshida, and N. Ebara, *Int. J. Mass Spectrom. Ion Phys.* **52,** 209 (1983).

111. P. C. Burgers, J. L. Holmes, A. A. Mommers, and J. K. Terlouw, *Chem. Phys. Lett.* **102,** 1 (1983).

112. P. C. Burgers, J. L. Holmes, A. A. Mommers, J. E. Szulejko, and J. K. Terlouw, *Org. Mass Spectrom.* **19,** 442 (1984).

113. H. M. Rosenstock, R. Stockbauer, and A. C. Parr, *J. Chim. Physique* **77,** 745 (1980).

114. J. H. D. Eland and H. Schulte, *J. Chem. Phys.* **62,** 3835 (1975).

115. W. J. Chesnavich and M. T. Bowers, *J. Am. Chem. Soc.* **99,** 1705 (1977).

116. R. C. Dunbar, in *Gas Phase Ion Chemistry*, M. T. Bowers, Ed., Academic, New York, 1979, Ch. 14.

117. R. L. Woodin and J. L. Beauchamp, *Chem. Phys.* **41,** 1 (1979).

118. J. P. Hanovich and R. C. Dunbar, *J. Am. Chem. Soc.* **104,** 6220 (1982).

119. R. C. Dunbar and J. H. Chen, *J. Phys. Chem.* **88,** 1401, (1984).

120. R. C. Dunbar, *Int. J. Mass Spectrom. Ion Processes* **54,** 109 (1983); R. C. Dunbar and J. P. Honovich, *Int. J. Mass Spectrom. Ion Processes* **58,** 25 (1984).

121. K. Kimura, S. Katsumata, Y. Achiba, T. Yamazaki, and S. Iwata, *Handbook of He I Photoelectron Spectra of Fundamental Organic Molecules*, Japan Scientific Soc. Press, Tokyo 1981.

122. H. M. Rosenstock, R. L. Stockbauer, and A. C. Parr, *J. Chem. Phys.* **71,** 3708 (1979).

123. F. W. McLafferty, *Interpretation of Mass Spectra*, 3rd ed., University Science Books, Mill Valley, California, 1980.

124. T. Baer, G. D. Willett, D. Smith, and J. S. Phillips, *J. Chem. Phys.* **70,** 4076 (1979).

125. A. C. Parr, A. J. Jason, and R. Stockbauer, *Int. J. Mass Spectrom. Ion Phys.* **26,** 23 (1978).

126. R. Stockbauer and H. M. Rosenstock, *Int. J. Mass Spectrom. Ion Phys.* **27,** 185 (1978).

127. A. C. Parr, A. Jason, R. Stockbauer, and K. E. McCulloh, *Int. J. Mass Spectrom. Ion Phys.* **30,** 319 (1979).

128. A. C. Parr, A. Jason, and R. Stockbauer, *Int. J. Mass Spectrom. Ion Phys.* **33,** 243 (1980).

129. A. S. Werner and T. Baer, *J. Chem. Phys.* **62,** 2900 (1975).

130. J. Dannacher, J. P. Flamme, J. P. Stadelmann, and J. Vogt, *Chem. Phys.* **51,** 189 (1980); R. Bombach, J. Dannacher, and J. P. Stadelmann, *J. Am. Chem. Soc.* **105,** 1824 (1983).

131. D. Smith, T. Baer, G. D. Willett, and R. C. Ormerod, *Int. J. Mass Spectrom. Ion Phys.* **30,** 155 (1979).

132. F. P. Lossing, *Can. J. Chem.* **50,** 3973 (1972).

133. H. M. Rosenstock, K. Draxl, B. W. Steiner, and J. T. Herron, "Energetics of Gaseous Ions," *J. Phys. Chem. Ref. Data* **6,** Suppl. 1 (1977).

134. Y. Ono and C. Y. Ng, *J. Chem. Phys.* **74,** 6985 (1981).

135. T. Baer, *J. Electr. Spectr. Rel. Phenomena* **15,** 225 (1979).

136. See H. M. Rosenstock, J. Dannacher, and J. F. Liebman, *Radiat. Phys. Chem.* **40,** 7 (1982), for an excellent and comprehensive review.

137. B. Andlauer and Ch. Ottinger, *J. Chem. Phys.* **55,** 1471 (1971).

138. B. Andlauer and Ch. Ottinger, *Z. Naturforsch.* **A27,** 293 (1972).

139. J. Dannacher, *Chem. Phys.* **29,** 339 (1978).

140. J. Dannacher, J. P. Stadelmann, and J. Vogt, *J. Chem. Phys.* **74,** 2094 (1981).

141. J. Dannacher, J. P. Stadelmann, and J. Vogt, *Int. J. Mass Spectrom. Ion Phys.* **38,** 69 (1981).

142. S. Leach, personal communcation.

143. G. D. Willett and T. Baer, *J. Am. Chem. Soc.* **102,** 6774 (1980).

144. J. J. Butler, T. Baer, and S. A. Evans, *J. Am. Chem. Soc.* **105,** 3451 (1983).

145. D. W. Davis, *Chem. Phys. Lett.* **28,** 520 (1974).

146. K. Osafune and K. Kimura, *Chem. Phys. Lett.* **25,** 47 (1974).

147. G. D. Willett and T. Baer, *J. Am. Chem. Soc.* **102,** 6769 (1980).

148. M. L. Fraser-Monteiro, L. Fraser-Monteiro, J. J. Butler, T. Baer, and J. R. Hass, *J. Phys. Chem.* **86,** 739 (1982).

149. L. Fraser-Monteiro, M. L. Fraser-Monteiro, J. J. Butler, and T. Baer, *J. Phys. Chem.* **86,** 752 (1982).

150. J. J. Butler, L. Fraser-Monteiro, M. L. Fraser-Monteiro, T. Baer, and J. R. Hass, *J. Phys. Chem.* **86,** 747 (1982).

151. D. J. McAdoo, D. N. Witiak, F. W. McLafferty, and J. D. Dill, *J. Am. Chem. Soc.* **100,** 6639 (1978).

152. W. J. Bouma, J. K. MacLeod, and L. Random, *J. Am. Chem. Soc.* **102,** 2246 (1980).

153. T. Baer, W. A. Brand, T. L. Bunn, and J. J. Butler, *Faraday Discuss,* **75,** 45 (1983).

154. W. A. Brand and T. Baer, *J. Am. Chem. Soc.* **106,** 3154 (1984).

155. P. N. Th. v. Velzen and W. J. v.d. Hart, *Chem. Phys.* **61,** 335 (1981).

156. T. Baer, D. Smith, B. P. Tsai, and A. S. Werner, *Adv. in Mass Spectrom.* **7A,** 56 (1977).

157. G. G. Meisels, M. Weiss, T. Hsieh, and G. M. L. Verboom, presented at the 26*th Ann. Conf. Mass Spectrom. Allied Topics,* St. Louis, May 28–June 2, 1978.

158. G. G. Meisels, T. Hsieh, and J. P. Gilman, *J. Chem. Phys.* **73,** 4126 (1980).

159. J. P. Gilman, T. Hsieh, and G. G. Meisels, *J. Chem. Phys.* **78,** 1174. (1983).

160. J. P. Gilman, T. Hsieh, and G. G. Meisels, *J. Chem. Phys.* **78,** 3767 (1983).

161. T. K. Ogden, N. Shaw, C. J. Danby, and I. Powis, *Int. J. Mass Spectrom. Ion Processes* **54,** 41 (1983).

162. M. L. Fraser-Monteiro, L. F. Monteiro, J. deWit, and T. Baer, *J. Phys. Chem.* **88,** 3622 (1984).

163. C. Lifshitz, M. Goldenberg, Y. Malinovich, and M. Peres, *Org. Mass Spectrom.* **17,** 453 (1982).

164. C. Lifshitz, S. Gefen, and R. Arakawa, *J. Phys. Chem.* **88,** 4242 (1984).

165. A. Maquestiau, R. Flammang, G. L. Glish, J. A. Laramee, and R. G. Cooks, *Org. Mass Spectrom.* **15,** 131 (1980).

166. D. H. Russell, M. L. Gross, and N. M. M. Nibbering, *J. Am. Chem. Soc.* **100,** 6133 (1978).

167. M. Panczel and T. Baer, *Int. J. Mass Spectrom. Ion Processes* **58,** 43 (1984).

168. J. P. Maier, in *Kinetics of Ion–Molecule Reactions,* P. Ausloos, Ed., Plenum, New York, 1979, p. 437.

169. T. Baer and J. R. Hass, *J. Phys. Chem.*, in press.

170. M. M. Bursey, J. R. Hass, D. J. Harvin, and C. E. Parker, *J. Am. Chem. Soc.* **101**, 5485 (1979).

171. W. M. Trott, N. C. Blais, and E. A. Walters, *J. Chem. Phys.* **69**, 3150 (1978).

172. C. Lifshitz and E. Tzidony, *Int. J. Mass Spectrom. Ion Phys.* **39**, 181 (1981).

173. J. L. Holmes and F. P. Lossing, *J. Am. Chem. Soc.* **102**, 3732 (1980); R. Weber and K. Levsen, *Org. Mass Spectrom.* **15**, 138 (1980).

174. R. Bombach, J. P. Stadelmann, and J. Vogt, *Chem. Phys.* **72**, 259 (1982).

175. D. M. Mintz and T. Baer, *Int. J. Mass Spectrom. Ion Phys.* **25**, 39 (1977).

176. C. Lifshitz, *Int. J. Mass Spectrom. Ion Phys.* **43**, 179 (1982).

177. G. Depke, C. Lifshitz, H. Schwartz, and E. Tzidony, *Angew. Chem. Int. Ed. Engl.* **20**, 792 (1981); R. C. Heyer and M. E. Russell, *Org. Mass Spectrom.* **16**, 236 (1981).

178. D. M. Mintz and T. Baer, *J. Chem. Phys.* **65**, 2407 (1976).

179. I. Powis, *Chem. Phys.* **74**, 421 (1983).

180. T. Baer, U. Buchler, and C. E. Klots, *J. Chim. Physique* **75**, 739 (1980).

181. W. A. Brand, T. Baer, and C. E. Klots, *Chem. Phys.* **76**, 111 (1983).

182. K. Johnson, I. Powis, and C. J. Danby, *Chem. Phys.* **63**, 1 (1981).

183. J. H. D. Eland, R. Frey, A. Kuestler, H. Schultz, and B. Brehm, *Int. J. Mass Spectrom. Ion Phys.* **22**, 155 (1976).

184. A. S. Werner, B. P. Tsai, and T. Baer, *J. Chem. Phys.* **60**, 3650 (1974).

185. K. Johnson, I. Powis, and C. J. Danby, *Chem. Phys.* **70**, 329 (1982).

186. C. E. Klots, D. M. Mintz, and T. Baer, *J. Chem. Phys.* **66**, 5100 (1977).

187. I. Powis, *J. Chem. Soc. Faraday Trans. II* **75**, 1294 (1979).

188. B. E. Miller and T. Baer, *Chem. Phys.* **85**, 39 (1984).

189. I. Powis, *J. Chem. Soc. Faraday Trans. II* **77**, 1433 (1981).

190. I. Powis, *Chem. Phys.* **68**, 251 (1982).

191. K. E. McCulloh, *Int. J. Mass Spectrom. Ion Phys.* **21**, 333 (1976).

192. T. Cvitas, H. Gusten, and L. Klasinc, *J. Chem. Phys.* **67**, 2687 (1977).

193. I. Powis, *Mol. Phys.* **39**, 311 (1980).

194. H. W. Jochims, W. Lohr, and H. Baumgartel, *Ber. Bunsenges. Phys. Chem.* **80**, 130 (1976).

195. S. J. Riley and K. R. Wilson, *Faraday Discuss. Chem. Soc.* **53**, 132 (1972).

196. P. J. Bassett and D. R. Lloyd, *J. Chem. Soc. Dalton Trans.* 248 (1972).

197. P. I. Mansell, C. J. Danby, and I. Powis, *J. Chem. Soc. Faraday Trans. II* **77**, 1449 (1981).

198. T. Baer, A. E. DePristo, and J. J. Hermans, *J. Chem. Phys.* **76**, 5917 (1982).

199. Y. Osamura, H. F. Schaefer II, M. Dupuis, and K. R. Newton, *J. Chem. Phys.* **75**, 5728 (1981).

200. G. Dujardin, S. Leach, O. Dutuit, P. M. Guyon, and M. Richard-Viard, *Chem. Phys.* **88**, 339 (1984).

THE FUNDAMENTALS OF SPONTANEOUS IGNITION OF GASEOUS HYDROCARBONS AND RELATED ORGANIC COMPOUNDS

JOHN F. GRIFFITHS

Department of Physical Chemistry
The University of Leeds, Leeds, England

CONTENTS

I. INTRODUCTION

A. Scope

The purpose of this Chapter is to describe the main experimental features of spontaneous ignition during the thermal oxidation of hydrocarbons and other organic substrates, to establish the physicochemical interactions that bring ignition about, and to discuss how the behavior is controlled by the chemical structure of the substrate.

Thermal ignition is associated traditionally with events occurring when a premixed fuel vapor and oxygen (or air) mixture reacts in a uniformly hot, closed vessel, or under flow through a hot tube. The current presentation rests very substantially on such idealized laboratory conditions, and on modern developments from them (the small-scale, well stirred flow reactor); accordingly it is possible for the respective roles of chemistry and physics, dominated by the rates of heat release and dissipation, to be investigated experimentally and expressed theoretically in their least complicated forms. It is important to acknowledge that far greater complexity is likely to be encountered in more general applications. On the one hand, spontaneous ignition hazards may be engendered in industrial environments where flammable gases are exposed to hot surfaces; the additional factors to be considered then include reactant mixing, the temperature gradients, and the free-convection or even turbulence that may be generated at the surface.[1] Evaporation of liquids from hot lagging or diffusion processes within the material are additional special topics.[2] On the other hand, in reciprocating engines the effects of droplet evaporation and mixing, of turbulence due to induction and exhaust, and of compression and expansion are added factors. The consequences of spark ignition, or similar influences, are not entirely thermal in origin and thus are not of direct concern here.[3] Aspects relating to ignition energies and flammability limits are discussed elsewhere.[4]

For the present purpose, spontaneous ignition may be classified as a sudden loss of stability in a system undergoing slow or even negligible reaction. It is exemplified admirably in a closed system by the sudden and violent generation of heat, light, and sound, that is, an explosion. None of the original material remains after the event. However, as we shall see, this phenomenon represents a wider realm, fashionably termed *criticality* or *parametric sensitivity*, whereby a marked change of behavior is brought about by a small change in control parameters. The effect needed not be so dramatic—or, superficially, so simple—and the concepts involved are those that give rise to stationary and oscillatory states and to criticality, dynamic instability, and reaction multiplicity.[5] This is the province of nonlinear dynamics and is of interest to chemical engineers and applied mathematicians.

This article is concerned with the physicochemical interactions underlying stability and criticality as they appear to the combustion chemist. This emphasis may be illustrated in the following way. At one extreme, oxidation of organic compounds may be carried out in a closed system such that no temperature change occurs despite exothermic reactions (i.e., the heat-dissipation rate far exceeds that of heat generation). The system is thus isothermal and any interest in it is of a strictly chemical kinetic and mechanistic nature. At the other extreme, oxidation may be carried out in a closed adiabatic system. Ignition is inevitable and all of the heat released then goes into raising the temperature of the reaction products;

the spotlight then falls on the state of thermodynamic equilibrium. Between these borders is the extremely fascinating region in which heat- and mass-transport processes vie with the heat release from the reaction: Both thermal and kinetic feedback mechanisms control the outcome.

From the mechanistic point of view, profound changes in the chemistry of hydrocarbon oxidation are brought about by changes in temperature. The present focus is on events that take place in the approximate temperature range 500–850 K. The chemistry of hydrocarbon combustion at higher temperatures is essentially that associated with hot flames and chemical equilibrium or pseudoequilibrium.[6] Cross reference to the high temperature realm is necessary but it is not a specific feature of this article. A very detailed understanding of processes in flames has begun to emerge in recent years.[7–9] Our preoccupation in the lower-temperature realm is with complex networks of elementary processes that, in general, are very far from equilibrium with their reverse steps; the system as a whole is also very far from thermodynamic equilibrium. Because the chemical nature of organic substrates is so diverse, the discussion will be directed to alkanes as the principal substrates. Generality will be maintained as far as possible, but for exemplifying kinetic detail, attention will be paid predominantly to n-butane and isobutane. These are the archetypes for illustrating how chemical structure may affect the behavior of higher hydrocarbons. However, alkenes are generated in very substantial yield at various stages of alkane oxidation, so aspects of their oxidation cannot be disregarded. Moreover, molecular intermediates, such as aldehydes, peroxides, and carbon monoxide, play very important parts in the development of spontaneous ignition and so these, too are singled out for special consideration. Very comprehensive reviews have appeared that discuss the kinetics and mechanistic details of hydrocarbon combustion[10–13] and that of other organic substrates.[10,14]

B. Historical Background

The development of theories for free-radical, branching-chain propagation by Semenov[15] in the Soviet Union and Hinshelwood[16] in England about 1930 put the interpretation of contemporary experimental studies of low-temperature hydrocarbon oxidation on a sound kinetic footing and stimulated much wider activity. The Soviet studies continue under the direction of Semenov and also his first research student, Nalbandyan.[17] Laffitte[18] was the founder of studies in France, with continuity of research maintained principally by Ben-Aim and Lucquin.[19] Hydrocarbon oxidation as a major research theme in many United Kingdom universities descended from Hinshelwood at Oxford and Norrish[20] at Cambridge; it reached its peak in the 1970s. The wider discipline of gas kinetics that disseminated from these schools remains intensely active. There seems never to have been similarly long-standing experimental activity in academic institutions in the United States, although studies by Pease[21] and co-workers at Princeton in the 1930s were particular landmarks. There has been ongoing interest at the Naval Research Laboratories,[22] and in the higher-temperature realm, hydrocarbon combustion studies by Glassman[23] and co-workers are now very well established and highly respected. There have been innumerable less sustained, but no less important, contributions over the decades from workers throughout the world. Currently there is a proliferation of experimental and theoretical activity aroused by concern over "knock" and pre-ignition in spark-ignition engines.[24]

There are many texts on combustion and kinetics that may be consulted to document the complete history (see, e.g., Minkoff and Tipper,[10] Shtern,[11] Lewis and von Elbe,[25] Benson,[26] and Mulcahy[27]), and the prevailing attitudes and progress in the field have been monitored since 1954 in the biennial proceedings of the International Symposia on Combustion. Contributions on this topic have pervaded the physical chemistry journals since 1930 and (since ~1960) the principal combustion-oriented journals, *Combustion and Flame, Combustion Science and Technology*, and *Combustion, Explosion and Shock Waves*.

A perusal of this literature would reveal that there is much more to hydrocarbon oxidation in the low-temperature realm than merely "slow reaction" or "ignition." The reader would be introduced to "cool flame phenomena," discovered by Davy[28] and investigated first by Perkin (1882),[29] multiple-stage ignitions,[30] the "piq d'ârret,"[31] and the region of negative temperature coefficient of reaction rate.[21] These phenomena are interlinked (see Section IV for their identities) and contribute to a very complex overall pattern of events, but they are also the basis of an immensely stimulating field for academic research that has only recently begun to yield to deep theoretical insight and understanding.

Until ~1970, exclusively "isothermal" interpretations of spontaneous ignition and the additional complexities of hydrocarbon oxidation predominated, despite the widespread acknowledgment that there were associated temperature changes.[32] However, one outstanding contribution that integrated the kinetic and thermal factors was that of Salnikov[33] in 1949; it had much in common with more recent developments in theories of chemical-reactor stability.[34] The novel concepts introduced by Salnikov[33] for dealing with the existence of oscillatory states in nonisothermal systems—the cool-flame phenomena—were not readily grasped by chemical kineticists, thwarting an early attack along fundamentally correct lines. (Interestingly, contemporary and rather similar developments in reactor theory[35,36] were not widely recognized by chemical engineers at the time and were "rediscovered" nearly 20 years later.[37]) The fundamentals of thermokinetic interactions are developed in a later section. The birth of modern interpretations for nonisothermal phenomena and spontaneous ignition in hydrocarbon oxidation was due to Gray and Yang.[38-42] They presented a unified interpretation of chain-thermal interactions based on the following skeleton kinetic scheme, in which X represents a reactive intermediate.

		Exothermicity	Rate constant	
initiation	Fuel $\rightarrow$ X	h_i	k_i	(1.1)
branching	X $\rightarrow$ 2X	h_b	k_b	(1.2)
termination	X $\rightarrow$ stable products	h_{t_1}	k_{t_1}	(1.3)
termination	X $\rightarrow$ stable products	h_{t_2}	k_{t_2}	(1.4)

The unifying link is the pair of conservation equations

$$\text{for mass:} \quad dx/dt = k_i - (k_{t_1} + k_{t_2} - k_b)x$$

$$\text{for energy:} \quad c_V dT/dt = k_1 h_1 + (k_{t_1} h_{t_1} + k_{t_2} h_{t_2} + k_b h_b) - l(T - T_0)$$

where x is concentration of X, t is time, c_V is a heat capacity per unit volume, and l is a Newtonian heat-transfer coefficient through the reactor surface. The term $(T - T_0)$ is the temperature excess of the reactants over that of the surface (T_0). The system is subject to the condition for activation energies $E_{t_1} < E_b < E_{t_2}$.

The implications of this scheme and its relationship to current interpretations of the mechanisms associated with hydrocarbon oxidation will be discussed later. That the complete understanding of thermokinetic interactions as applied to hydrocarbon combustion has roots in nonlinear dynamics is beyond dispute, and the problems that exercise the minds of combustion chemists are the problems that occupy not only engineers and applied mathematicians but also scientists in many other disciplines [43,44] One main feature of the current state of knowledge of spontaneous-ignition phenomena is that such topics do not yield to unequivocal qualitative description, but demand rigorous mathematical accounts.

C. Presentation

The unifying theme throughout this review is the portrayal of events in the context of well-stirred flowing conditions (the continuously stirred tank reactor, or c.s.t.r.). This experimental technique confers spatial uniformity of temperature and concentration throughout the vessel, and either stationary or oscillatory states are sustained indefinitely. It offers the only sound basis for (i) an unambiguous interpretation of the nature of thermokinetic phenomena, (ii) the distinctions between them and the conditions at which changes occur, and (iii) fully quantitative links between experiment and theory. From the theoretical point of view, in stationary conditions algebraic equations replace the differential or partial differential expressions that are representative of other systems. Thus some aspects of the interpretation are greatly simplified and may facilitate the application of stability theorems in nonlinear mechanics and of numerical methods. However, experimental investigations in well-stirred flow vessels have been initiated in earnest only during the last decade and so a very extensive review of physical and chemical observations that have been made in the traditional unstirred, closed vessels is also necessary here to present a complete picture.

The review opens with a detailed account of how (i) simple, irreversible, and nonisothermal and (ii) isothermal autocatalytic reactions may be expected to behave in a c.s.t.r. Links to the more complicated hydrocarbon oxidations are forged after the elaboration of their principal physical and chemical features. The theoretical background is thus followed by a description and critical appraisal of the experimental techniques that have been employed and, in a separate section, the phenomenological aspects of hydrocarbon combustion and the significant overall kinetic features. Detailed kinetics and mechanisms are then discussed. The fundamental theory is drawn together with the main experimental facts, and aspects of numerical interpretation are presented, in the final section. Distinctions among the behaviors of different organic substrates are given in the conclusion.

II. THEORETICAL BACKGROUND

A. Stability of Exothermic Reactions in a Continuously Stirred Tank Reactor (c.s.t.r.): Adiabatic Operation

Although chemical reaction in a closed system proceeds to the unique point of thermodynamic equilibrium, in an open system permanent stationary states far from equilibrium and sustained oscillatory states are both possible. Moreover, in open systems, multiple stationary states may also be attained; which one is realized depends on the path by which it is approached. Conditions for changes of multiplicity are the conditions for critical phenomena such as ignitions and extinctions: discontinuous jumps occur in response to a continuous change in control parameters. We need to establish these conditions and to understand the factors that govern the stability of stationary states and those that give rise to oscillations.

For simple reactions (irreversible, exothermic, first-order reactions obeying the Arrhenius law) these features have been the subject of extensive study by chemical-reactor engineers. The foundations were laid in very distinguished papers by Zel'dovich and Zysin[35,36] (1941), Salnikov[33] (1949), and Longwell[45] (1953), but the papers that stimulated the extraordinarily intensive activity over the past three decades were those of Bilous and Amundson[46] (1955). The major codifying papers of Uppal, Ray, and Poore[47] did not appear until the mid 1970s.

1.   Criticality and Reaction Multiplicity

Consider a first-order, single-step, exothermic reaction in an adiabatic c.s.t.r. There is no heat loss via the vessel walls. The starting points are the equations for mass and energy conservation.

Mass balance:

$$\frac{dx}{dt} \text{ (mol/s)} = \underbrace{\frac{f}{V}x_0}_{\substack{\text{(inflow} \\ \text{rate)}}} - \underbrace{\frac{f}{V}x}_{\substack{\text{(outflow} \\ \text{rate)}}} - \underbrace{kx}_{\substack{\text{(chemical-reaction} \\ \text{rate)}}} \tag{2.1}$$

(rate of change of concentration)

Energy balance:

$$\sigma V c_p \frac{dT}{dt} \text{ (J/s)} = \underbrace{(-\Delta H)Vkx}_{\substack{\text{(rate of heat} \\ \text{output)}}} - \underbrace{\sigma f c_p (T - T_0)}_{\substack{\text{(rate of heat uptake by} \\ \text{incoming reactants} \\ \text{at } T_0)}} \tag{2.2}$$

Note that f is a total volume flow rate of reactants at temperature T in the vessel, as described in Fig. 1. All terms and symbols are defined in the Appendix. We may express these equations in a more convenient, unifying form in the following way:

$$\frac{d\gamma}{dt} = \frac{(1-\gamma)}{t_{ch}} \exp(\theta/1 + \varepsilon\theta) - \frac{\gamma}{t_{res}} \tag{2.3}$$

$$\frac{d\theta}{dt} = \frac{B(1-\gamma)}{t_{ch}} \exp(\theta/1 + \varepsilon\theta) - \frac{\theta}{t_{res}} \tag{2.4}$$

The various terms are defined as follows:

fractional extent of conversion, $\gamma = 1 - x/x_0$
dimensionless temperature excess, $\theta = (T - T_0)E/RT_0^2$
dimensionless adiabatic temperature excess, $B = (T_{ad} - T_0)E/RT_0^2 = (-\Delta H)x_0 E/\sum c_p RT_0^2$

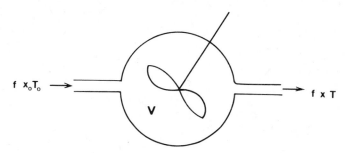

Figure 1.   Representation of a c.s.t.r. of volume V. Reactants enter the vessel at concentration x_0, volume flow rate f, and temperature T_0. Products leave at concentration x, volume flow rate f, and temperature T. In adiabatic operation the vessel temperature is raised to T and no heat is dissipated through the vessel walls. In nonadiabatic operation the walls are thermostated at temperature T_a, and heat is transported through them. The "figure eight" in the drawing represents the stirrer.

mean residence time (in seconds), $t_{res} = V/f$
characteristic chemical time (in seconds), $t_{ch} = 1/k(T_0)$

Since t_{ch} is defined at the inlet temperature T_0, t'_{ch} the chemical time at reactant temperature T, is $1/k(T)$. Thus $t_{ch} = t'_{ch} \exp(\theta/1 + \varepsilon\theta)$, from Frank-Kamenetskii's[48] identity $k(T) = k(T_0) \exp(\theta/1 + \varepsilon\theta)$, where $\varepsilon = RT_0/E$. T_0 is used as the reference temperature throughout these expressions because it is a control parameter. T is a variable of the system and is able to float to its own level. In circumstances where T is being measured directly it may be convenient to circumvent the Frank-Kamenetskii term by defining t'_{ch} in the expressions.

In a stationary state, $d\gamma/dt = d\theta/dt = 0$, from which

$$\gamma_{ss} = \frac{\exp(\theta/1 + \varepsilon\theta)}{t_{ch}} \left[\frac{1}{t_{res}} + \frac{\exp(\theta/1 + \varepsilon\theta)}{t_{ch}} \right]^{-1} \tag{2.5}$$

and

$$\gamma_{ss} = \frac{\theta_{ss}}{B} \tag{2.6}$$

where ss stands for stationary state. At low degrees of conversion, the temperature excess (expressed here in the dimensionless form θ) is also low. At complete conversion $\theta_{ss}/B = 1$ and the adiabatic temperature rise is measured. Equation (2.6) is a very important identity, because it reveals that the extent of conversion and the temperature are not independent variables under adiabatic operation. The consequences of this are discussed below. A single expression of the condition for stationary states is

$$\frac{\theta_{ss}}{t_{res}} = \frac{B}{t_{ch}}\left[1 - \frac{\theta_{ss}}{B}\right]\exp(\theta_{ss}/1 + \varepsilon\theta_{ss}) \qquad (2.7)$$

("heat-loss term" L)   ("heat-release term" R)

We can map some of the properties of this equation.

Figure 2 displays the separate representations of the two terms of Eq. (2.7) in the forms $R(\theta)$ and $L(\theta)$; the intersections between them are the stationary-state values for θ. The fan of lines representing $L(\theta)$ arises because the heat-loss rate depends on the mean residence time. Shallow gradients signify long residence times; steep gradients signify short residence times or very fast flows. There is only one $R(\theta)$ at constant T_0 and it may take either the form depicted in Fig. 2a or that of Fig. 2b with respect to the heat-loss lines.

The qualitative distinctions illustrated by Fig. 2 show that there may be either one or three values of θ at a given residence time (Fig. 2a) or θ may be only single-valued regardless of t_{res} (Fig. 2b). The scaling of $R(\theta)$ as T_0 is changed emerges from the magnitude of B; it is inversely proportional to T_0^2. The maximum value that θ can take is B and the

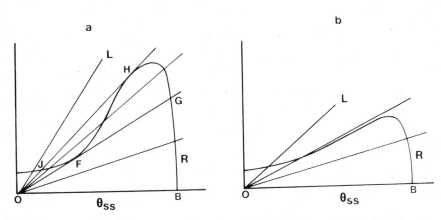

Figure 2.   Representation of rates of heat release and dissipation in the adiabatic c.s.t.r. as a function of θ [Eq. (2.7)]. Points of intersection (e.g., J, F, H, G in a) between R and L signify the stationary states of the system.

maximum of $R(\theta)$ occurs at $B-1$. Since in general, B is very much greater than unity, the R curves of Fig. 2a and b are distortions of their quantitative forms that do not properly reflect the sharpness of the peak and the precipitous drop following it. As the "tangency conditions" at points F and H in Fig. 2a show, we may expect to see "critical ignition" $(F \rightarrow G)$ as t_{res} is raised and "critical extinction" $(H \rightarrow J)$ as t_{res} is diminished. Between the two is a region of hysteresis and multiple stationary states.

2. Criteria for Ignition and Extinction and the Existence of Multiplicity

To establish the analytical criteria for ignition and extinction and the existence of multiplicity with reference to T_0, Eqs. (2.3) and (2.4) must be expressed in simplified form by invoking the Frank-Kamenetskii approximation $\varepsilon \rightarrow 0$, $E \rightarrow \infty$. (Conditions for the validity of this approximation are discussed below.) Thus

$$\frac{d\gamma}{dt} = \frac{1-\gamma}{t_{ch}} \exp \theta - \frac{\gamma}{t_{res}} \qquad (2.3')$$

$$\frac{d\theta}{dt} = \frac{B(1-\gamma)}{t_{ch}} \exp \theta - \frac{\theta}{t_{res}} \qquad (2.4')$$

The analytical conditions for the critical transitions representing ignition and extinction are

$$\frac{dR(\theta)}{d\theta} = \frac{dL(\theta)}{d\theta} \quad \text{and} \quad R(\theta) = L(\theta) \qquad (2.8)$$

The critical values of θ, $\theta_{\pm}$, are given by the solutions of the quadratic equation

$$\theta^2 - B\theta - B = 0 \qquad (2.9)$$

from which

$$\theta_{\pm} = \frac{B}{2} \left[1 \pm \left(1 - \frac{4}{B} \right)^{1/2} \right] \qquad (2.10)$$

Expanding the binomial and taking the two roots, which correspond to

θ_{ign} (θ_-) and θ_{ext} (θ_+), we obtain

$$\theta_{ign} = \frac{B}{2}\left[1 - \left(1 - \frac{2}{B} - \frac{2}{B^2} - \cdots\right)\right] = 1 + \frac{1}{B} + \cdots \tag{2.11}$$

$$\theta_{ext} = B - 1 - \frac{1}{B} \tag{2.12}$$

For large values of B, $\theta_{ign} \simeq 1$, and the temperature excess at "ignition" $\Delta T_{cr} \simeq RT_0^2/E$. The value of θ_{ext} is very close to that of B itself, and only marginally less than that at the maximum in $R(\theta)$, for which $\theta = B - 1$. These solutions are exact in the hypothetical limit $E = \infty$. The reality of the exponential approximation is that $\varepsilon \ll 1$. This means that Eq. (2.11) is a very satisfactory solution for θ_{ign}, since $\varepsilon\theta_{ign} \ll 1$. However, $\varepsilon\theta_{ext} \simeq \varepsilon B$, and for $B \gg 1$, the product $\varepsilon\theta_{ext}$ is not negligible and may even exceed unity.

Exact solutions can be obtained using the same mathematical treatment but retaining $k(T)$. For ignition and extinction, they take the respective forms

$$T_{ign} - T_0 = \frac{RT_0^2}{E} + \cdots \tag{2.11'}$$

which is very similar to Eq. (2.11), and

$$T_{ad} - T_{ext} = \frac{RT_{ad}^2}{E} + \cdots \tag{2.12'}$$

The latter takes a form in which T_{ad} has replaced T_0 as the reference temperature. The respective extents to which Eqs. (2.11) and (2.11') and Eqs. (2.12) and (2.12') may differ are demonstrated in an example in the next subsection.

Criticality just ceases to exist when

$$\frac{dR(\theta)}{d\theta} \leqslant \frac{dL(\theta)}{d\theta}, \quad \frac{d^2R(\theta)}{d\theta^2} \leqslant \frac{d^2L(\theta)}{d\theta^2}, \quad \text{and} \quad R(\theta) = L(\theta) \tag{2.13}$$

or, more simply, when $\theta_+ = \theta_-$ and the discriminant of Eq. (2.9) goes to zero. Hence $\theta_+ = \theta_-$ when

$$1 - \frac{4}{B} = 0, \quad \text{that is, when} \quad B = 4. \tag{2.14}$$

Thus critical ignition and extinction are possible only when

$$B > 4 \qquad (2.15)$$

or in full analytical form,

$$B > \frac{4}{1 - 4\varepsilon} \qquad (2.15')$$

3. A Practical Example of Combustion in the Adiabatic c.s.t.r.

There have been very few experimental studies of exothermic reactions in the c.s.t.r. Jenkins et al,[49] for example, investigated aspects of hydrogen combustion under adiabatic conditions, but their goals were kinetically oriented rather than directed toward reactor dynamics. There is currently research activity on adiabatic turbulent combustion in recirculating flows relating to gas-turbine studies. Some liquid-phase systems[50,51] have been studied in the *nonadiabatic* c.s.t.r., but the only gaseous studies directed toward quantitative dynamic results have been those by Bush[52] and by Gray et al.[53] The latter study involved the first-order exothermic decomposition of di-*t*-butyl peroxide (DTBP), which serves as a useful model system for how it might be expected to behave in an *adiabatic* c.s.t.r. The relevant kinetic[54] and thermal data are as follows:

$$(CH_3)_3COOC(CH_3) \rightarrow 2(CH_3)_2CO + C_2H_6$$
$$\Delta H^0_{298} = -170 \text{ kJ/mol}; \qquad E = 152 \text{ kJ/mol} \qquad (2.16)$$

The value of $\sum_i (nc_p)_i$ per mole DTBP reacted is 203 J/K at 300 K, rising to 427 J/K at 900 K. Here n is the number of moles of each product formed.

Let $T_0 = 300$ K, $\Delta T_{ad} = 557$ K, $B = 113.2$, and $\varepsilon = 0.0164$. From Eq. (2.11), $\theta_{ign} = 1 + 1/113.2 = 1.008$, corresponding to a critical temperature excess of 4.9 K at ignition. This value is virtually identical to that derived directly from RT_0^2/E [Eq. (2.11)]. At ignition $\varepsilon\theta = 0.0165$.

We may note that the criterion for ignition in the adiabatic c.s.t.r. is very similar to that for the marginal loss of stability derived from Semenov's quasi-stationary interpretation of a criterion for thermal ignition in a closed system.[48]

From (2.12), at extinction $\theta_{ext} = 112.2 - 1/113.2 = 112.192$. This yields $\Delta T_{ext} = \theta_{ext} RT_0^2/E = 552.3$ K, from which $T_{ext} = 852.3$ K.

Taking the critical criterion from analysis without invoking the exponential approximation (2.12),

$$T_{ad} - T_{ext} = 40.2 \text{ K}$$

from which

$$T_{ext} = 816.8 \text{ K}$$

Thus, whereas a precise prediction of T_{cr} and ΔT_{cr} at ignition is obtained using the simplest expressions, calculation of T_{cr} at extinction leads to a marked discrepancy when approximations are made. The error arises because at extinction $\varepsilon\theta = 1.84$, which value is no longer negligible, even though ε itself is very small.

Clearly $B \gg 1$ even for this modestly exothermic reaction in which the reactant and product compositions have quite large heat capacities. The criterion for failure of criticality to occur ($B < 4$) seems very remote, and to achieve it we would have to carry out reaction either in extremely dilute conditions, so that $(-\Delta H)/c_p$ per mole of mixture would be much reduced, or at very high inlet temperatures T_0.

4. Patterns of Stable and Unstable States

Rearranging Eq. (2.7) yields, in reduced form,

$$t_{res} = \frac{\theta_{ss} \exp(-\theta_{ss}) t_{ch}}{B(1 - \theta_{ss}/B)} \tag{2.17}$$

The behavior of θ_{ss} and its counterpart γ_{ss} as t_{res} varies are displayed graphically in Figs. 3a and 3b, respectively. There may be one or three

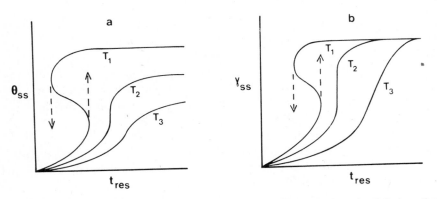

Figure 3. Relationships between (a) θ_{ss} and t_{res} and (b) γ_{ss} and t_{res} for inflow to the adiabatic c.s.t.r. at increasing temperatures of the reactants ($T_3 > T_2 > T_1$). The curve at T_2 corresponds to the condition $B = 4(1 - 4\varepsilon)$.

values of θ at a given t_{res}. Each of the curves relates to a different inflow reactant temperature T_0; this value controls the magnitudes of t_{ch} and B. In the limit $t_{res} \to \infty$, $\theta \to B$; θ_{ss} diminishes as T_0 is raised, since $B(T_0)$ diminishes. Regardless of T_0, $\gamma_{ss} \to 1$ as $t_{res} \to \infty$. The different forms of the overall patterns of behavior, namely monostability without criticality (curves T_3), the margin of criticality (curves T_2), and the pattern in which multiple singularities can coexist (curves T_1), are discernable in Fig. 3. The ignition and extinction transitions via changes of residence time are also shown.

What is the nature of the stationary states along the curves in Fig. 3, and are oscillatory states possible? The answers to these questions are obtained by investigating the local stability of the singularities in the $\gamma-\theta$ phase plane.[55] This is equivalent to studying mathematically the time-dependent response to a perturbation from the singularity. An experimental analogy to this is brought about in the c.s.t.r. by the physical disturbance of the system from its stationary state.

At the stationary state, $d\gamma/dt = d\theta/dt = 0$ and the singularities that are the solutions to this expression are points in the $\gamma-\theta$ plane. The time-independent expression $d\gamma/d\theta$, obtained by dividing Eq. (2.3) by Eq. (2.4), is indeterminate at the singularities. The mathematical procedure for determining the character of these points is to take each pair γ_{ss}, θ_{ss} and to express the time-dependent system in the new coordinates

$$\gamma' = \gamma - \gamma_{ss}, \qquad \theta' = \theta - \theta_{ss}.$$

At the coordinates γ', θ' ($\neq 0, 0$) the system is displaced from the singularity and therefore must exhibit a time-dependent response. Expressing $d\gamma'/dt$ and $d\theta'/dt$ in term of $d\gamma/dt$ and $d\theta/dt$ by the Taylor expansion, but linearizing by taking only first-order terms (thereby limiting the study to local stability, i.e., marginal displacement from γ, θ) leads to the expressions

$$\frac{d\gamma'}{dt} = \underbrace{\left(\frac{\partial(d\gamma/dt)}{\partial\gamma}\right)_\theta}_{A'} \gamma' + \underbrace{\left(\frac{\partial(d\gamma/dt)}{\partial\theta}\right)_\gamma}_{B'} \theta' \qquad (2.18)$$

$$\frac{d\theta'}{dt} = \underbrace{\left(\frac{\partial(d\theta/dt)}{\partial\gamma}\right)_\theta}_{C'} \gamma' + \underbrace{\left(\frac{\partial(d\theta/dt)}{\partial\theta}\right)_\gamma}_{D'} \theta' \qquad (2.19)$$

where the partial differential terms are assigned the identities A', B', C', and D'. The stability of the singularity is determined by the roots of the characteristic equation

$$\begin{bmatrix} A'-\lambda & B' \\ C' & D'-\lambda \end{bmatrix} = 0 \qquad (2.20)$$

This is a quadratic in λ, from which

$$2\lambda_\pm = (A'+D') \pm [(A'+D')^2 - 4(A'D'-B'C')]^{1/2}$$

The roots are linear combinations of the form

$$\gamma(t) = a \exp(\lambda_1 t) + b \exp(\lambda_2 t) \qquad (2.21)$$

$$\theta(t) = c \exp(\lambda_1 t) + d \exp(\lambda_2 t) \qquad (2.22)$$

Positive values for λ_1 and λ_2 signify growth and departure from the singularity (instability). Negative values for λ_1 and λ_2 guarantee decay and return to the singularity (stability). These and other significant features are given in Table I, and their implications portrayed in an X–Y plane are shown in Fig. 4.

TABLE I

The Nature of Singularities in the Two-Dimensional Phase Plane and Their Local Stabilities Determined from the Solutions to the Characteristic Equation

Sign and character of roots λ_1 and λ_2	Condition	Nature of phase-plane singularity	Time-dependent response to perturbation
Both real and negative	$A'+D'<0$, $A'D'-B'C'>0$	Stable node	Monotonic decay
Both real and positive	$A'+D'>0$, $A'D'-B'C'>0$	Unstable node	Monotonic growth
Opposite sign	$A'D'-B'C'<0$	Unstable saddle point	Not accessible
Complex conjugates with negative real parts	$A'+D'<0$, $(A'+D')^2-4(A'D' -B'C')<0$	Stable focus	Oscillatory decay
Complex conjugates with positive real parts	$A'+D'>0$, $(A'+D')^2-4(A'D' -B'C')<0$	Unstable focus	Oscillatory growth
Pure imaginary	$A'+D'=0$	Unstable singularity surrounded by stable limit cycle	Growth or decay to sustained oscillatory state

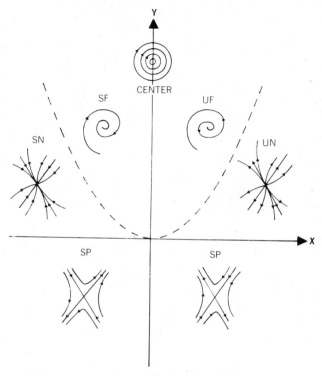

Figure 4. Representation of the trajectories following local disturbances from singularities in terms of X and Y, where $X = A' + D'$ and $Y = A'D' - B'C'$. See also Table I. SF, stable focus; SN, stable node; SP, saddle point; UF, unstable focus; UN unstable node.

The partial differentials of Eqs. (2.18) and (2.19) are as follows:

$$A' = -\frac{\exp \theta}{t_{ch}} - \frac{1}{t_{res}}, \qquad B' = \frac{(1-\gamma)}{t_{ch}} \exp \theta$$

$$C' = -\frac{B \exp \theta}{t_{ch}}, \qquad D' = \frac{B(1-\gamma)}{t_{ch}} \exp \theta - \frac{1}{t_{res}}$$

For a saddle point to exist, the condition $A'D' - B'C' < 0$ must be met. Hence

$$\left[-\frac{\exp \theta}{t_{ch}} - \frac{1}{t_{res}} \right]\left[-\frac{B(1-\gamma)}{t_{ch}} \exp \theta - \frac{1}{t_{res}} \right] - \left[-\frac{B \exp \theta}{t_{ch}} \right]\left[\frac{(1-\gamma)}{t_{ch}} \exp \theta \right] < 0$$

$$(2.23)$$

Substituting $\gamma = \theta/B$ and resolving the terms gives

$$\frac{1}{t_{res}t_{ch}}\left[\frac{t_{ch}}{t_{res}}+\exp\theta - B\exp\theta + \theta\exp\theta\right]<0 \qquad (2.24)$$

Returning to $\theta(t_{res})$, we may note that in Fig. 3, curve T_1, the middle branch of the region of multiplicity is opposite in gradient (negative) with respect to all other parts of the curve. A positive gradient exists when only unique solutions are possible (Fig. 3, curve T_2 or T_3).

Rearranging Eq. (2.17) and differentiating with respect to θ yields

$$\frac{\theta t_{ch}}{t_{res}^2}\frac{dt_{res}}{d\theta}=\frac{t_{ch}}{t_{res}}+\theta\exp\theta+\exp\theta - B\exp\theta \qquad (2.25)$$

Thus

$$A'D'-B'C'=\frac{\theta}{t_{res}^3}\frac{dt_{res}}{d\theta} \qquad (2.26)$$

and the saddle point exists when

$$\frac{\theta}{t_{res}^3}\frac{dt_{res}}{d\theta}<0 \qquad (2.27)$$

This condition is possible only in the center branch of $\theta(t_{res})$ and must extend throughout the whole range between the ignition and extinction points (at which $dt_{res}/d\theta = 0$).

The criteria for stable states are

$$A'+D'<0, \qquad A'D'-B'C'>0,$$

$$A'+D'=-\frac{\exp\theta}{t_{ch}}-\frac{1}{t_{res}}+\frac{B(1-\gamma)}{t_{ch}}\exp\theta-\frac{1}{t_{res}} \qquad (2.28)$$

Thus when γ is eliminated, for stable states to exist, the condition

$$\frac{B\exp\theta}{t_{ch}}-\frac{\theta\exp\theta}{t_{ch}}-\frac{\exp\theta}{t_{ch}}-\frac{2}{t_{res}}<0 \qquad (2.29)$$

must be met. But

$$\frac{B\exp\theta}{t_{ch}}-\frac{\theta\exp\theta}{t_{ch}}-\frac{\exp\theta}{t_{ch}}-\frac{1}{t_{res}}=-\frac{\theta}{t_{res}^2}\frac{dt_{res}}{d\theta} \qquad (2.25)$$

Hence

$$A' + D' = -\frac{\theta}{t_{res}^2}\frac{dt_{res}}{d\theta} - \frac{1}{t_{res}} \tag{2.30}$$

Stable states exist throughout the entire range when $dt_{res}/d\theta$ is positive, and not surprisingly, they are excluded from the realm in which saddle points occur. Moreover, there is no possibility of oscillatory states, since either small perturbations or the transitions at ignition or extinction must restore the system to a stationary state. In fact, we may have anticipated this outcome, since it is very well established that a system having only one independent variable cannot give rise to oscillatory behavior. The link between γ and θ [Eq. (2.6)] reduces the system to one variable. Whether or not there is a spiral approach to the stationary state is determined by the condition for the existence of an imaginary part to λ_1 and λ_2.

For complex roots $(A'+D')^2-4(A'D'-B'C')<0$ or $(A'-D')^2+4B'C'<0$,

$$(A'+D')^2+4B'C' = \left[-\frac{\exp\theta}{t_{ch}} - \frac{B(1-\gamma)}{t_{ch}}\exp\theta\right]^2 - \frac{4B(1-\gamma)}{t_{ch}^2}\exp(2\theta) \tag{2.31}$$

Multiplying out the terms in (2.31) and simplifying leads to

$$(A'-D')^2+4B'C' = \left[\frac{\exp\theta}{t_{ch}} - \frac{B(1-\gamma)}{t_{ch}}\exp\theta\right]^2 \tag{2.32}$$

This quadratic term can never be negative and so the condition $(A'-D')^2+4B'C'<0$ is not satisfied under adiabatic operation of the c.s.t.r. Thus the singularities are either stable nodes or saddle points, and where multiple states can exist, the saddle points occupy the centre branch exclusively. A $\gamma-\theta$ phase portrait for the region of three singularities is displayed in qualitative form in Fig. 5. The trajectories indicate the paths that would be taken from one or the other of the states according to the magnitude of disturbance to the system. In principle, there is a strong analogy to induced ignition: to bring about ignition, a spark or other stimulus applied to reactants in a state of low reactivity (γ_1, θ_1) would have to be sufficiently strong to perturb the system beyond the separatrices so that it would converge on the "ignition" state (γ_2, θ_2). The dynamics of the local dissipation of energy are not taken into account in

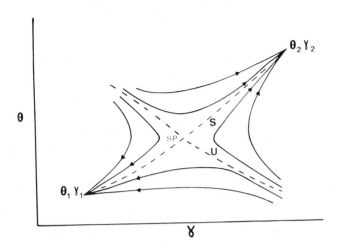

Figure 5.   Representation of the trajectories associated with local disturbances from the singularities that exist in the region of multiplicity in the adiabatic c.s.t.r. The broken lines represent the separatrices of the saddle point (SP); S is the stable manifold U is the unstable manifold. The system converges on one or the other of the stable nodal singularities (θ_1, γ_1 or θ_2, γ_2).

this simplified description. Levy and co-workers[56] have investigated related characteristics of the (more complicated) oxidation of methane in nonadiabatic closed conditions.

There is a saddle-node bifurcation at ignition or extinction and the only singularity that remains in the phase plane is a stable node.

B.   Stability of Exothermic Reactions in a c.s.t.r.: Nonadiabatic Operation

1.   Criticality and reaction multiplicity

The starting point of the treatment for nonadiabatic operation of the c.s.t.r. is the same as in the foregoing sections. The primary distinction is that there is an additional term to represent heat transport via the vessel walls. The energy-conservation equation now takes the form

$$\sigma V c_p \frac{dT}{dt} = (-\Delta H)Vkx \; - \; \sigma f c_p (T-T_0) \; - \; \chi S(T-T_a) \tag{2.33}$$

$$\underset{\substack{\text{(rate of heat} \\ \text{output)}}}{} \quad \underset{\substack{\text{(heat-release} \\ \text{rate)}}}{} \quad \underset{\substack{\text{(rate of heat} \\ \text{uptake by} \\ \text{incoming reactants)}}}{} \quad \underset{\substack{\text{(rate of heat loss} \\ \text{via walls)}}}{}$$

where χS is the specific heat transfer rate through surface area S. The mass-conservation equation remains as

$$\frac{dx}{dt} = \frac{f}{V} x_0 - \frac{f}{V} x - kx \qquad (2.34)$$

(rate of change of   (inflow   (outflow   (chemical-
    concentration)   rate)   rate)   reaction rate)

There are now two possible reference temperatures, T_0 and T_a, but they may be rationalized to a common base as[34]

$$T_* = \frac{\chi S T_a + \sigma f c_p T_0}{\chi S + \sigma f c_p} \qquad (2.35)$$

T_* is the temperature to which nonreactive gases are raised from temperature T_0 when they flow at rate f into the vessel at surface temperature T_a. We may define the characteristic (Newtonian) heat-transfer time t_N as $\sigma c_p V/\chi S$; then

$$T_* = \frac{T_a/t_N + T_0/t_{res}}{1/t_N + 1/t_{res}} \qquad (2.36)$$

Following (2.3′) and (2.4′), the mass- and energy-balance equations become

$$\frac{d\gamma}{dt} = \frac{(1-\gamma)}{t_{ch}} \exp \theta - \frac{\gamma}{t_{res}} \qquad (2.37)$$

$$\frac{d\theta}{dt} = \frac{B(1-\gamma)}{t_{ch}} \exp \theta - \theta\left(\frac{1}{t_{res}} + \frac{1}{t_N}\right) \qquad (2.38)$$

In a stationary state,

$$\gamma_{ss} = \frac{\exp \theta}{t_{ch}}\left[\frac{1}{t_{res}} + \frac{\exp \theta}{t_{ch}}\right]^{-1} \qquad (2.5′)$$

and

$$\gamma_{ss} = \frac{\theta_{ss}}{B}\left(1 + \frac{t_{res}}{t_N}\right) \qquad (2.39)$$

No longer is γ_{ss} tied directly to θ_{ss}, since now the ratio of the mean residence time to the characteristic cooling time is relevant. Equation

(2.39) may be written in the simplified form

$$\gamma_{ss} = \frac{\theta_{ss}}{B^*} \qquad (2.40)$$

where

$$B^* = B/(1 + t_{res}/t_N). \qquad (2.41)$$

When $t_{res} \ll t_N$ (fast-flow limit) $B^* \simeq B$. When $t_{res} \gg t_N$, $B^* \to 0$.
The reduction to B^* leads to an expression analogous to (2.7) for the
stationary states, namely

$$\frac{\theta_{ss}}{t_{res}} = \frac{B^*}{t_{ch}} \exp \theta_{ss}\left[1 - \frac{\theta_{ss}}{B^*}\right] \qquad (2.42)$$

There can still be only one or three values of θ_{ss} for any values of t_{res} and
B and the forms of $R(\theta)$ and $L(\theta)$ are similar to those shown in Fig. 2 for
adiabatic operation, but there are deeper ramifications. In qualitative
terms we may note that whereas for adiabatic operation a change of
residence time affects only the gradient of $L(\theta)$, in nonadiabatic operation
$R(\theta)$ also "collapses" as the ratio t_{res}/t_N is raised. If the overall diminution of $R(\theta)$ is more sensitive to the change of t_{res} than is the changing
gradient of $L(\theta)$, there will be an additional extinction from the ignited
state. This will be associated with long residence times ($t_{res}/t_N \gg 1$).
Retreating from long residence times then gives reignition, followed,
ultimately, by extinction in the fast-flow limit ($t_{res} \to 0$). Thus there may
be two separate regions of reaction multiplicity in the $\theta(t_{res})$ diagram.
This, however, is not the only new pattern of behavior (see below).

There are also additional constraints on the existence of multiplicity.
Critical ignition and extinction points are located at

$$\theta_\pm = \frac{B^*}{2}\left[1 \pm \left(1 - \frac{4}{B^*}\right)^{1/2}\right] \qquad (2.43)$$

[cf. Eq. (2.10)] and critical phenomena exist only as long as

$$B^* > \frac{4}{1 - 4\varepsilon} \qquad (2.44$$

In the limit $\varepsilon \to 0$, by combining (2.41) and (2.44) and rearranging, w find that the criterion under which criticality *ceases* to exist is

$$\frac{t_{res}}{t_N} > \frac{B}{4} - 1 \qquad (2.45)$$

A value of $B < 4$ thus eliminates any possibility of criticality, as in the adiabatic c.s.t.r. [Eq. (2.14)]; this result is not flow-rate dependent. However, even at $B > 4$ criticality ceases to be possible if t_{res} grows too large, and we thus distinguish between the losses of critical phenomena in the "fast"- and "slow"-flow limits in a nonadiabatic c.s.t.r.:

$$t_{res} > t_{res,cr} = t_N\left(\frac{B}{4} - 1\right) \qquad (2.46)$$

Thus, for example, whereas "ignition" is inevitable in the decomposition of DTBP in an adiabatic c.s.t.r. at all moderate conditions, under nonadiabatic operation ($t_N = 0.5$ s, say) criticality and ignition at $T_* = 500$ K become impossible at $t_{res} = 4.5$ s. This is in good agreement with experimental measurements.[53]

2. *Patterns of Stable and Unstable States and Oscillatory Reaction*

Rearrangement of Eq. (2.42) leads to a quadratic equation in t_{res} of the form

$$t_{res}^2 + [t_N + t_{ch} \exp(-\theta_{ss}) - Bt_N/\theta_{ss}]t_{res} + t_{ch} \exp(-\theta_{ss})t_N = 0 \qquad (2.47)$$

Solutions of this equation define the behavior in the (θ_{ss}, t_{res}) plane at constant T_* and t_N. The expression may take zero or two values in t_{res}. It may, however, take one or three values in θ_{ss}. In all, five patterns of multiplicity are possible[47,57,58] for Eq. (2.47); these are shown in Fig. 6.

The properties (at $t_N < 1$ s) of the present generation of mechanically stirred flow vessels[59–61] being used to study gaseous organic oxidation systems (see Section III) and the typical residence times studied ($t_{res} = 1$–30 s; see Section IV) are such that the accessible realms lie to the right-hand side of each $\theta(t_{res})$ curve (Fig. 6). It is possible that the upper branch of the "breaking wave" (Fig. 6b) to its maximum in θ may be observed and the right-hand tip of the "isola" (Fig. 6d) may yet be found. Ignition and extinction at "long" residence times (criticality at the right-hand side of the "mushroom" (Fig. 6c)) can be established.[53] All other features are accessible only at much faster flows ($t_{res} \ll 1$ s) or in reactors that operate much closer to adiabatic conditions ($t_N > 1$ s).

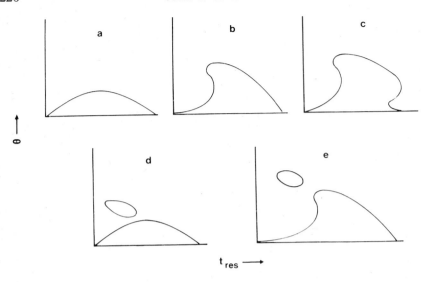

Figure 6.   The patterns of multiplicity that may exist in the θ_{ss}–t_{res} plane for the nonadiabatic c.s.t.r. (cf. Fig. 3a). t_{res} may take zero or two values; θ_{ss} may take one or three values. (a) Curve (b) "Breaking wave." (c) "Mushroom." (d) "Isola." (e) "Breaking wave" + "isola."

For a given vessel and reactant composition (hence a preordained t_N), which of the patterns in Fig. 6 is observed depends on the magnitude of T_*. A very great variety of bifurcation phenomena is associated with each of these forms, and the route to their determination is, in part, via the stability theorems that have been applied to the adiabatic c.s.t.r. However, oscillatory reaction also becomes a feature and the link to it in stability analysis is the existence of a limit cycle in the γ–θ plane. Although its existence may be inferred (a stable limit cycle is expected to surround an unstable singularity, for example), theorems for local stability are insufficient to identify the limit cycle,[62] and commonly its birth or death is bound up in complex phenomena such as the Hopf bifurcation.[62] Analytical routes are used for the identification of limit cycles, but at present substantial recourse is made to numerical methods. Uppal, Ray, and Poore[47] were the first to present a comprehensive account of the sequences of singularities that are possible along the trajectories shown in Fig. 6 and of the birth, growth, and death of limit cycles associated with them.

The criteria for the nature of singularities are those described in Table I, and the partial differential terms derived from Eqs. (2.37) and (2.38) for the nonadiabatic c.s.t.r. are as follows:

$$A' = -\frac{\exp \theta}{t_{ch}} - \frac{1}{t_{res}}, \qquad B' = \frac{(1-\gamma)}{t_{ch}} \exp \theta,$$

$$C' = -\frac{B \exp \theta}{t_{ch}}, \qquad D' = \frac{B(1-\gamma)}{t_{ch}} \exp \theta - \left(\frac{1}{t_N} + \frac{1}{t_{res}}\right) \qquad (2.48)$$

For a saddle point to exist, $A'D' - B'C' < 0$ must hold; hence

$$\frac{1}{t_{ch}t_{res}}\left[\left(\exp \theta + \theta \exp \theta + \frac{t_{ch}}{t_{res}}\right)\left(1 + \frac{t_{res}}{t_N}\right) - B \exp \theta\right] < 0$$

or

$$\frac{dt_{res}}{d\theta}\left[\frac{t_{ch}}{t_{res}^2} - \frac{\exp \theta}{t_N}\right] < 0 \qquad (2.49)$$

Following the procedure adopted for the adiabatic c.s.t.r., it can be shown that the saddle points exist exclusively in *each* of the center branches where multiplicity occurs in $\theta(t_{res})$. That is, at short residence times (the S-shaped curve), $dt_{res}/d\theta < 0$, but $[t_{ch}/t_{res}^2 - \exp \theta/t_N]$ must be positive. This is the only common ground between the nonadiabatic and adiabatic c.s.t.r. At long residence times (the Z-shaped curve), $dt_{res}/d\theta > 0$, but $[t_{ch}/t_{res}^2 - \exp \theta/t_N]$ must be negative, since $[t_{ch}/t_{res}^2 - \exp \theta/t_N]$ changes sign once at the apex $d\theta/dt_{res} = 0$.

For stable singularities to exist, $A' + D' < 0$ must hold; hence

$$\frac{B \exp \theta}{t_{ch}} - \frac{\exp \theta}{t_{ch}} - \frac{\theta \exp \theta}{t_{ch}}\left(1 + \frac{t_{res}}{t_N}\right) - \frac{2}{t_{res}} - \frac{1}{t_N} < 0 \qquad (2.50)$$

or

$$-\frac{1}{t_{ch}}\frac{dt_{res}}{d\theta}\left[\frac{t_{ch}}{t_{res}^2} - \frac{\exp \theta}{t_N}\right] - \frac{1}{t_{res}} + \frac{t_{res}}{t_{ch}t_N}\exp \theta < 0 \qquad (2.50')$$

For complex roots $(A' + D')^2 - 4(A'D' - B'C') < 0$; hence

$$\frac{\exp 2\theta}{t_{ch}^2}\left(\left[\theta\left(1 + \frac{t_{res}}{t_N}\right) - B\right]^2 + 2\theta\left(1 + \frac{t_{res}}{t_N}\right) - 2B + 1\right)$$

$$+ \frac{2 \exp \theta}{t_{ch}t_N}\left[\theta\left(1 + \frac{t_{res}}{t_N}\right) - B - 1\right] + \frac{1}{t_N^2} < 0 \qquad (2.51)$$

or

$$\left(-\frac{1}{t_{ch}}\frac{dt_{res}}{d\theta}\left[\frac{t_{ch}}{t_{res}^2}-\frac{\exp\theta}{t_N}\right]-\frac{1}{t_{res}}+\frac{t_{res}\exp\theta}{t_{ch}t_N}\right)^2-\frac{4}{t_{res}t_{ch}}\frac{dt_{res}}{d\theta}\left[\frac{t_{ch}}{t_{res}^2}-\frac{\exp\theta}{t_N}\right]<0$$

$$(2.51')$$

Neither of these criteria can be related in such a simple fashion to points on the $\theta(t_{res})$ trajectory as is achieved for the adiabatic c.s.t.r. However, for the simple first-order reaction under nonadiabatic conditions the axes displayed in Fig. 4 are

$$Y=\frac{1}{t_{res}t_{ch}}\frac{dt_{res}}{d\theta}\left[\frac{t_{ch}}{t_{res}^2}-\frac{\exp}{t_N}\right]$$

$$X=\frac{1}{t_{ch}}\frac{dt_{res}}{d\theta}\left[\frac{t_{ch}}{t_{res}^2}-\frac{\exp\theta}{t_N}\right]-\frac{1}{t_{res}}+\frac{t_{res}}{t_{ch}t_N}\exp\theta$$

The locus of the parabolic boundary between nodal and focal singularities is given by

$$\left(-\frac{1}{t_{ch}}\frac{dt_{res}}{d\theta}\left[\frac{t_{ch}}{t_{res}^2}-\frac{\exp\theta}{t_N}\right]-\frac{1}{t_{res}}+\frac{t_{res}\exp\theta}{t_{ch}t_N}\right)^2-\frac{4}{t_{res}t_{ch}}\frac{dt_{res}}{d\theta}\left[\frac{t_{ch}}{t_{res}^2}-\frac{\exp\theta}{t_N}\right]=0$$

$$(2.52)$$

Because there is a common term in X and Y, $X\neq0$ when $Y=0$ and so the system cannot pass through the origin. Only limited sequences of singularities can occur. Thus, moving clockwise from top left in Fig. 4, the states SN, SF, UF, UN, SP are possible. For $Y\not<0$ (no saddle point) only the upper half of the figure is traversed, yielding SN, SF, UF, SF, SN. For $X\not>0$, $Y\not<0$ (stable states only) the system remains in the upper left-hand quadrant, giving the sequence SN, SF, SN.

3. Hopf Bifurcation Phenomena

Solutions to the characteristic equation are pure imaginary at the SF–UF bifurcation; under general conditions, the Hopf bifurcation theorem[62] assures the existence of a limit cycle in the neighborhood of the point given by $A'+D'=0$. A unique solution guarantees the existence of a stable limit cycle surrounding the unstable singularity. When multiple singularities can coexist, the limit cycle may be either stable or unstable. The stability of the limit cycle cannot be established from a (local) linear analysis, and the properties of higher-order terms of the characteristic equation must then be investigated, yielding a global stability analysis via the Floquet exponent or Poincaré index.[63]

The significance of the nature of these transitions in relation to observed oscillatory behavior is as follows. When the system undergoes a transition from a stable focus to an unstable focus surrounded by a stable limit cycle [the supercritical Hopf bifurcation (Fig. 7a)] there is a soft excitation, namely the birth of oscillations the growth of whose amplitude initially follows a parabolic law. Their death, brought about by retracing the bifurcation parameter, follows a reverse path.

When the system undergoes a transition from a stable focus surrounded by an unstable limit cycle to an unstable focus surrounded by a stable limit cycle [the subcritical Hopf bifurcation (Fig. 7b)] there is a hard excitation, namely the instantaneous appearance of finite amplitude oscillations. Progression in the same direction ultimately leads to a "soft" death at a second UF–SF bifurcation. Retracing the pathway from this point gives rise to a "mirror image", soft excitation, but the range of stable oscillatory behavior extends beyond the initial subcritical Hopf bifurcation to extinction at the confluence of the stable-limit-cycle–unstable-limit-cycle envelopes. The subcritical Hopf bifurcation is not readily distinguished from criticality due to a saddle-node bifurcation into a stable limit cycle, since they each exhibit both a hard excitation and

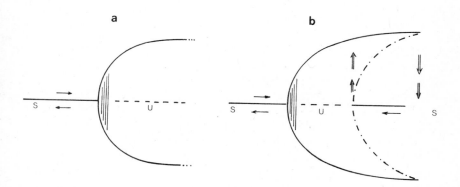

Figure 7.   Diagrammatic representations of the (a) supercritical and (b) subcritical Hopf bifurcations. The bifurcation point is at the stable (S, —) to unstable (U, ---) singularity transition. The solid envelopes represent the limit cycles surrounding an unstable singularity. The size of the limit cycle determines the amplitude of the time-dependent oscillations (signified by the lengthening vertical lines). When multiplicity exists, an unstable limit cycle (–·–·–) may surround a stable singularity. Hysteresis (open vertical arrows) may then be observed when the bifurcation parameter is varied.

hysteresis before extinction of oscillations. One quantitative distinguishing feature of the subcritical Hopf bifurcation is that the "jump" transition is brought about without critical tangency (as displayed in Fig. 2a, F–G, for example).

C. P–T_a Ignition Boundaries in a c.s.t.r.

In combustion studies, a common experimental basis for establishing under closed-vessel conditions the regimes in which different reaction modes exist is to portray events in a P–T_a ignition diagram for a given reactant composition. The corresponding representation under well-stirred flowing conditions is achieved by maintaining, in addition, a constant mean residence time. For the irreversible, first-order, exothermic reaction under *nonadiabatic conditions*, oscillatory ignition is distinguished in the P–T_a diagram from the stationary state of low reactivity as follows.

Critical ignition occurs at

$$\theta_- = \frac{B^*}{2}\left[1 - \left(1 - \frac{4}{B^*}\right)^{1/2}\right] \simeq 1 + \frac{1}{B^*} \tag{2.53}$$

But

$$\theta = \frac{Bt_{\text{res}}(1-\gamma)\exp(1+1/B^*)}{t_{\text{ch}}(1+t_{\text{res}}/t_{\text{N}})} = \frac{Bt_{\text{N}}(1-\gamma)\exp(1+1/B^*)}{t_{\text{ch}}(1+t_{\text{N}}/t_{\text{res}})} \tag{2.54}$$

Hence

$$1 + \frac{1}{B^*} = \frac{Bt_{\text{N}}(1-\gamma)\exp(1+1/B^*)}{t_{\text{ch}}(1+t_{\text{N}}/t_{\text{res}})} \tag{2.55}$$

For the present purpose, let us suppose $B^* \to \infty$, so that $1+1/B^* \simeq 1$. Then

$$e^{-1} \simeq \frac{Bt_{\text{N}}(1-\gamma)}{t_{\text{ch}}(1+t_{\text{N}}/t_{\text{res}})} \tag{2.56}$$

In exact form for a given reactant composition, $1+1/B^* < 1$; its magnitude would vary marginally with change of T_*. The term $Bt_{\text{N}}/t_{\text{ch}}$ corresponds to the Semenov dimensionless heat release term (Se or ψ) familiar in the context of stationary-state, thermal explosion theory,[48] and is given by

$$Se = \frac{qVP_0Ek(T_*)}{\chi SR^2T_*^3} \qquad (2.57)$$

in which the ideal-gas relationship,

$$x_0 = \frac{P_0}{RT_*}$$

has been incorporated. At criticality $Se_{cr} = e^{-1}$.

For $B \gg 1$, the extent of reactant consumption is not very significant at criticality (typically less than 10% of the reactant has been consumed[64]) so that $1 - \gamma = (1 - \gamma_{cr})$ is approximately constant. Thus from (2.56) the limiting ignition criterion is

$$\frac{1}{e} \simeq \frac{qVP_{0,cr}Ek(T_*)(1 - \gamma_{cr})}{\chi SR^2T_{*,cr}^3(1 + t_N/t_{res})} \qquad (2.58)$$

or

$$\frac{1}{e} \simeq \frac{qVP_{0,cr}EA \exp(-E/Rt_*)(1 - \gamma_{cr})}{\chi SR^2T_{*,cr}^3(1 + t_N/t_{res})} \qquad (2.59)$$

Taking logarithms and rearranging (2.59)

$$\ln\left(\frac{P_0}{T_*^3}\right)_{cr} = \ln\left[\frac{\chi SR^2(1 + t_N/t_{res})}{eqVEA(1 - \gamma_{cr})}\right] + \frac{E}{RT_{*,cr}} \qquad (2.60)$$

from which, for a given flow rate,

$$\ln\left(\frac{P_0}{T_*^3}\right)_{cr} \propto \frac{1}{T_{*,cr}} \qquad (2.61)$$

and the constant of proportionality is E/R (Fig. 8b). From (2.59),

$$P_{0,cr} = \frac{\chi SR^2T_{*,cr}^3(1 + t_N/t_{res})}{eqVEA \exp(-E/RT_*)(1 - \gamma_{cr})} \qquad (2.62)$$

Although $P_{0,cr}$ is proportional to $T_{*,cr}^3$, it is inversely proportional to $\exp(-E/RT_*)$. In consequence, provided that $E \gg 0$, $P_{0,cr}$ diminishes as T_* is raised (Fig. 8a).

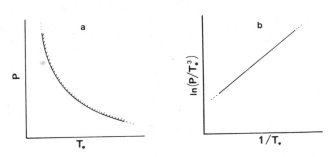

Figure 8. Criticality in the P–T_a ignition diagram for a first-order exothermic reaction under nonadiabatic conditions in a c.s.t.r. (a) P versus T_*. (b) $\ln(P/T_a^3)$ versus $1/T_*$.

In adiabatic conditions ($t_N \to \infty$) the relationship that corresponds to (2.56) is

$$Ze(1-\gamma) = e^{-1} \tag{2.63}$$

where the Zel'dovich number $Ze = Bt_{res}/t_{ch}$. Thus for a first-order reaction,

$$\frac{1}{e} = \frac{q\sigma VEk(T_*)(1-\gamma)}{x_0 c_p fRT_*^2} \tag{2.64}$$

A critical criterion for ignition is controlled only by the mean residence time (V/f) or the mole fraction of reactant (x_0/σ). Changes in total concentration or pressure do not bring about a transition from one mode of reaction to another. Thus for a first-order exothermic reaction under *adiabatic conditions* there is no critical P–T_a ignition boundary. This is a consequence of the linear dependence of heat-release and heat-transport rates on reactant concentration. Higher-order reactions may be expected to exhibit a P–T_a ignition boundary even under adiabatic conditions.

In summary, a critical P–T_* ignition boundary may be identified in the *nonadiabatic* c.s.t.r. even for a first-order exothermic reaction and it bears a striking resemblance to that observed for low-pressure gases in closed vessels. As we shall see (Section IV), the P–T_a ignition diagrams for hydrocarbon oxidation in closed and open vessels are qualitatively similar. These diagrams display much richer patterns of behavior, however, arising from thermally coupled kinetic mechanisms of much greater complexity than that already considered.

D. Stability of Isothermal Reactions in a c.s.t.r.

1. Nonautocatalytic Reactions

Isothermal reactions that exhibit no form of kinetic feedback cannot show any novel responses or exotic patterns of behavior in the c.s.t.r. Regardless of the reaction order, $\gamma(t_{res})$ increases monotonically throughout the range of t_{res} and can be only single valued at any point. The mass-conservation equations and their solutions are trivial.

2. Autocatalytic Reactions

Although particular forms of isothermal autocatalysis have been studied very extensively,[65-67] with special reference to the formal representation of the oscillatory Belousov–Zhabotinsky[68] reaction and other liquid-phase redox systems, only since 1980 has there been a comprehensive investigation of the most basic autocatalytic rate laws, by Gray and Scott.[69-74] They may be expressed with reference to the mechanistic forms

$$A + B \rightarrow 2B \tag{2.65}$$

$$A + 2B \rightarrow 3B \tag{2.66}$$

either of which may be accompanied by (a) decay of the catalyst, (b) addition of catalyst to the flow, or (c) reversibility of the autocatalytic step. The rate law associated with (2.65) and (2.66) takes the general form

$$\text{rate} \propto [A][B]^n$$

where $n = 1$ or 2 in the respective cases. The overall dependence on concentration is thus quadratic (when $n = 1$) or cubic (when $n = 2$), and these are the classifications that have been assigned by Gray and Scott to each of the cases.

The mass-balance equation when there is no addition of catalyst to the inflow or chemical decay is thus

$$\frac{db}{dt} = kab^n - \frac{b}{t_{res}} \tag{2.67}$$

where $a = [A]$ and $b = [B]$. In terms of fractional conversion, following the convention of the previous section,

$$\frac{d\gamma}{dt} = \frac{\gamma^n(\gamma - 1)}{t_{ch}} - \frac{\gamma}{t_{res}} \tag{2.68}$$

This equation is bound by the simplifying constraint

$$a + b = a_0$$

Here t_{ch} is the reciprocal of the pseudo-first-order rate constant ka_0 (quadratic) or ka_0^2 (cubic). Equation (2.68) is not a complete nondimensional form. Gray and Scott[72] add three elements to make it so. First, they scale time to the chemical time as $\tau = t/t_{ch}$; second, they scale residence time to chemical time as the Damköhler number $Da = t_{res}/t_{ch}$; third, they normalize the chemical-reaction rate to a maximum value of unity. The effect of this final component is to add a numerical factor to the chemical-rate term in the overall expression for $d\gamma/d\tau$. For quadratic autocatalysis the normalizing factor is 4; for cubic autocatalysis it is 27/4. Thus the fully nondimensional expressions become

$$\frac{d\gamma}{d\tau} = 4\gamma(1 - \gamma) - \frac{\gamma}{Da} \qquad \text{(quadratic)} \qquad (2.69)$$

$$\frac{d\gamma}{d\tau} = \left(\frac{27}{4}\right)\gamma^2(1 - \gamma) - \frac{\gamma}{Da} \qquad \text{(cubic)} \qquad (2.70)$$

Since these equations do not have obvious relevance to nonisothermal combustion of hydrocarbons, it is not the intention to analyze them or derivatives from them in detail in the context of reaction multiplicity and stability of the singularities. It may yet be the case that where strong autocatalysis accompanies nonisothermal reaction such that criticality is dominated by the branching component, schemes of the kind studied by Gray and Scott have a part to play in interpretation of the observed phenomena.

There are very clear, entirely analytical accounts by Gray and Scott[69-74] that elaborate the criteria for multiplicity, the variety of forms of $\gamma(t_{res})$ that can be found, and the stabilities of the singularities in the $\gamma(t_{res})$ diagram. Numerical interpretations of the birth and death of oscillatory events are also straightforward, and there has even been progress in their analytical interpretation. Moreover, there are remarkable parallels in qualitative form and in some aspects of quantitative detail to the single-step, nonisothermal reaction. In particular the formal scheme

$$A + 2B \rightarrow 3B \qquad (2.66)$$

shows all of the properties of the adiabatic c.s.t.r. that were presented in the previous section. The sequence

$$A + 2B \rightarrow 3B \tag{2.66}$$

$$B \rightarrow inert \tag{2.71}$$

has the attributes of the nonadiabatic c.s.t.r. The decay of the catalyst (B) is equivalent to the transport of heat via the vessel walls; Eq. (2.71) introduces the additional characteristic time t_B, just as wall losses introduce the characteristic Newtonian time t_N. The equations involved are free of the complications introduced by the exponential dependence of rate on temperature under nonisothermal conditions; thus, guided by the work of Gray and Scott,[69–74] the reader who is inspired to learn more about the fascinating properties of nonlinear kinetic systems may be well advised to study the isothermal autocatalytic systems first.

The dependences of $d\gamma/d\tau$ on γ for the quadratic and cubic cases are shown in Fig. 9. The analogy of cubic autocatalysis to the single-step exothermic reaction in the adiabatic c.s.t.r. is clear. In each case, $d\gamma/d\tau = 0$ at $\gamma = 0$ and 1. When some of the catalyst is added to the flow, $d\gamma/d\tau > 0$ at $\gamma = 0$. When either the catalyst can decay [Eq. (2.71)] or the autocatalytic step is reversible, $d\gamma/d\tau = 0$ at $\gamma < 1$. Quadratic autocatalysis, even with catalyst decay, shows a limited variety of patterns (stable node, stable focus). Cubic autocatalysis with a stable catalyst exhibits the constraints of the adiabatic c.s.t.r. (multiplicity in certain conditions, with its associated stable nodes and saddle points). With catalyst decay, cubic autocatalysis gives rise to "mushrooms" and "isolas" and the same sequences of the

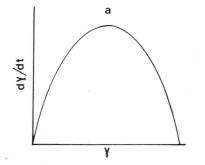 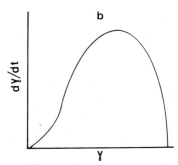

Figure 9.   Relationships between $d\gamma/d\tau$ and γ for (a) quadratic and (b) cubic isothermal autocatalysis.

changing characters of the singularities are observed as in the nonadiabatic c.s.t.r.

E.   Nonisothermal Oxidation of Hydrocarbons in a Nonadiabatic c.s.t.r.

The quantitative interpretation of nonisothermal phenomena associated with hydrocarbon oxidation must follow the guidelines laid down
in the foregoing sections, but at present the formal background is far less
developed and some of the mathematical tools needed for work on
systems involving more than two variables have yet to be provided. At
this stage we may locate the following cornerstones:

1.   Either kinetic or thermal feedback is the key to criticality and
multiple stabilities.
2.   Unstable states and oscillatory phenomena are introduced only by
an additional independent variable (a time constant that is independent of
flow control), thus creating at least a two-dimensional frame of reference.
3.   Highly nonlinear feedback terms are also a prerequisite for oscillatory phenomena (and the Arrhenius rate dependence on temperature is
particularly powerful).
4.   Deep insights have already been obtained into the behavior of
nonlinear two-dimensional systems.
5.   The kinetic forms so far discussed display some, but not all, of the
thermokinetic phenomena associated with hydrocarbon oxidation (see
Section IV); greater kinetic complexity is certainly involved in these
systems and two-dimensional models are probably not adequate even for
qualitative descriptions.

Progress has been made in the understanding of spontaneous ignition,
and will be discussed following the presentation of the main physical and
chemical features.

III.   EXPERIMENTAL METHODS

A.   Well-Stirred Flow Vessels

Progress in spontaneous-ignition research during the 1980s is being
dominated by the application of the c.s.t.r. to experimental studies of
hydrocarbon oxidations (Table II).[59-61,75-82] The mode of operation leads
to spatial homogeneity in temperature and concentration; gradients are
smoothed out and measurements at one point are representative of the
whole. Its strength as a technique lies also in the ability it gives the
investigator to establish true stationary states, to sustain oscillatory events
as long as reactants continue to be fed to the reactor, and to locate

TABLE II
Experimental Investigations of Gas-Phase Oxidations in the c.s.t.r.

Reactant composition	Pressure range (atm)	Mean residence time (s)	Reference
$C_2H_6 + O_2 \rightarrow 8C_2H_6 + O_2$	1.0	15–120	75
$C_3H_8 + O_2$	1.0	10–25	76
$C_3H_8 + 1.25O_2 \rightarrow 2C_3H_8 + O_2$	0.5–1.0	60	77
$n\text{-}C_4H_{10} + O_2$ } $i\text{-}C_4H_{10} + O_2$ }	1	7.5–29	59
$c\text{-}(C_5H_{12})CH_3 + 6O_2$	1	2.8–4.3	78
$n\text{-}C_7H_{16} + 2O_2 + N_2$	1	8–21	79
$i\text{-}C_8H_{18} + 2O_2 + N_2$	1	8–21	79
$n\text{-}C_7H_{16} + 2O_2 + N_2$	8–12	0.1	80
$i\text{-}C_8H_{18} + 2O_2 + N_2$	8–12	0.1	80
$CH_3CHO + O_2$	0.1–0.5	3–10	60
$CH_3CHO + O_2$	0.1	3–21	81
$C_2H_5OC_2H_5 + O_2$	0.1–0.6	3	82
$(CH_3)_2C(H)OC(H)(CH_3)_2 + O_2$	0.1–0.6	3	82
$CH_2\!\!-\!\!CH_2 + O_2$ (O epoxide)	0.1–0.5	25–36	61

precisely the conditions under which switches occur from one mode of behavior to another. Clearly, it is possible to control the system at the verge of criticality or dynamic instability and to make sufficiently detailed physical and chemical measurements to yield an interpretation of the criteria necessary to bring about the change. These are important advantages over classical, unstirred, closed-vessel studies, in which all events are ephemeral, but they are not the only ones. As is shown in the discussion of the theoretical background, there is a need to understand the nature of stationary states. The experimental analog to the local stability analysis is to create a disturbance, such as a momentary perturbation of flow, and to observe the manner in which the system responds to the effect. Monotonic responses reflect a nodal character; oscillatory responses reflect a focal singularity.

Modern experiments using the c.s.t.r. began after 1970; so far, small glass vessels[59-61] (300–1000 cm^3, cylindrical or spherical) have been employed (Fig. 10). A mechanical stirrer is incorporated to ensure spatially uniform temperatures and concentrations, since the flow rates are not sufficiently rapid (commonly $t_{res} = 3$–30 s) for effective mixing by jet injection to be achieved, as in the Longwell[83] type of stirred vessel. Previous, isolated investigations of organic oxidations in the gas phase

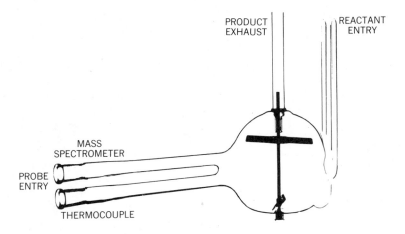

Figure 10. Mechanically stirred, 500 cm³ flowing reactor. The rotor is driven by a
magnetic couple.

were carried out by Gervart and Frank-Kamenetskii[84] and by Dutton.[85]
Bush studied thermokinetic oscillations in the chlorination of methyl
chloride.[52] The control parameters are the vessel-wall temperature (T_a)
and the reactant composition (ϕ), pressure (P), and mean residence time
(t_{res}).

As Eq. (2.35) shows, we may make allowance for differences between
the reactant entry temperature and the vessel-wall temperature. However,
if mixing of reactants is brought about at the point of entry, then at
moderate flow rates, it is very easy to ensure preheating of the separate
components en route to the vessel. Extremely careful and sustained
control of the reactant flows is essential; critical flow orifices are usually
used. These permit the variation of composition via the independent
change of the molar flow rate of each of the components. Thermostating
is usually done in a recirculating air oven. Control to ± 1 K has been
satisfactory for immediate needs, but as theories for criticality and bifurcation develop, still more precise experimental control will be sought at
the margins of stability. The measured parameters are reactant temperature (T), reactant and product concentrations in the vessel (c_i), and,
where appropriate, light output (I). The methods used are discussed
below. Most experiments have been confined to a working pressure of
1 atm or less. A jet-stirred c.s.t.r. has also been developed recently;[80] it is
operated at pressures in the 8–12 atm range at mean residence times of
~ 0.1 s.

B.   Unstirred Closed Vessels

Set against modern theoretical interpretations, the best that unstirred closed-vessel studies can achieve is a rudimentary qualitative representation of predicted events. Their shortcomings reside principally in the inherent transitory nature of the behavior: reactants have to be dumped into a heated evacuated vessel, invariably from room temperature and sometimes without premixing. Reaction then proceeds at a varying rate and terminates at complete consumption of reactant. The only clear distinction emerges when ignition takes place; the margins of other reaction modes are blurred by the effects of reactant consumption. Moreover, during nonisothermal reaction, very marked asymmetric temperature gradients may persist or may wax and wane as a consequence of natural convection. These spatial and temporal variations lead to formidable mathematical complexities in interpretation and are worth pursuing only to satisfy a specific need (as in the more simple thermal ignition of low-pressure vapors, in which conduction is the principal mode of heat transport,[54] leading to symmetric spatial temperature profiles).

A quasi-stationary state of reaction can be discerned experimentally in a closed vessel if events do not occur too rapidly, and this state may be construed as a qualitative analog of the stationary state in the c.s.t.r. The initial behavior could be regarded as a parallel to the response in the c.s.t.r. to the effect of perturbations, since the experimentalist is observing events when a reactant system is brought to a particular pressure, composition, and temperature from remote conditions. The quasi-stationary behavior may be eclipsed by this transitory behavior, especially when reaction rates are high.

The foregoing qualifications are directed particularly at the use of closed systems to study nonisothermal combustion systems. They are still the mainstay of isothermal gas kinetic studies and quantitative measurements of rate data are made very successfully in them. Moreover, these systems still have a useful part to play in present developments: they offer a convenient and economical method for initial exploration, and a simple and safe starting point for the recognition of the exotic nonisothermal features associated with hydrocarbon oxidation. In what follows there is extensive reference to closed-vessel studies, because decades of research activity using them has provided the wealth of vital kinetic and mechanistic data on which present interpretations are based, and has given rise to very widespread familiarity with the main experimental features. It is folly to expect profound quantitative insights to emerge from the experimental measurements: that rates of pressure change in nonisothermal reaction have a simple relationship to the rate of chemical change[86] and that irregular temperature records have deep mechanistic significance[87] are just two examples of the misleading interpretations to be found in the literature.

C. Flow Tubes and Flat-Flame Burners

The bulk of experimental investigation has focused on the closed vessel, and the major quantitative developments relate to the c.s.t.r. But an important part is also played by the (isothermal and nonisothermal) flow tube in providing a thermal and chemical history.[88–90] Some of the earliest observations exploited this technique.[91] At temperatures above 1000 K the turbulent-flow adiabatic reactor has proved to be an important source of kinetic and mechanistic data.[23] In the circumstances of present interest, laminar flow conditions have been used to stabilize the complex combustion phenomena associated with hydrocarbon oxidation; flow tubes[88–90] and flat-flame burners[92,93] have served this purpose very successfully. The merit of these systems is that the continuous supply of reactant ensures stabilization of a stationary state. Kinetic and thermal histories are resolved spatially downstream from the point of entry. Careful sampling via probes going to chemical analyzers[89] or direct thermocouple measurements[89] yield very great detail with high spatial (and hence temporal) resolution. Similar remarks may be made about controlled reactant flows across a hot surface, although quantitative interpretation is more difficult in that case, because marked temperature and concentration gradients are involved.[94]

D. Other Experimental Methods

Complementary experimental methods used to study aspects of hydrocarbon oxidation include the use of shock tubes,[95] rapid compression ahead of a mechanical piston,[96,97] and the use of spark-ignited or motored engines.[98] Shock tubes lend themselves to high-temperature pyrolysis or oxidation studies; in general T exceeds 1000 K behind the incident shock. Reaction intervals fall in the microsecond to millisecond range. They may be used to derive elementary rate data or global information such as ignition delay times. Spectroscopic methods based on absorption or emission are normally used to obtain kinetic measurements. The techniques involving rapid compression machines or motored engines come much closer to applied aspects of hydrocarbon combustion than do any of the other techniques relating to spontaneous ignition in engines. Cylinder pressures may be tens of atmospheres and times to ignition are on the order of milliseconds. While such experiments are very important technically, detailed chemical analyses are extremely difficult to make, and the technique is thus better directed toward specific goals than toward the provision of general information.

E. Methods of Chemical Analysis

The accumulation of detailed knowledge about the chemistry of hydrocarbon oxidation has advanced most significantly in parallel with the development of gas-chromatographic techniques. The present-day refinement of these makes possible the quantitative identification of components in extremely complex product compositions, even when there are very close chemical similarities such as are found among the isomeric forms of hydrocarbons and their derivatives. Present interpretations of kinetic data and mechanisms for hydrocarbon oxidation rely strongly on a background of data acquired during the flourishing of such methods during the 1960s and 1970s.

The virtually exclusive use of closed vessels for experimental investigations means that most studies have relied on batch sampling and transfer to the gas chromatograph following quenching of the reaction. One qualification in the use of these techniques is that reactant temperatures may not always be as carefully defined as is required. This is especially true of analyses associated with markedly nonisothermal reactions. Probe sampling and direct transfer to the chromatograph injection system has also been exploited, especially in association with flow-tube and flat-flame-burner studies.

Mass spectrometry is an inherently attractive technique for quantitative chemical analysis. It makes possible continuous monitoring of chemical changes, and the short response times mean that sharp changes in concentration can be followed. The technique has been exploited successfully for measurement of reactant, molecular-intermediate, and final-product concentrations during the oxidation of some of the simpler organic oxidation systems,[99] for which overlap of the mass-spectral cracking-pattern data for the components of the system is not too complicated. Very careful calibration is required if quantitative data are to be obtained, since there is a transition from viscous to molecular flow en route to the ionization chamber, which is operated at pressures of $<10^{-3}$ torr. Selective pumping rates are then incurred that are inversely proportional to the square root of the relative molecular mass, and the quantitative sensitivity to each of the components becomes dependent on the overall viscosity of the mixture entering the probe.

Spectroscopic analyses (UV, visible, or IR) based on absorption or emission have not been much exploited because the resolution of unique spectral features among the structurally similar reactants and products of hydrocarbon oxidation is not easy.[100] Experimental complications are also associated with light paths and sensitivity in the conventional apparatus used to investigate hydrocarbon combustion. Such techniques are inherently attractive because the instrumental responses are fast and there is no perturbation to the combustion system; it would be desirable to tap the available resources and expertise in spectroscopy to establish novel applications in the realm of low-temperature combustion of hydrocarbons. The spectroscopy of flames is, of course, extremely well established;[101] the emissions are a rich source of spectroscopic data and much has learned about excited states from flame spectra. Studies of the weak chemiluminescent emissions from excited formaldehyde[102] that accompany oscillatory cool-flame phenomena have yielded some information about the radical–radical interactions involved in the generation of this molecular intermediate.

All of the foregoing pertains to the detection and measurement of the reactants, the molecular intermediate, and the final products of oxidation. Molecular beam sampling to a mass spectrometer has been used successfully for the direct measurement of radical and atom concentrations in hydrocarbon flames,[103,104] but there has yet to be a successful application to low-temperature oxidation. Such methods are technically very difficult and even in flames the concentrations of the reactive intermediates do not exceed the limits of detection by very much. The lower concentrations associated with low-temperature combustion offer little promise for extensions to this realm in the immediate future.

An alternative attack on the measurement of free radicals involves the continuous trapping of reactive species in an inert matrix at the temperature of liquid nitrogen within the cavity of an electron spin resonance spectrometer.[105] By repetitive sweeps through a range of frequencies it is possible to discern the gradual accumulation of radical species and to identify characteristic spectra. Nalbandyan and co-workers[17,105,106] developed this technique in Armenia to study kinetics involving alkyl, alkylperoxy, and hydroperoxy radicals. Sochet[107] has also utilized the technique in Lille.

F. Temperature Measurements and Other Physical Observations

The application of the analytical methods discussed so far contributes to the understanding of the mechanisms and kinetic data of elementary processes involved in hydrocarbon oxidation at prescribed conditions. It is the association with temperature change that confers the exotic behavior on combustion processes, and thus direct temperature records are at least as important as the monitoring of chemical change. In the simplest mode of criticality, thermal explosion,[108,109] it is the thermal history that controls events, and the development of sensitive and responsive methods for measuring temperatures directly using very fine Pt–Pt/Rh thermocouples has evolved to meet the need for such data. These techniques have been applied to studies of the nonisothermal oxidation of hydrocarbons[110,111] since about 1965. Moreover, if one calibrates the c.s.t.r. to obtain a value for its heat-transfer coefficient,[59] one can use the direct measurements of temperature excess for the assessment of an overall heat-release rate for stationary reaction.[60,75] This is an extremely important advance toward the quantitative interpretation of nonisothermal behavior in hydrocarbon oxidation.

Light output accompanies hydrocarbon oxidations not only from normal ignition but also from cool-flame phenomena (CH_2O^*).[102] The intensity of the cool-flame emission (wavelength 370–480 nm) is commonly monitored by photomultiplier to provide an additional means of delineation of different reaction modes and to provide a "timing mark" during oscillatory events.[60] There has yet to be a quantitative study in which radical concentrations are monitored from the intensity of emission.

IV. EXPERIMENTAL OBSERVATIONS

The purpose of this section is to describe the overall features of alkane and other organic oxidations. Detailed mechanistic and kinetic aspects are to be discussed separately. While the main thrust is toward the features that are revealed under a well-stirred flow, the knowledge from this source is still too limited for us to confine our attention solely to the c.s.t.r. Data must be drawn also from the very extensive history of unstirred, closed-vessel studies and comparisons and contrasts must be made between the two techniques.

The natural link to theory is the portrayal of events in the terms R or ΔT versus T_0; we shall come to this later. In closed-vessel combustion studies, the control parameters are pressure, vessel temperature, and reactant composition; in flow studies, the mean residence time is added to these. Events in closed vessels are represented most commonly in terms of the two-dimensional P–T_a ignition diagram for a given reactant composition.[13,14] The simplest experimental procedure in the c.s.t.r. is to fix the composition and mean residence time and to sweep a range of vessel temperatures at successive reactant pressures. This provides a common P–T_a basis for comparisons between the two methods and it is a natural starting point in the present account. We begin with a description of events accompanying the oxidation of acetaldehyde.

A. *P–T*$_a$ Ignition Diagram for Acetaldehyde Oxidation in a c.s.t.r.

Acetaldehyde is the only organic substrate for which a complete $P-T_a$ ignition diagram has been obtained in the c.s.t.r.[60] (Carbon monoxide oxidation has been studied comprehensively,[112] and results are now emerging for n-heptane and isooctane oxidations in the jet-stirred high-pressure flow reactor.[80]) A description of the acetaldehyde plus oxygen ignition diagram is relevant also for other reasons. One is that despite its structural simplicity, acetaldehyde exhibits a richness of nonisothermal behavior surpassing those of many more complex substrates and including features closely resembling characteristics of butane oxidation.[30] Another is that under certain conditions, acetaldehyde and its higher homologs are molecular intermediates of the oxidations of alkanes from propane up,[10,11] and particular facets of their structure and reactivity have an influential role on the overall course of oxidation (see Section V).

The $P-T_a$ ignition diagram for $CH_3CHO + O_2$ at a mean residence time of 5 s in a c.s.t.r. (500 cm^3, spherical) incorporates five distinct regions[60a] (Fig. 11). Two of these (regions I and V) represent stationary reaction, and the remaining three (regions II, III, and IV) represent different oscillatory modes. Although the boundaries between regions are

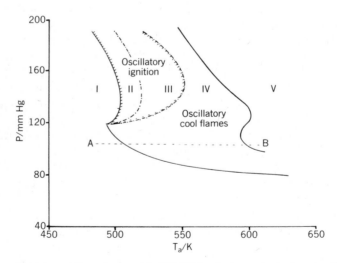

Figure 11. $P-T_a$ ignition diagram for $CH_3CHO + O_2$ in a spherical c.s.t.r. (500 cm^3) at a mean residence time of 5 s. I and V, stationary reaction; II, oscillatory two-stage ignition; III, complex oscillatory ignitions; IV, oscillatory cool flames. Shaded lines distinguish the $P-T_a$ regions in which ignition phenomena are observed.

sharply defined, only that between I and II and that between I and IV exhibit criticality.

1. *Stationary States*

In region I stationary reaction occurs at a modest rate, which increases as the vessel temperature is raised: there is a positive temperature coefficient for the heat-release rate. The highest temperature excesses are reached at the boundary with region II or IV; they do not exceed about 30 K, and this corresponds to a heat-release rate of about 20 W/dm^3 [see Eq. (2.33)]. If the stationary state is disturbed by, for example, momentarily interrupting the flow, there is a sharp diminution in ΔT, which is restored by a monotonic trajectory (Fig. 12, curve a); this is characteristic of a stable nodal point in the two-dimensional phase plane (or multidimensional phase space).

Stationary reaction also exists in region V, but there are marked differences from that observed in region I. First, the reaction rate is greatly enhanced, and the temperature excess may exceed 80 K. Second, ΔT is highest at the boundary with region IV and diminishes as the vessel temperature is raised: there is a negative temperature coefficient for the heat-release rate. Third, a very weak blue chemiluminescence accompanies reaction in region V. This originates from excited formaldehyde molecules. Finally, a momentary interruption to the reactant flow causes a sharp diminution in ΔT, which is restored by a damped oscillatory sequence (Fig. 12, curve b); this characterizes a stable focus in phase space.

It is appropriate to emphasize the existence of the negative temperature coefficient before moving on to a description of the oscillatory reaction modes. The transition from a positive to a negative temperature

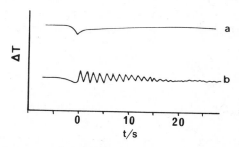

Figure 12.   Responses to a disturbance of flow in the stationary state as measured by the temperature excess in region I (curve a) and in region V (curve b).

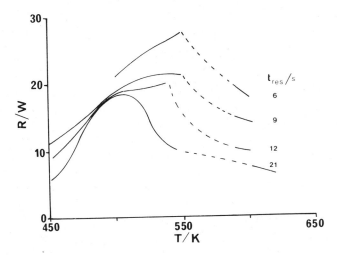

Figure 13. Dependence of the rate of heat release on temperature for $CH_3CHO + O_2$ ($P = 86$ mm Hg) at various residence times in a c.s.t.r. (1000 cm³). The broken lines represent the region (IV) in which oscillatory cool flames are observed. Across this region, the heat-release rate shows a negative temperature dependence.

dependence is displayed in Fig. 13, which shows results measured during the oxidation of acetaldehyde in a c.s.t.r. (1000 cm³).[81]

2. Oscillatory States

Sustained oscillatory "cool flames" occur in region IV. These are so named because the temperature excursions are rarely found to exceed 200 K and they may be so small as to be barely perceptible. The largest amplitudes occur at the boundary with region III; the smallest, at the boundary with region V. The frequencies vary in an inverse manner. The transition from IV to V is a smooth one; the temperature excess about which the smallest detectable (<1 K) oscillations occur coincides with the stationary state at the marginally higher vessel temperature. Examples of such sequences are displayed in Fig. 14. An extremely important additional feature of the oscillatory cool flames is that even at their most intense, only a portion (<50%) of the fuel and oxygen is consumed and the final products of combustion (carbon oxides and water) are formed only in a very minor yield. Thus only a small proportion of the available enthalpy is released; the reaction rate has a "self-quenching" mechanism.

The behavior in region II is in marked contrast. It too is an oscillatory realm, but the successive events in it are very violent, they give rise to

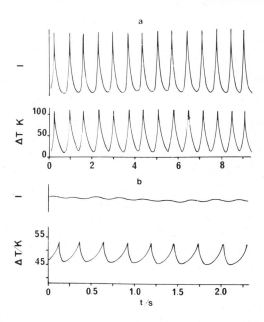

Figure 14.   Temperature excess (ΔT) time records and light output (I) from oscillatory cool flames of $CH_3CHO + O_2$ in the c.s.t.r. (500 cm^3; $t_{res} = 3$ s). (a) $P = 100$ mm Hg; $T_a = 570$ K. (b) $P = 100$ mm Hg; $T_a = 590$ K.

very substantial temperature excursions (>800 K), and they are accompanied by bright flashes of light. All of the fuel is consumed at each event, and carbon oxides and water are the major (if not the sole) products. The period of oscillation is greater than the mean residence time. Superficially these oscillations resemble the oscillatory states that are brought about at criticality, when

$$\theta_- = \frac{B^*}{2}\left[1 - \left(1 - \frac{4}{B^*}\right)^{1/2}\right] \tag{2.43}$$

The realm is entered via a discontinuity at the I–II boundary, and it is exited via a sharp extinction when temperature is reduced below the "critical ignition" temperature; that is, there is a realm of hysteresis in T_a. More detailed scrutiny reveals a finer structure to the oscillatory ignitions. They exhibit a two-stage character (Fig. 15a), the first stage of which resembles the cool flame in all physical and chemical respects.

Region III is the most complicated oscillatory sequence of all; the behavior in it is a composite of those found in regions II and IV, and the

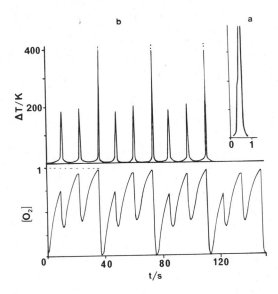

Figure 15. (a) Two-stage and (b) multiple-stage oscillatory ignitions of CH_3CHO+O_2 in a c.s.t.r. The change of concentration of oxygen in complex, four-stage ignition, as monitored continuously by mass spectrometry, is also shown. Two-stage oscillatory ignitions occur in region II and the multiple-stage oscillatory ignitions in region III of the $P-T_a$ ignition diagram.

"relative contributions" are controlled by the $P-T_a$ location within the region. Adjacent to the boundary with region II the cycle of events is

$$\cdots + \text{cool flame} + \text{ignition} + \text{cool flame} + \text{ignition} + \text{cool flame} + \cdots$$

Since each of the ignition pulses is itself a two-stage event, we may term this sequence "repetitive three-stage ignition," with each unit being

$$\cdots + \text{cool flame} + \text{two-stage ignition} + \cdots$$

If T_a is raised, there is a transition from this cycle into the sequence (Fig. 15b)

$$\cdots + \text{cool flame} + \text{cool flame} + \text{ignition} + \text{cool flame}$$
$$+ \text{cool flame} + \text{ignition} + \cdots$$

Again the ignitions show a two-stage character and this particular sequence is thus termed "repetitive four-stage ignition." So far, the greatest multiplicity observed has been five cool flames interspersed by one ignition pulse; the location of the boundary with region IV is determined by the marginal failure of hot ignition ever to occur. In all chemical respects, too, the characteristics of region III are a composite of those in regions II and IV.

Although these "hybrid" oscillations are, in part, an artifact of the flow system, they are not unique to it. They bear striking parallels to the multiple-stage ignitions observed in closed-vessel studies (see below). The existence of these hybrid events was one of the most important discoveries obtained from nonisothermal oxidation of organic compounds as far as contributing to the understanding of two-stage spontaneous-ignition phenomena is concerned. The way in which advances have been obtained based on them is elaborated in Section VI.

The same five regions exist in the P–T_a ignition diagram when experiments are carried out at faster flow rates ($t_{res} = 3$ s). There are quantitative differences in the locations of the boundaries, but the main distinction is that at the faster rates the total pressures need not be so high to achieve the same ends. This is consistent with the controlling influence of the quotient $B/(1 + t_{res}/t_N)$ on events in the nonadiabatic c.s.t.r. (see Section II.B.1).

Just as there is direct identification of the negative temperature coefficient of the heat-release rate during stationary reaction in region V, the falling reactivity through the oscillatory realm (II, III, and IV) as T_a is raised is also a manifestation of an overall negative dependence of rate on temperature. That ignition can fail to occur on *raising* the temperature is the most striking manifestation of this unusual phenomenon.

B. P–T_a Ignition Diagram for Acetaldehyde Oxidation in an Unstirred, Closed Vessel

The main features of $CH_3CHO + O_2$ reaction measured in an unstirred, closed vessel (500 cm^3, spherical) are shown in Fig. 16.[113] This ignition diagram is in accord with results obtained by Chamboux and Lucquin[114] using a very small (30 cm^3) cylindrical vessel. The major distinction between the two closed-vessel studies is the displacement of the common boundaries to higher temperatures ($\Delta T_a \simeq 35$ K) and pressures ($\Delta P \approx 20$ mm Hg) in the smaller vessel. This is in semiquantitative agreement with the effect of enhanced heat-transfer rates via the magnitude of $\chi S/V$ [see Eq. (2.33)]. The relevance of vessel size and shape to the location of the boundaries in the P–T_a ignition diagram is not a feature that has been stressed in earlier work.

Figure 16 is a complicated ignition diagram showing a peculiar fragmentation of the multiple-stage ignition and cool-flame zones. Because

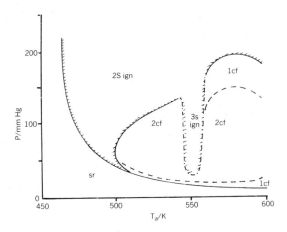

Figure 16. P–T_a ignition diagram for $CH_3CHO + O_2$ in an unstirred closed vessel (500 cm³). Slow reaction (sr), one cool flame (1cf), two cool flames (2cf), two-stage ignition (2s ign), and three-stage ignition (3s ign) are distinguished.

the results displayed in Fig. 16 were obtained in Leeds in the late 1960s, before well-stirred flow studies had begun, the more ordered structure of the (later) c.s.t.r. results was a surprising and welcome discovery. Moreover, the much closer qualitative resemblance of the ignition diagram for acetaldehyde oxidation under well-stirred flow to that of hydrocarbons in closed vessels[30] testified to deeper common links between the properties of organic oxidations than might have been perceived from Fig. 16 (see below). The reasons why this closed-vessel ignition diagram exhibits unusual structural features are still uncertain, and establishing such reasons is no longer important. They are connected, in part, with the extraordinary reactivity of acetaldehyde, which may give rise to two perturbing effects. The first is that substantial reaction may occur even during the course of injection of reactants into an evacuated, closed vessel; there is no measurable induction time to the nonisothermal events (see Fig. 17). The second is that the rate and extent of temperature change are sufficient to produce peculiar spatial inhomogeneities and gas motion due to strong natural convection. Even mechanical stirring is sometimes inadequate to break up the stronger currents (for example, see Fig. 5 in reference 60a).

The unifying features of this particular closed-vessel ignition diagram in relation to that of the c.s.t.r. are (i) the two-stage ignition promontory that invades the slow reaction zone, (ii) the cool-flame zone bounding the

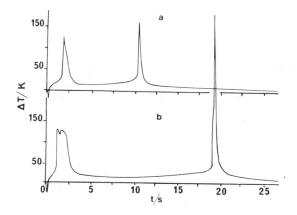

Figure 17. Temperature–time records displaying (a) two cool flames ($P = 23$ mm Hg) (b) three-stage or "delayed" ignition ($P = 32$ mm Hg) in the reaction of $CH_3CHO + O_2$ in an unstirred, closed vessel (500 cm^3); $T_a = 550$ K. Time zero is the time of admission of premixed reactants to the evacuated vessel. The thermocouple registers the entry of the cold reactants.

promontory at its high-temperature side, and (iii) the occurrence of three-stage ignition and multiple cool flames (Fig. 17).

C.   Cool-Flame Multiplicity in Closed Vessels

One important contrast between acetaldehyde oxidation in closed vessels and that of hydrocarbons is the limitation in the former of cool-flame multiplicity to two successive events. This is a consequence of the reactivity of acetaldehyde. Nearly 50% of the reactant is consumed in the first cool flame;[115] in propane oxidation, by contrast, less than 15% of the alkane is consumed, and in consequence, up to eight successive cool flames may be observed in a closed vessel before all of the reactant disappears. Cool-flame multiplicity in closed vessels has been stressed in the past and elaborate diagrams have been drawn separating regions showing different numbers of events. That the multiple cool flames exist in closed vessels is extremely important, but the numbers of them detected are of no great account, since we are witnessing merely a quasi-oscillatory state perturbed by reactant consumption. The form of the closed regions representing different numbers of cool flames arises from the shifting balance between chemical reactivity and the nature of the stability of the system. Some of these aspects were clarified by Gray et al.[87] using a closed vessel into which a mechanical stirrer had been incorporated. Figure 18 shows that in propane oxidation there is a

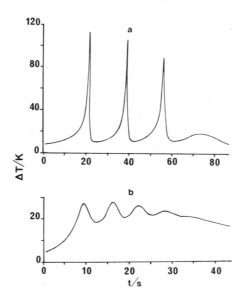

Figure 18. Temperature-time records during oscillatory cool flames of $C_3H_8 + O_2$ in a mechanically stirred closed vessel. $P = 240$ mm Hg. (a) $T = 570$ K. (b) $T = 600$ K.

transition from a quasi-oscillatory state to a damped oscillatory approach to a quasi-stationary state when the vessel temperature is raised. A higher oscillatory damping factor (enhanced by raising the temperature) leads to an apparent diminution in cool flame multiplicity.

The observation that nonisothermal oscillatory (or quasi-oscillatory) cool flames can occur in closed vessels was one of the most important contributions to interpretations of the behavior of complex kinetic systems accompanied by heat release. Their existence distinguishes a class of reactions among all of the types of reactions that are described by two variables, namely one-step kinetics with thermal feedback or isothermal branching schemes (see Section II). Oscillatory phenomena in either of these systems are not possible if the system is not open. As we shall see, Gray and Yang[38–42] were able to describe oscillatory cool-flame phenomena in terms of a two-variable (temperature–concentration) model that was not confined to the open system. The fundamental prerequisites of their simple skeleton scheme (see Section VI) included nonisothermal chain branching coupled to a negative temperature dependence of the overall kinetics.

D. $P-T_a$ Ignition Diagrams for Normal Butane and Isobutane in Closed Vessels

From the structural point of view, normal and isobutane differ in the form of their carbon skeletons and in the numbers of primary, secondary, and tertiary C–H bonds in each molecule. These are the simplest chemical features that might be expected to affect chemical kinetics and the form of the ignition diagram, and the butane isomers are the simplest hydrocarbons that exhibit them. In Figs. 19 and 20 are displayed the $P-T_a$ ignition diagrams for normal[116] and isobutane[117] oxidations, respectively, in unstirred closed vessels (500 cm^3, spherical) at the compositions $C_4H_{10}+2O_2$. Normal butane is the more reactive of the two. Not only does it exhibit an ignition boundary and promontory at pressures below 250 torr, but multiple-stage ignitions are also observed (cf. CH_3CHO+O_2, Fig. 17b). Oscillatory cool flames are observed below 560 K and at pressures down to 50 torr. By contrast, isobutane oxidation fails to give rise to spontaneous ignition at pressures up to 300 torr and temperatures up to 670 K. Oscillatory cool flames do not occur below 570 K or at pressures below 170 torr. These distinctions may be explained from a qualitative mechanistic point of view, as outlined in Section V. We cannot expect to be able to establish the distinctions in unstirred, closed vessels on a quantitative basis.

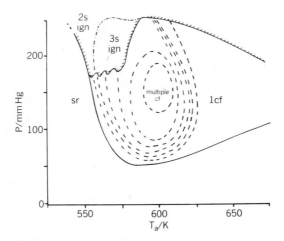

Figure 19. $P-T_a$ ignition diagram for n-$C_4H_{10}+2O_2$ in an unstirred, closed vessel (500 cm^3, spherical). See legend to Fig. 16 for identification of regions. The broken lines distinguish the regions in which different numbers of consecutive cool flames are observed. Shaded lines distinguish spontaneous ignition from slow reaction or cool flames.

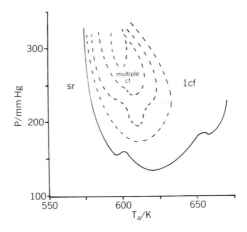

Figure 20. $P-T_a$ ignition diagram for $i\text{-}C_4H_{10}+2O_2$ in an unstirred closed vessel ($500\ cm^3$, spherical). Details as for Fig. 19.

E. Summary of Principal Features

The occurrence of ignition (whether of simple or complex form) is determined by the pressure and temperature to which a specific reactant composition is subjected in a particular vessel. The fundamental properties that control events are the rate of heat release (ordained by thermokinetic evolution) and the rate of heat dissipation from the reaction vessel. The pressure at which ignition can be brought about may be regarded as a measure of the reactivity of a particular composition.

The changing temperature dependence of the reaction rate is an overall feature of complex kinetics involving free-radical branching chains (see Section V); it may be observed even under isothermal conditions. Both the ignition peninsula that penetrates the stationary-reaction and cool-flame zones of the $P-T_a$ diagram and the oscillatory cool flames themselves are born in these complex kinetic interactions, but their existence also depends strongly on heat release and temperature change. An additional feature not described in detail here, the *pic d'ârret*, would seem also to have nonisothermal origins. It is a phenomenon associated with the final stages of reaction in closed vessels. There have been extensive investigations of it by Lucquin and co-workers.[31]

V. KINETICS AND MECHANISMS

Hydrocarbon oxidation starts from a position very far from equilibrium, and there is a complex network of reaction pathways by which the system may progress toward it. Temperature is a major influence in determining the main routes and the extent of progress made toward equilibrium or, in the c.s.t.r., the extent of conversion at the stationary state. Whether or not criticality and ignition are possible has to do with kinetic autocatalysis and thermal feedback, as exemplified in Section II.

From the complex forms of the critical boundaries of the $P-T_a$ ignition diagrams described in Section IV, we may anticipate that the kinetic interactions involved in hydrocarbon oxidation at temperatures below 850 K are far from simple. A natural division falls at about 850 K, because beyond this temperature, chain branching augments the overall reaction rate and heat-release rate so strongly that without special control, the system is swept on to the high temperatures associated with flames. The predominant reaction mode at a particular temperature is a consequence of the relative rates of the elementary processes involved, and the gradual transition from one to another brought about when the temperature is changed is determined by their activation energies.

A. Stoichiometry, Thermochemistry, and Completeness of Combustion

The complete oxidation of alkanes is associated with the stoichiometry

$$C_nH_{2n+2} + \left(\frac{3n+1}{2}\right)O_2 \rightarrow nCO_2 + (n+1)H_2O$$

n-Butane, for example, gives the overall stoichiometry

$$n\text{-}C_4H_{10} + 6.5O_2 \rightarrow 4CO_2 + 5H_2O \tag{5.1}$$

and the enthalpy change associated with it is $\Delta H^\circ_{298} = -2654$ kJ/mol of n-butane, -408 kJ/mol of oxygen, or -393 kJ/mol of the reactant mixture. The theoretical adiabatic temperature is in excess of 5000 K. At the stoichiometric composition in air $(C_4H_{10} + 6.5O_2 + 24.5N_2)$, the heat release per mole of reactant mixture (ΔH°_{298}) is -79 kJ/mol. The overall stoichiometry is the same for isobutane, but there is a marginal change in enthalpy due to the different heat of formation of the substrate. The equilibrium composition for a stoichiometric butane–oxygen mixture at the measured adiabatic flame temperature ($\sim$3000 K) differs from the stoichiometry given in (5.1) because substantial dissociation of the products is brought about above $\sim$2000 K, and the energy required for such processes moderates the temperature that can be achieved. Atoms and radicals constitute more than 20 mol % of the equilibrium composition. It is inevitable that spontaneous ignition in adiabatic conditions produces this final state, and it is implicit that isothermal oxidation of butane at 3000 K, as in a shock tube, for example, will also lead to the adiabatic flame composition, provided that the reaction time is sufficiently long to permit complete equilibration. Compositions that differ from that at thermodynamic equilibrium may be measured in flames, especially at low pressures, because termolecular recombination reactions to do not have sufficient time to become equilibrated. "Super equilibrium" concentrations of atoms and radicals (e.g., O, OH, and H) then exist in the reaction zone; the enhancement in their concentrations may be substantial.[118]

The temperatures and flame speeds of freely propagating (near adiabatic) hydrocarbon–oxygen or hydrocarbon–air flames in stoichiometric mixtures are rather similar to each other.[25] This is partly because the heat release per mole of reactant mixture does not vary very much from hydrocarbon to hydrocarbon, and also because at high temperatures, the predominant elementary processes involving the fuel are degradative in kind. The body of elementary steps that controls the rate of reaction is then virtually independent of the substrate[7,8] (see Section V.F).

Matters are quite different in the low-temperature realm ($T <$ 850 K), where the development of spontaneous ignition begins. There are markedly different stoichiometries and exothermicities from those measured at high temperatures. For example, the final composition of the isothermal oxidation of isobutane at 550 K is broadly consistent with the stoichiometry[119]

$$i\text{-}C_4H_{10} + 2.5O_2 \rightarrow 0.35(CH_3)_2CO + 0.27CH_3OH + 0.18CH_2O + 0.15C_4H_8$$
$$+ 0.12C_3H_6 + 0.88CO + 0.33CO_2 + 2.64H_2O \quad (5.2)$$

The associated enthalpy change, ΔH°_{298}, is -877 kJ/mol of $i\text{-}C_4H_{10}$. Only 150 K higher, at 700 K, the oxidation is accounted for more satisfactorily by the overall stoichiometry[119]

$$i\text{-}C_4H_{10} + 2.0O_2 \rightarrow 0.14(CH_3)_2CO + 0.06CH_3OH + 0.12CH_2O + 0.26C_4H_8$$
$$+ 0.24C_3H_6 + 0.06C_2H_6 + 0.84CO$$
$$+ 0.20CO_2 + 2.44H_2O \quad (5.3)$$

from which $\Delta H^\circ_{298} = -689$ kJ/mol of $i\text{-}C_4H_{10}$.

Although these stoichiometries do not show perfect elemental balances and, more importantly, do not reflect changes of stoichiometry throughout the course of reaction, they do serve to illustrate a number of important points. First, oxidation is far from complete at low temperatures. Second, the predominant chemical features change as the temperature is increased. Third, a relatively small proportion of the total available enthalpy is released.

Except for carbon dioxide and water, each of the molecular products constitutes a reserve of energy at an intermediate stage of spontaneous ignition, and must eventually contribute to the exothermicity of reaction in the complete combustion. Elementary processes that may be regarded as subsidiary to the main kinetic pathway are, thus, not without effect on the overall reaction exothermicity and associated temperature change. The different stoichiometries at 550 and 700 K also imply that overall kinetic parameters and dependences on reactant concentrations are unlikely to be applicable at temperatures outside the experimental range in which they are measured.

B. The Molecular Products of Isothermal or Near-Isothermal Oxidation at Low Temperatures

The advent of gas chromatography in the late 1950s and technical advances since then have made it possible to detect and measure the molecular constituents of low-temperature oxidation processes in remarkable detail. Chemical analysis throughout the course of reaction in closed

TABLE III

Analyses of Intermediate and Final Product from the Low-Temperature Oxidation of Hydrocarbons in Closed Vessels

Hydrocarbon	Fuel–oxygen composition	Total pressure (mm Hg)	Vessel temperature (K)	Number of products identified	Reference
Propane	1:1	300	570	11	120
	2:1	150	708	9	121
n-Butane	1:3.5	160	588	15	122
i-Butane	1:1	300	590	21	123
n-Pentane	1:1	150	430	19	110
	3:4	20–200	523	16	124
Neopentane	1:2	170	473–503	20	125
	1:1	300	553	21	126
n-Hexane	1.25:1	20–200	473–723	8[b]	127
c-Hexane	2:1–1:2	50–100	533	18	128
2-Methyl pentane	1:2	40–220	503–583	26(46)[a]	129
3-Methyl pentane	1:2	80–150	568–678	48(86)[a]	130
2,3-Dimethyl butane	1:2	325	591	30	131
n-Heptane	1:1	150	523	26(48)[a]	132
3-Ethyl pentane	1:2	80–150	368	39	133
n-Decane	2:1:7 (RH:O_2:N_2)	510	483	31	134

[a] Number including unidentified components is given in parentheses.
[b] Study confined specifically to certain θ-heterocyclic products.

TABLE IV

Product Compositions from the Extensive Slow Oxidation of C$_4$ Hydrocarbons: Moles per Mole Reactant (%)

Composition	Temp. (K)	Ref.	x (mol% yield)			
			x > 30	30 > x > 10	10 > x > 2	2 > x > 0.1
3.5n-C$_4$H$_{10}$+O$_2$	588	122	Butene-1	Butene-2	Propene, ethene	Tetrahydrofuran, 2-methyloxetan, 2.3-dimethyloxiran, 2-ethyloxiran, methyl ethyl ketone, propionaldehyde, propene oxide, acrolein, acetone, ethene oxide, acetaldehyde
i-C$_4$H$_{10}$+O$_2$	579	123	Isobutene, carbon monoxide, water	Isobutyraldehyde, propene, carbon dioxide	Methyloxiran, t-butanol, methanol, formaldehyde	2.2-Dimethyloxiran, s-butanol, methacrolein, isopropanol, propionaldehyde, ethene, methane
trans-C$_4$H$_8$+O$_2$	540	135	Carbon dioxide	2.3-Epoxybutane, acetaldehyde, methanol, carbon monoxide	2.3-Epoxybutane, formaldehyde, methane	but-3-en-2-one, butan-2.3-dione, but-2-ene-1-al, 3-hydroxybutene-1, 3.4-epoxybutane-1, cis-butene-2, propene

258

TABLE V

Initial Products of Radical Attack on Butanes

Composition	Temp. (K)	Ref.	Products			
i-$C_4H_{10} + 2O_2$ (at 1% conversion)	583	124	t-Butyl hydroperoxide, isobutene	Propionaldehyde, propene, acetone		Methane, formaldehyde, carbon monoxide, carbon dioxide
$2H_2 + O_2 + 4N_2 + 0.07\,i$-$C_4H_{10}$	753	138b	Isobutene, 3-methyloxetan, isobutene oxide, isobutyraldehyde, 2-methylprop-2-en-1-al, 2-methylprop-2-en-1-ol	Propene, propane, propene oxide, propionaldehyde, acetone, prop-2-en-1-ol	Ethene, ethane, ethene oxide, acetaldehyde	
$2H_2 + O_2 + 4N_2 + 0.07\,n$-$C_4H_{10}$	753	139	But-1-ene, trans-but-2-ene, cis-but-2-ene, butadiene, 1,2-epoxybutane, trans-2,3-epoxybutane, cis-2,3-epoxybutane	2-Methyloxetan, tetrahydrofuran, methyl ethyl ketone, but-3-en-2-one, but-2-en-3-ol, 1,2-epoxybut-3-ene	Propene, propene oxide, acetone, propionaldehyde, prop-2-en-1-al, prop-2-en-1-ol	Ethene, ethane, ethene oxide, acetaldehyde, methane, formaldehyde, carbon monoxide, carbon dioxide

vessels, normally by repetitive batch operation over different time intervals, has yielded concentration–time profiles for the oxidations of hydrocarbons in the range C_3–C_{10}. An indication of the extent of this activity, the conditions employed, and the complexity of the analyses is given in Table III. Many products are present in only very minor yield.

Most workers have relied on the thermostating-oven temperature as a monitor of the conditions. Rarely has an internal thermocouple been used to measure the reactant temperature. Thus if reaction is nonisothermal, then the measured composition may be associated with temperatures in excess of those deemed to prevail. The effect can be significant, especially when cool flames accompany reaction (e.g., see Figs. 17 and 18).

Selected results for compositions at the late stages of the oxidations of C_4 hydrocarbons are given in Table IV. These reflect the consequences of secondary reactions involving the molecular products of the initial oxidation of the substrate. There seems to be virtually no direct information on product compositions at temperatures above 700 K. To investigate the primary products of radical attack on the substrate, it is essential either to curtail reaction when only a very small proportion ($<1\%$) of the reactant has been consumed, so that secondary processes are suppressed, or to control the radical and atom environment in a very careful manner. The latter goal has been achieved by Baldwin and co-workers by means of the addition of small amounts of hydrocarbons to slowly reacting hydrogen–oxygen mixtures (e.g., see references 136–140). Application of this technique has clarified much mechanistic detail and provided extremely important kinetic data in the vicinity of 750 K for elementary reactions involved in hydrocarbon combustion. Some results from analytical studies of initial products are given in Table V.

C. A Kinetic Framework for the Low–Temperature Oxidation of Hydrocarbons

To proceed to a kinetic interpretation of spontaneous ignition, it is most economical to formulate a possible basic mechanism and to analyze its constituent parts. Appropriate to the formulation of a simple mechanism reactions in the range to ~750 K (see Table VI) are discussed in this section. With the exception of 6b and 8, the reactions considered in Table VI involve only the parent hydrocarbon or oxygen as molecular reactants. This gross simplification denies the possibility of radical attack on all other molecular intermediates. Some of these additional factors are considered in Section V.E and the processes that occur at higher temperature are reviewed in Section V.F.

TABLE VI
Mechanistic Foundation for the Oxidation of Hydrocarbons at
Temperatures Below 750 K

Reaction no.	Reaction
1	$RH + O_2 \rightarrow R + HO_2$
2	$R + O_2 \rightarrow AB + HO_2$
3	$R + O_2 \rightarrow RO_2$
4	$RO_2 \rightarrow R + O_2$
5a	$RO_2 \rightarrow QOOH \rightarrow R'CHO + P_1 + OH$
5b	$\rightarrow AO + P_2 + OH$
6a	$RO_2 + RH \rightarrow ROOH + R$
6b	$RO_2 + R'CHO \rightarrow ROOH + R'CO$
7a	$\nearrow P_3$
7b	$RO_2 + RO_2 \rightarrow RO + RO + O_2$
8	$ROOH \rightarrow RO + OH$
9	$RO + O_2 \rightarrow R'CHO + HO_2$
10	$RO + RH \rightarrow ROH + R$
11	$RO + RO \rightarrow ROH + R'CHO$
12	$OH + RH \rightarrow H_2O + R$
13	$HO_2 + RH \rightarrow H_2O_2 + R$
14	$HO_2 + HO_2 \rightarrow H_2O_2 + O_2$
15	$R'CO \rightarrow R' + CO$
16	$R' \rightarrow$ new subset of reactions

1. A Simplified Overview

The initial source of alkyl radicals is almost certainly reaction 1, and propagation by attack of molecular oxygen follows. Except for methyl radicals, there are two possibilities. The first is the abstraction of a hydrogen atom to yield the conjugate alkene (AB in reaction 2); an activation energy in excess of 20 kJ/mol may be inferred for this reaction from present data. The second is an association (reaction 3), and this occurs with no activation energy.

Alkylperoxy radicals formed via reaction 3 may undergo various reactions; these are designated 4–7. First, there is the decomposition 4, which is the reverse of 3. This is the only equilibrium between forward and backward steps that has to be considered in oxidation at low temperatures; it is displaced toward dissociation as the temperature is raised (see below). Secondly, intramolecular hydrogen-atom transfer may occur (reaction 5) to yield an alkylhydroperoxy radical (QOOH); this decomposes by initial fission of the peroxide linkage. [Addition of molecular oxygen to QOOH is also possible[141] (see below).] There can be multiplicity of products from reaction 5, and it is the major source of the very great variety of partially oxygenated products (aldehydes, ketones, ethers) produced, as exemplified in Table IV. Their types are governed by the location from which the internal H atom is drawn and their proportions are governed by the "strain energy" that has to be surmounted in the ring transition state for completion of the transfer. A particular distinction is made here between the route to aldehydes (reaction 5a) and the route to other oxygenated products AO (reaction 5b). "P" signifies any other molecular product of the decomposition, and the overall process is a propagation generating hydroxyl radicals. Third, intermolecular abstraction of hydrogen atoms (reaction 6) may occur to yield an alkyl hydroperoxide; the activation energy varies between ~40 and ~80 kJ/mol. E is lowest when abstraction occurs from the acyl moiety of aldehydes (reaction 6b); it is highest when abstraction occurs from primary C–H bonds in an alkyl moiety (reaction 6a). Finally, radical–radical interaction of the alkylperoxy radical may occur (reaction 7).

The strength of the peroxide linkage in the alkyl hydroperoxide is sufficiently low ($E \simeq 140$–180 kJ/mol) that secondary initiation of additional reaction chains ("degenerate branching") may result from it. Alkoxyl radicals so formed may undergo various steps, as illustrated in 9–11. Hydrogen-atom abstraction (reaction 10) is not strongly favored, since the activation energy exceeds 10 kJ/mol even when a tertiary C–H bond is involved; it rises to nearly 30 kJ/mol at a primary C–H bond.

The predominant chain-propagation steps that restore the alkyl radical involve the hydroxyl (reaction 12) and hydroperoxy (reaction 13) radicals. There is a complete substructure involving the alkyl radical (R') derived from the intermediate aldehydes that has much in common with the reactions in Table VI. The formyl radical (HCO) derived from formaldehyde is a special case, and discussion of it is reserved for Section V.E. Secondary sequences have purposely been disregarded. These include the further oxidation of the conjugate alkene (AB), the oxygenated products AO, the alcohol ROH, hydrogen peroxide, and all other products P. These are all susceptible to further reaction, especially through H-atom abstraction by the hydroxyl radical. Activation energies for reaction 12 are small ($0 < E < 7$ kJ/mol) and preexponential factors do not vary much across the entire range.

Before discussing in detail some of these kinetic features, it may be instructive to make (simplified) assertions in the context of isothermal oxidation to illustrate the concerted effects of the elementary processes in Table VI.

(a)   At temperatures below 400 K, the sequence 1, 3, 6a, 7a predominates, yielding the organic peroxide as a final product. It is an unbranched, long-chain process and is the core of liquid-phase oxidation of hydrocarbons.[142,143]

(b)   At temperatures above 650 K, the sequence 1, 2, 13, 14 predominates. It is an unbranched chain, propagated by HO_2 radicals and yields the conjugate alkene (AB) and hydrogen peroxide. Chain lengths are very short, because the rate of propagation (reaction 13) is slow ($E \approx 60$–80 kJ/mol). The balance is thrown onto this particular sequence because the equilibrium ratio $[R]/[RO_2]$ in reactions 3 and 4 is displaced in favor of the alkyl radical. (Hydrogen peroxide decomposition may become an important secondary source of OH radicals above ~ 800 K.)

(c)   Alkylperoxy (RO_2) is the predominant radical at intermediate temperatures and its isomerization leads to fast and unselective propagation by OH radicals. Secondary initiation through the organic peroxide becomes a significant feature. Moreover, the formation of ROOH is augmented by the aldehydes derived from the isomerization (reaction 5a), because the abstraction of the acyl H atom (reaction 6b) is so much more favorable than abstraction from primary, secondary, or tertiary C–H bonds in hydrocarbons (reaction 6a). Interactions between the elementary steps are thus at their most complex in the intermediate-temperature range.

We may now anticipate the effect of such changes in mechanism on the overall rate of isothermal reaction. A realm of negatively temperature-dependent reaction rate becomes a feature, dominated by the shifting balance from RO_2 to R in the equilibrium of reactions 3 and 4 and causing the transition from (c) to (b) above as the temperature is raised. In practice, the secondary oxidation of molecular intermediates must also be influential; the final stoichiometries (5.2) and (5.3) show that the overall chemistry is much more complicated than the simple form described here.

D.   Detailed Kinetic Interpretations

We now focus our attention quantitatively on five features of Table VI. These are (1) initiation; (2) propagation by OH attack, and by RO_2 or HO_2 attack; (3) the RO_2–R equilibrium; (4) alkylperoxy-radical isomerization; and (5) conjugate alkene formation. The differences brought about by structural features of the hydrocarbon molecule are illustrated by reference to the isomers normal butane and isobutane. Preexponential factors for the rate parameters of reactions involving intermolecular hydrogen-atom abstraction are expressed in terms of the contribution per (primary, secondary, or tertiary) C–H bond.[144,145] This is an important, unifying link among the homologous series of hydrocarbons. The activation energies, expressed here in kilojoules per mole, depend solely on the type of C–H bond; they are independent of carbon-chain length or structural form.

1. *Initiation*

Initiation is strongly endothermic and has a high activation energy. Hydrogen-atom abstraction by molecular oxygen from normal butane can yield either normal butyl or secondary butyl radicals. Isobutane can yield either isobutyl or tertiary butyl radicals. Their probabilities of formation are not the same. Rate constants for initiation (m^3/mol/s) at four temperatures and their relative magnitudes are given in Table VII.[144] Rate constants at the high-pressure limit for unimolecular decomposition of each of the butanes are also given for comparison.[144] There are two main points to be observed. The first is that there is a strongly dominant formation of the secondary butyl and tertiary butyl radical as a result of reaction of n-butane and i-butane, respectively, with oxygen. The degree of this predominance diminishes as the temperature is raised. The second concerns the relative rates of initiation by the two modes, given (for normal butane) by $k_1[O_2]/k_{17}$ (the subscripts denote the reaction numbers).

Oxidative initiation (reaction 1) is very greatly favored over pyrolysis of the hydrocarbon in the low-temperature range, $T < 850$ K, except at an extreme condition of exceptionally low oxygen concentration. This throws the basis for the mechanism of oxidation at low temperatures onto the reactions of the conjugate alkyl (i.e., the butyl) radical. This is not the case at higher temperatures. As the relative magnitudes of the rate constants for oxidation and pyrolysis at 1000 K show, the balance shifts strongly toward pyrolytic processes at higher temperatures, and we may thus anticipate that the principal routes to oxidation by-pass reactions of butyl and concentrate on those of lower alkyl radicals.

2. *Propagation*

The rate constants and relative rates of propagation involving the formation of the different butyl radicals are presented in Table VIII for hydroxyl-radical attack (reaction 12) on the butanes and in the Table IX for alkylperoxy- (reaction 6a) or hydroperoxy- (reaction 13) radical attack on the butanes.[144] Such kinetic data as are available indicate that attacks

TABLE VII
Relative Rates of Initiation for Normal and Isobutane Oxidation

	$C_4H_{10} + O_2 \rightarrow C_4H_9 + HO_2$ (Reaction 1)			
	n-Butane		*i*-Butane	
	$\underset{\underset{\mathrm{H}}{\|}}{CH_2CH_2CH_2CH_3}$	$\underset{\underset{\mathrm{H}}{\|}}{CH_3CHCH_2CH_3}$	$\underset{\underset{\mathrm{H}}{\|}}{\overset{\overset{CH_3}{\|}}{CH_2-CH-CH_3}}$	$\underset{\underset{\mathrm{H}}{\|}}{\overset{\overset{CH_3}{\|}}{CH_3-C-CH_3}}$
D_{R-H}(kJ/mol)	413	395	398	383
R	$\underset{(n\text{-}C_4H_9)}{CH_2CH_2CH_2CH_3}$	$\underset{(s\text{-}C_4H_9)}{CH_3CHCH_3CH_3}$	$\underset{(i\text{-}C_4H_9)}{\overset{\overset{CH_3}{\|}}{CH_2-CH-CH_3}}$	$\underset{(t\text{-}C_4H_9)}{\overset{\overset{CH_3}{\|}}{CH_3-C-CH_3}}$
k_1(m³/mol/s) (per C–H bond)	6.6×10^6 exp $(-213/RT)$	1×10^7 exp $(-199/RT)$	4.5×10^6 exp $(-213/RT)$	4×10^7 exp $(-184/RT)$
k				
at 550 K	2.35×10^{-13} $(1)^a$	5.03×10^{-12} (21)	2.35×10^{-13} (1)	1.34×10^{-10} (553)
at 650 K	3.05×10^{-10} (1)	4.07×10^{-9} (13)	3.05×10^{-10} (1)	6.50×10^{-8} (213)
at 750 K	5.85×10^{-8} (1)	5.52×10^{-7} (9.5)	5.84×10^{-8} (1)	6.20×10^{-6} (106)
at 1000 K	2.99×10^{-4} (1)	1.61×10^{-3} (5.4)	2.99×10^{-4} (1)	9.78×10^{-3} (33)

	$n\text{-}C_4H_{10} \rightarrow C_2H_5 + C_2H_5$ (Reaction 17)	$i\text{-}C_4H_{10} \rightarrow CH_3 + C_3H_7$ (Reaction 18)
$D_{R-R'}$(kJ/mol)	342	345
k (s⁻¹)	1.9×10^{14} exp$(-342/RT)$	6.3×10^{14} exp$(-345/RT)$
at 550 K	6.3×10^{-19}	1.1×10^{-18}
at 650 K	6.2×10^{-14}	1.2×10^{-13}
at 750 K	2.9×10^{-10}	5.9×10^{-10}
at 1000 K	2.6×10^{-4}	6.0×10^{-4}

[a] Numbers in parenthesis are relative magnitudes of rate constants.

of HO_2 and RO_2 on hydrocarbons have comparable preexponential factors and activation energies.[144]

Hydroxyl radicals are unselective in abstraction because the activation energies are never very large. Where significant differences in activation energy do prevail (cf. *i*-butyl and *t*-butyl formation from isobutane), there may be a compensatory effect of the number of centres from which H-atom abstraction can occur; thus in isobutane there are nine primary C–H bonds but only one tertiary C–H bond. By contrast, there is a much greater selectivity of abstraction by RO_2 and HO_2. This means that if

TABLE VIII

Relative Rates of Hydroxyl Radical Attack on Normal and Isobutane

	$C_4H_{10} + OH \rightarrow C_4H_9 + H_2O$ (Reaction 12)			
	n-Butane		i-Butane	
	$R = n\text{-}C_4H_9$	$R = s\text{-}C_4H_9$	$R = i\text{-}C_4H_9$	$R = t\text{-}C_4H_9$
k_{12} (m^3/mol/s) (per C–H bond)	6.15×10^5 exp $(-6.9/RT)$	1.4×10^6 exp $(-3.6/RT)$	6.15×10^5 exp $(-6.9/RT)$	1.25×10^6 exp $(0.8/RT)$
k (m^3/mol/s)				
at 550 K	8.16×10^5 (1)^a	2.55×10^6 (3.1)	1.22×10^6 (1.5)	1.45×10^6 (1.53)
at 650 K	1.02×10^6 (1)	2.87×10^6 (2.8)	1.54×10^6 (1.5)	1.45×10^6 (1.2)
at 750 K	1.22×10^6 (1)	3.14×10^6 (2.6)	1.83×10^6 (1.5)	1.45×10^6 (1.05)

^{a} Numbers in parentheses have same meaning as in Table VII.

HO$_2$-radical propagation is able to predominate over OH-radical propagation, there is a corresponding shift toward a simpler mechanism involving one type of butyl radical as a major component.

3. $R + O_2 \rightleftharpoons RO_2$ Equilibrium

Equilibrium constants for the association of oxygen with each of the four butyl radicals derived from normal and isobutane at four temperatures are given in Table X. A comparison is also made with the equilibrium

$$H + O_2 + M \rightleftharpoons HO_2 + M$$

TABLE IX

Relative Rates of hydroperoxy- and Alkylperoxy-Radical Attack on Normal and Isobutane

	$C_4H_{10} + RO_2 \rightarrow C_4H_9 + RO_2H$ (Reaction 6a) or $C_4H_{10} + HO_2 \rightarrow C_4H_9 + H_2O_2$ (Reaction 13)			
	n-Butane		i-Butane	
	$R = n\text{-}C_4H_9$	$R = s\text{-}C_4H_9$	$R = i\text{-}C_4H_9$	$R = t\text{-}C_4H_9$
k_{6a} or k_{13} (m^3/mol/s) (per C–H bond)	1×10^6 exp $(-81.0/RT)$	1×10^6 exp $(-71.3/RT)$	1×10^6 exp $(-81.0/RT)$	1×10^6 exp $(-60.3/RT)$
k (m^3/mol/s)				
at 550 K	0.121 (1)^a	0.671 (5.6)	0.182 (1.5)	1.875 (15.4)
at 650 K	1.856 (1)	7.449 (4)	2.785 (1.5)	14.258 (7.7)
at 750 K	14.262 (1)	43.262 (3)	21.394 (1.5)	63.122 (4.4)

^{a} Numbers in parentheses have same meaning as in Table VII.

TABLE X

Constants for Equilibria Between C_4H_9 and $C_4H_9O_2$ and Between H and HO_2[a]

| | $C_4H_9 + O_2 \underset{4}{\overset{3}{\rightleftharpoons}} C_4H_9O_2$ | | | | |
	$n\text{-}C_4H_9$	$s\text{-}C_4H_9$	$i\text{-}C_4H_9$	$t\text{-}C_4H_9$	$H + O_2 \underset{4}{\overset{3}{\rightleftharpoons}} HO_2$
$\log_{10}(K_p/(\text{mm Hg})^{-1})$					
at 300 K	9.85	10.5	9.6	10.1	26.6
at 550 K	0.85	1.1	0.7	0.7	11.3
at 650 K	−0.80	−0.65	−1.05	−1.15	8.6
at 750 K	−2.01	−1.87	−2.33	−2.53	6.3
Ceiling temperature[b] (K)					
	750	770	720	710	1600

[a] Data calculated from Arrhenius parameters for reactions 3 and 4.[144]

[b] Temperature at which $[R] = [RO_2]$ when $P(O_2) = 100$ mm Hg.

The additional data, the "ceiling temperatures," are a convenient way of distinguishing among the behaviors of different radicals on the basis of the temperature required to establish equal concentrations of alkyl and alkylperoxy radicals at a prescribed concentration of oxygen.[146] Thus the displacement of the equilibrium in favor of the butyl radical as temperature is raised occurs in the order of increasing temperature: $t\text{-}C_4H_9 < i\text{-}C_4H_9 < n\text{-}C_4H_9 < s\text{-}C_4H_9$. There is a concomitant distinction in the shifting balance from oxygenated-product formation and predominant hydroxyl-radical propagation to alkene formation and predominant hydroperoxy-radical propagation. The ranges of temperatures at which alkyl and alkylperoxy radicals are balanced are quite different from those at which hydrogen atoms and hydroperoxy radicals achieve equal proportions.

4. Alkylperoxy-Radical Isomerization and Decomposition

The kinetic factors that determine the activation energy required to bring about alkylperoxy-radical isomerization and decomposition are (1) the nature of the C–H bond that has to be broken and (2) the number of atoms associated with the ring transition state for intramolecular hydrogen-atom transfer.[13,147-151] The least strained transition state is an eight-membered ring,[151] as, for example, in the reaction

$$CH_3CH_2CH_2CH_2CH_2OO$$

$$\longrightarrow CH_2CH_2CH_2CH_2CH_2OOH \qquad (5.4)$$

$$\downarrow$$

molecular products $+$ OH

Activation energies for this rearrangement are believed to vary between 82 and 48 kJ/mol for the intramolecular transfer from a primary, secondary, or tertiary C–H bond. Seven-, six-, and five-membered ring transition structures are increasingly more strained; the activation energies are estimated to fall in the range 98–145 kJ/mol when abstraction occurs from a primary C–H bond. Comparison between calculated and measured yields of O-heterocyclic compounds in very carefully controlled conditions suggests that the overall rate constants were estimated previously from activation energies that were too low.[13,149,150] Although we may regard this qualitative basis, dating from ~1965–1970, for the interpretation of mechanistic routes to the multiplicity of oxygenated products formed as being very sound, it is only very recently that a satisfactory quantitative footing has begun to emerge.

From the nature of other products found, it seems likely that internal rearrangement of the alkylperoxy radicals by methyl-radical transfer is also possible.[13] Moreover, the detection of dihydroperoxides during the low-temperature oxidation of long-chain hydrocarbons is evidence for the addition of an oxygen molecule to the rearranged species:[141]

$$QOOH \xrightarrow{O_2} O_2QOOH \xrightarrow{RH} HOOQOOH \qquad (5.5)$$

The qualitative differences among the types of products that may be formed from the alkylperoxy radicals derived from normal and isobutane are displayed in Table XI. The formation of 2-methyloxetan and of propene and formaldehyde as complementary products to it is common to both of the butanes. The major distinction between the two is that normal butane is able also to yield acetaldehyde via isomerization and decomposition of the secondary butylperoxy radical; there seems to be no primary route to acetaldehyde by alkylperoxy-radical rearrangement during isobutane oxidation. It is because of mechanistic differences such as these that oscillatory cool flames and spontaneous ignition are brought about more easily in normal-butane oxidation.

TABLE XI

Internal Isomerisation Reactions of $C_4H_9O_2$: $C_4H_9O_2 \rightarrow [C_4H_8OOH] \rightarrow$ Products $+$ OH
(Reaction 5)

$C_4H_9O_2$	C_4H_8OOH	O-Heterocycles	Other products	Products of CH_3-radical transfer
		Products of normal butane oxidation		
$\underset{\overset{\mid}{O\text{—}O}}{CH_2CH_2CH_2CH_3}$	$\underset{\overset{\mid}{O\text{—}OH}}{CH_2CH_2CH_2CH_2}$	(oxolane ring: $CH_2\text{—}CH_2$, CH_2, CH_2, O)	$\underset{+C_2H_4}{CH_2\text{—}CH_2 \, O}$	CH_2O+ $C_2H_4+CH_3O$
$(n\text{-}C_4H_9O_2)$	$\underset{\overset{\mid}{O\text{—}OH}}{CH_2CH_2CHCH_3}$	$\underset{O\text{——}CHCH_3}{CH_2\text{—}CH_2}$	CH_2O $+CH_2CHCH_3$	$\underset{O\text{——}CH_2+CH_3O}{CH_2\text{—}CH_2}$
	$\underset{\overset{\mid}{O\text{—}OH}}{CH_2CHCH_2CH_3}$	$\underset{O}{CH_2\text{—}CHCH_2CH_3}$		
$\underset{\overset{\mid}{O\text{—}O}}{CH_3CHCH_2CH_3}$	$\underset{\overset{\mid}{O\text{—}OH}}{CH_3CHCH_2CH_2}$	$\underset{O\text{—}CH_2}{CH_3CH\text{—}CH_2}$	CH_3CHO $+C_2H_4$	$\underset{+CH_3O}{CH_3CH\text{—}CH_2 \, O}$
$(s\text{-}C_4H_9O_2)$	$\underset{\overset{\mid}{O\text{—}OH}}{CH_3CHCHCH_3}$	$\underset{O}{CH_3CH\text{—}CHCH_3}$		
		Products of isobutane oxidation		
$\underset{\overset{\mid}{O\text{—}O}}{\overset{\overset{CH_3}{\mid}}{CH_3CHCH_2}}$	$\underset{HO\text{—}O}{\overset{\overset{CH_3}{\mid}}{CH_3\text{—}C\text{—}CH_2}}$	$\underset{CH_3}{\overset{CH_3}{>}}C\text{—}CH_2 \, O$	None	$\underset{+CH_3O}{CH_3\text{—}CH\text{—}CH_2 \, O}$
$(i\text{-}C_4H_9O_2)$	$\underset{HO\text{—}O}{\overset{\overset{CH_3}{\mid}}{CH_2\text{—}CH\text{—}CH_2}}$	$\underset{CH_2\text{—}O}{CH_3CH\text{—}CH_2}$	CH_2CHCH_3 $+CH_2O$	
$\underset{\overset{\mid}{O\text{—}O}}{\overset{\overset{CH_3}{\mid}}{CH_3\text{—}C\text{—}CH_3}}$	$\underset{O\text{—}OH}{\overset{\overset{CH_3}{\mid}}{CH_3\text{—}C\text{—}CH_2}}$	$\underset{CH_3}{\overset{CH_3}{>}}C\text{—}CH_2 \, O$	$\underset{\overset{\parallel}{O}}{CH_3\text{—}C\text{—}C_2H_5}$	$(CH_3)_2CO$ $+CH_3O$
$(t\text{-}C_4H_9O_2)$				

5. Conjugate Alkene Formation

During isobutane oxidation the only alkene that may be formed by hydrogen-atom abstraction (reaction 2) from the butyl radical is isobutene:

$$CH_3\text{—}CH\text{—}CH_3 + X \longrightarrow$$
$$\overset{|}{\underset{CH_3}{}}$$

$$\left\{ \begin{array}{c} CH_3\text{—}C\text{—}CH_3 \\ \overset{|}{\underset{CH_3}{}} \\ \\ CH_2\text{—}CH\text{—}CH_3 \\ \overset{|}{\underset{CH_3}{}} \end{array} + HX \right\} \xrightarrow{O_2} CH_2\text{=}C\text{-}CH_3 + HO_2 \quad (5.6)$$
$$\underset{CH_3}{\overset{|}{}}$$

Oxidation of normal butane leads to the alternative isomers 1-butene or cis- and trans-2-butene:

$$CH_3CH_2CH_2CH_3 + X \longrightarrow$$

$$\left\{ \begin{array}{l} CH_3CHCH_2CH_3 \xrightarrow{O_2} \left\{ \begin{array}{l} CH_2\text{=}CHCH_2CH_3 + HO_2 \\ CH_3CH\text{=}CHCH_3 + HO_2 \end{array} \right. \\ \\ \qquad + HX \qquad\qquad\qquad\qquad\qquad (5.7) \\ \\ CH_2CH_2CH_2CH_3 \xrightarrow{O_2} CH_2\text{=}CHCH_2CH_3 + HO_2 \end{array} \right.$$

Rate constants for hydrogen-atom abstraction have been obtained experimentally at 753 K from the addition of the butanes to slowly reacting hydrogen–oxygen mixtures. There are insufficient data available to allow the Arrhenius parameters to be assessed with certainty. However, combining[145] the rate constant at 753 K[140] for isobutene formation from isobutane with that obtained at 313 K[152] suggests an activation energy of 24.3 kJ/mol and a preexponential factor of 4×10^6 m^3/mol/s. Data for ethene formation from ethane at 753 and 896 K lead to the values $E = 23.5$ kJ/mol and $A = 2.3 \times 10^6$ m^3/mol/s. These are useful indications of the magnitudes of the activation energies associated with this type of hydrogen-atom abstraction.

E. Oxidation of Molecular Intermediates

Each of the molecular compounds formed during the oxidation of alkanes is itself susceptible to radical attack, especially by addition to unsaturated carbon–carbon linkages or abstraction at carbon–hydrogen bonds. Except in circumstances where adjacent functional groups modify the reactivity, the rate constants for hydrogen-atom abstraction fit into the patterns described in Section V.D. The overall rates of oxidation of intermediates by this route are governed by the concentrations of the molecular intermediates relative to that of the parent alkane.

The principal mode of reaction of aldehydes is a sequence of the type 6b, 15, 16. Abstraction of the hydrogen atom at the acyl moiety, by any radical, occurs extremely readily since the bond strength is low ($D_{RCO-H} \approx$ 368 kJ mol). Aldehydes are much more reactive than the parent alkane on this account (cf. D_{R-H} in Table VII), and their important contribution to secondary initiation or degenerate branching (Table VI, reactions 6b and 8) is one consequence.

Rather less is known about the reactivity of cyclic ethers. The oxidation of ethylene oxide does not occur very readily, in marked contrast to that of its isomer acetaldehyde,[61] and special reactivity of higher members of the homologous series is unlikely to be conferred by the oxiran ring structure.

Alkenes are susceptible to radical addition at the double bond as well as hydrogen-atom abstraction.[153] Their oxidation is extremely complex and, as yet, not especially well characterized. They are found to be precursors of formaldehyde and carbon monoxide, especially at temperatures in the vicinity of 750 K; this may be of particular importance to the evolution of hot ignition (see Section V.F). Ethene, for example, added to slowly reacting hydrogen–oxygen mixtures at 770 K gives rise to the approximate proportions (in mole percentages)[154] 70% carbon monoxide, 10% ethylene oxide, 8% carbon dioxide, 6% formaldehyde, and 5% methane. Methanol, propane, ethane, and acetylene are also formed, in very minor yield. These overall proportions result from primary, secondary, or tertiary stages of reaction. Reactions of the following kinds may be important:

$$C_2H_4 + OH \longrightarrow C_2H_3 \xrightarrow{O_2} CH_2O + HCO \qquad (5.8)$$
$$+$$
$$H_2O$$

$$C_2H_4 + OH \longrightarrow C_2H_4OH \begin{cases} \xrightarrow{O_2} C_2H_4O + HO_2 \\ \xrightarrow{O_2} 2CH_2O + OH \\ \longrightarrow CH_2O + CH_3 \end{cases} \qquad (5.9)$$

$$C_2H_4 + HO_2 \longrightarrow C_2H_4OOH \longrightarrow C_2H_4O + OH \quad (5.10)$$

The reactions of higher members of the series of alkenes are more complex, but may follow similar patterns. For example, the following reactions may be associated with isobutene oxidation:

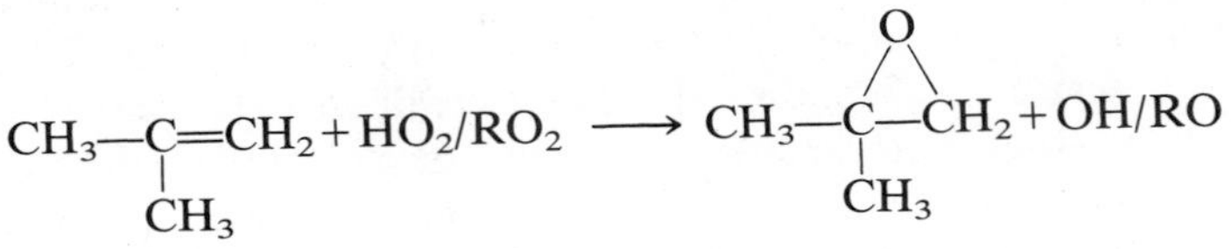

$$CH_3\!-\!\underset{\underset{CH_3}{|}}{C}\!\!=\!\!CH_2 + HO_2/RO_2 \longrightarrow CH_3\!-\!\underset{\underset{CH_3}{|}}{\overset{\overset{\displaystyle O}{\diagup\!\diagdown}}{C}\!-\!CH_2} + OH/RO \quad (5.11)$$

$$CH_3\!-\!\underset{\underset{CH_3}{|}}{C}\!\!=\!\!CH_2 + X \longrightarrow \underset{\underset{\overset{+}{HX}}{CH_3}}{CH_2\!-\!\underset{}{C}\!\!=\!\!CH_2} \longrightarrow CH_2\!\!=\!\!C\!\!=\!\!CH_2 + CH_3$$

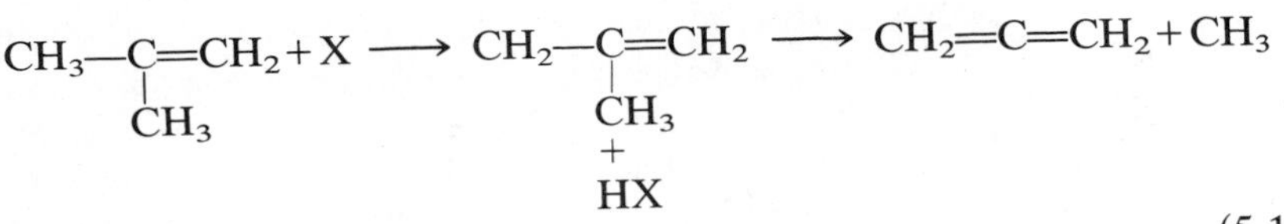

$$(5.12)$$

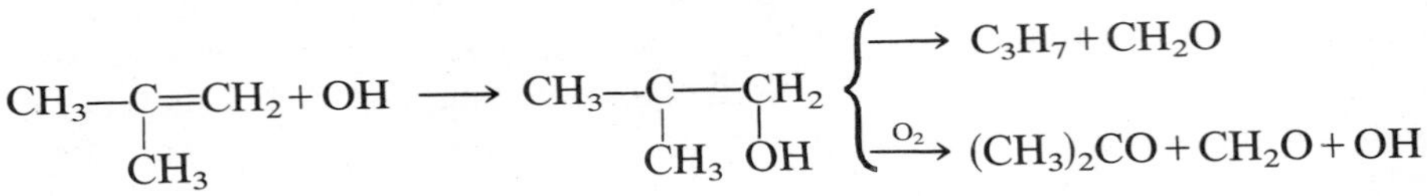

$$CH_3\!-\!\underset{\underset{CH_3}{|}}{C}\!\!=\!\!CH_2 + OH \longrightarrow CH_3\!-\!\underset{\underset{CH_3}{|}}{C}\!-\!\underset{\underset{OH}{|}}{CH_2} \begin{cases} \longrightarrow C_3H_7 + CH_2O \\ \overset{O_2}{\longrightarrow} (CH_3)_2CO + CH_2O + OH \end{cases}$$

$$(5.13)$$

$$CH_3\!-\!\underset{\underset{CH_3}{|}}{C}\!\!=\!\!CH_2 + OH \longrightarrow CH_3\!-\!\underset{\underset{CH_3}{|}}{\overset{\overset{OH}{|}}{C}}\!-\!CH_2 \longrightarrow (CH_3)_2CO + CH_3$$

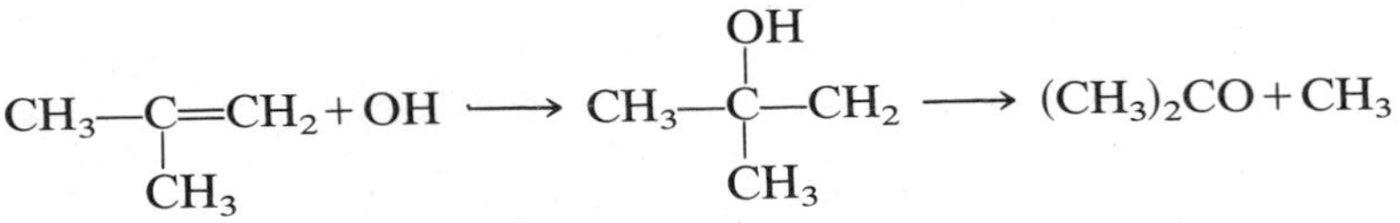

$$(5.14)$$

One important feature of these types of elementary reactions is that they contribute to a stepwise breakdown of the carbon-chain structure of the parent alkane; in consequence, a complete description of the mechanisms for oxidation of higher-molecular-weight alkanes is extremely complicated. Although many elementary reactions are required also to describe events at very high temperatures, in general, the overall behavior is dominated by the reactions of hydrocarbon fragments containing only one or two carbon atoms.[8,9]

F. Oxidation at Temperatures Above 750 K

At 750–850 K and higher temperatures, there is a marked transition in the mode of oxidation. It becomes dominated by the chain-branching processes associated with hydrogen and oxygen atoms:

$$H + O_2 \rightarrow OH + O \qquad (5.15)$$

$$O + XH \rightarrow X + OH \qquad (5.16)$$

$$OH + XH \rightarrow X + H_2O \qquad (5.17)$$

where XH may be any molecular species, including the principal reactant, from which a hydrogen atom may be abstracted. Autocatalysis will occur if hydrogen atoms are regenerated. In hydrocarbon oxidation, the two molecular intermediates that are capable of furnishing hydrogen atoms at temperatures below ~850 K are formaldehyde and carbon monoxide, as follows:

$$CH_2O + X \rightarrow HCO + HX \qquad (5.18)$$

$$HCO + M \rightarrow H + CO + M \qquad (5.19)$$

$$CO + OH \rightarrow CO_2 + H \qquad (5.20)$$

The importance of (5.20) as the major route to carbon dioxide as a final product of combustion has long been recognized. The special significance of this propagation in combustion at moderate temperatures, in which hydroxyl radicals are replaced by hydrogen atoms, has not been stressed previously. The part played by hydrogenous impurities in carbon monoxide oxidation is also due substantially to this propagation.[155]

To maintain an emphasis on the kinetics and mechanisms of the transition realm at 750–850 K, we may raise four further questions:

1. *To what extent does carbon monoxide owe its existence to formaldehyde and formyl radicals?* Within the temperature range of particular interest in this section, they are virtually exclusive precursors. The reaction chain also includes the important step

$$HCO + O_2 \rightarrow CO + HO_2 \qquad (5.21)$$

At lower temperatures, where RO_2 isomerizations (reaction 5a) may give rise to higher aldehydes, the acyl radicals derived from them may also yield carbon monoxide, as follows:

$$RCO + M \rightarrow R + CO + M \qquad (5.22)$$

At rather higher temperatures, especially as adiabatic flame temperatures are approached, there is a multiplicity of pyrolytic processes and radical–radical or radical–atom interactions that are believed to yield formyl radicals or carbon monoxide directly.

2. *How is formaldehyde formed?* Methyl radicals are one source, via processes of the type

$$CH_3 + O_2 + M \rightleftharpoons CH_3O_2 + M$$

$$CH_3O_2 + CH_3O_2 \rightarrow CH_3O + CH_3O + O_2$$

$$CH_3O + CH_3O \rightarrow CH_3OH + CH_2O$$

$$CH_3O + O_2 \rightarrow CH_2O + HO_2$$

Oxidation of ethene of the lower alkenes may also be very important sources (see Section V.E).

3. *Do any other molecular intermediates contribute to chain branching?* Hydrogen peroxide is formed from hydroperoxy radicals (reactions 13 and 14). Its decomposition via

$$H_2O_2 + M \rightarrow 2OH + M \tag{5.23}$$

for which $k = (1.9 \pm 1.5) \times 10^{11} \exp(-196.5/RT)$ m^3/mol/s in the range 650–900 K, may strongly augment chain branching, but only at temperatures above ~850 K, where its half-life is sufficiently short.

4. *Are routes to hydrogen atoms other than via formyl radicals or carbon monoxide possible?* Homolytic fission of C–H bonds in alkyl radicals can occur; for example,

$$C_4H_9 \rightarrow C_4H_8 + H \tag{5.24}$$

However, activation energies are high ($E > 130$ kJ/mol), so processes of this type can compete successfully with abstraction of hydrogen atoms by oxygen (reaction 2) only at temperatures in excess of 1000 K. Dissociation of alkenes may occur via, for example,

$$C_4H_8 \rightarrow C_4H_7 + H \tag{5.25}$$

but the associated activation energies are extremely high and such processes are unlikely to occur much below flame temperatures. Unimolecular dissociation of alkanes occurs virtually exclusively by C–C bond fission, since $D_{C-C} < D_{C-H}$.

VI. THEORETICAL INTERPRETATIONS OF OSCILLATORY COOL FLAMES AND SPONTANEOUS IGNITION

To understand the origins and interrelationships among the complex nonisothermal events associated with hydrocarbon oxidation, it is necessary to marry the basic elements of the mechanisms described from an essentially isothermal standpoint in Section V to thermal feedback as portrayed for the single-step, first-order process in Section II. This is far from simple, since formal analytical interpretations involving multistep kinetic structures in the nonadiabatic c.s.t.r. go far beyond the present state of the art in theories of chemical-reactor stability. Nevertheless, there has been substantial progress, beginning with the remarkable, if rudimentary, insight by Salnikov[33] (1949), which predates virtually all of the theoretical studies of reactor stability, and followed by the outstanding unifying interpretation of chain–thermal interactions by Gray and Yang[38–42] (1965–1970). A description of this work forms the basis of the first part of the present section. The mathematical tools used are the stability theorems for nonlinear systems exemplified in Section II.

The understanding of thermokinetic phenomena has also benefitted from the advancement of computational techniques and the ability of large computers to perform extremely rapid and accurate numerical integrations of extensive networks of stiff, nonlinear differential equations. However, the supplementary requirement for successful implementation of numerical methods is complete confidence in the mechanistic foundation to be exploited and in the quality of the kinetic, thermochemical, and physical parameters associated with the "real" system. The advances made through numerical studies are discussed later in this section.

A. Salnikov's Analysis (1949)

Virtually all contemporary experimental studies were under closed-vessel conditions when Salnikov[33] first introduced the concept of an oscillatory state to represent multiple cool flames. He argued that thermokinetic oscillations could arise even in a closed vessel when the reaction system was no more complicated than

$$A \xrightarrow{\text{I}} X \xrightarrow{\text{II}} B$$

provided that the heat output was associated with reaction II and that $E_{II} > E_I$. Using the simplifying assumptions of pseudo-zero-order kinetics for reaction I, that is, considering the rate of depletion of A (a) to be slow compared with the rate of change of concentration of X (x), he investigated three kinetic possibilities. The first was an autocatalytic generation of X taking the rate law $v_I = k_I ax$. There is a precedent here for a nonisothermal complement to the isothermal studies by Gray and Scott[68-74] relating to

$$A + B \rightarrow 2B \qquad (2.65)$$

$$B \rightarrow \text{inert}$$

The second was the pseudo-zero-order generation of X taking the rate law $v_I = k_I a_0$. The third was an invariant rate of supply of X, given by $v_I = \text{constant}$. Salnikov interpreted the third case as a limiting condition of the second, and he recognized that it simulated the single-step, first-order, exothermic reaction under open conditions.[156]

Although there are no obvious unifying links between this work and the striking facets of nonisothermal oxidation of hydrocarbons, the important achievements were threefold. Salnikov recognized that in the $T-[x]$ phase plane there may be one or three singularities; he determined the nature of these singularities, drawing for the first time on stability theorems for nonlinear dynamic systems; and he established the conditions for the existence of limit cycles. The most fruitful developments in the interpretation of oscillatory cool flames follow these principles.

B. Gray and Yang's Analysis: Thermally Coupled Chain-Branching Theories (1965–1970)

The skeleton scheme adopted by Gray and Yang[38-42] is set out in Section I. Its principal elements are the generation of a reactive (molecular or radical) intermediate X. Only linear terms in the concentration of X were considered, and this facilitated a tractable, comprehensive analysis. To simplify the algebra, only the rate-determining branching step of the chain was retained and dependences on molecular concentrations were subsumed into the rate constants. As in Salnikov's treatment, the concentration of the reactant was assumed constant, thus confining the analysis to a two-dimensional (T, x) system under closed-vessel conditions.

To preserve the present emphasis on c.s.t.r. studies and to highlight important features additional to those associated with the single-step, exothermic reaction, Gray and Yang's scheme is cast here in modified form:

		Rate constant	Activation energy	Heat output	
$A_0 \rightarrow A$	(inflow/outflow)	k_f	0	0	(6.1)
$A \rightarrow P$	(nonchain conversion)	k_p	E_p	h_p	(6.2)

A → X	(initiation)	k_i	0	h_i	(6.3)
A + X → 2X	(branching)	k_b	E_b	h_b	(6.4)
X → inert	(outflow)	k_f	0	0	(6.5)
X → inert	(termination)	k_t	E_t	h_t	(6.6)

This system is described by the three conservation equations

$$\frac{da}{dt} = k_f(a_0 - a) - k_p a - k_i a - k_b a x \tag{6.7}$$

$$\frac{dx}{dt} = k_i a - (k_f + k_t - k_b)x \tag{6.8}$$

$$V\sigma c_p \frac{dT}{dt} = V k_p h_p a + V k_i h_i a + V(k_b h_b + k_t h_t)x$$
$$- (\chi S + f\sigma c_p)(T - T_a) \tag{6.9}$$

At this point only general comments can be made about the types of singularities that may exist. Analytical routes to the detailed interpretation of three or more equations are not sufficiently advanced to go beyond the Hurwitz criteria for stability.[157]

Let us dissect the system into two parts to establish not only the links to Section II but also additional properties of interest. Consider steps (6.1)–(6.3). These constitute the system of two parallel, first-order reactions in the c.s.t.r., one of which [(6.3)] is deemed here to be insensitive to temperature change. The conservation equations representing the system may be deduced from (6.7) and (6.9) in modified form, and, following the procedure in Section II, they take the reduced forms

$$\frac{d\gamma}{dt} = (1 - \gamma)\left(\frac{\exp \theta}{t_{chp}} + \frac{1}{t_{chi}}\right) - \frac{\gamma}{t_{res}} \tag{6.10}$$

$$\frac{d\theta}{dt} = B(1 - \gamma)\left(\frac{\exp \theta}{t_{chp}} + \frac{1}{t_{chi}}\right) - \theta\left(\frac{1}{t_{res}} + \frac{1}{t_N}\right) \tag{6.11}$$

where t_{chp} and t_{chi} represent the characteristic chemical times at T_* for (6.2) and (6.3); t_{chi} is assumed to be independent of temperature in order to allow us to retain only one activation energy (E_P) and to define θ in terms of it.

Solving (6.10) and (6.11) at the stationary states leads to an expression analogous to (2.42):

$$t_{\text{res}}^2 + \left(\frac{1}{\exp \theta_{\text{ss}}/t_{\text{chp}} + 1/t_{\text{chi}}} + t_{\text{N}} - \frac{Bt_{\text{N}}}{\theta_{\text{ss}}}\right)t_{\text{res}} + \frac{t_{\text{N}}}{\exp \theta_{\text{ss}}/t_{\text{chp}} + 1/t_{\text{chi}}} = 0$$

(6.12)

We should expect similar qualitative forms for $t_{\text{res}}(\theta_{\text{ss}})$ to those displayed in Fig. 6 and the same sequences of stability for the singularities in each of these patterns of multiplicity. The treatment of a pair of parallel first-order reactions, both of which are dependent on temperature, is much more complex and its forms depend on the relative magnitudes of $E_{6.2}$ and $E_{6.3}$.

Now consider a simplified representation for the sequence of reactions (6.3)–(6.6), as follows:

		Rate constant	Activation energy	Heat output	
$A \rightarrow X$	(constant initiation rate v_i)			h_i	(6.3′)
$X \rightarrow 2X$	(branching)	k_b	E_b	h_b	(6.4′)
$X \rightarrow$ inert	(outflow)	k_f	0	0	(6.5′)
$X \rightarrow$ inert	(termination)	k_t	E_t	h_t	(6.6′)

The singularities are solutions to the pair of equations

$$\frac{dx}{dt} = v_i - (k_f + k_t - k_b)x = 0 \qquad (6.13)$$

$$\sigma c_p \frac{dT}{dt} = R_i + (k_b h_b + k_f h_f)x - \left(\frac{\chi S + f\sigma c_p}{V}\right)(T - T_0) = 0 \qquad (6.14)$$

or, with simplifying substitutions,

$$\frac{dx}{dt} = v_i - \phi x = 0 \qquad (6.13′)$$

$$\frac{dT}{dt} = R_i + \psi x - l(T - T_0) = 0 \qquad (6.14′)$$

from which

$$x_{\text{ss}} = \frac{v_i}{\phi} \quad \text{and} \quad (T - T_0)_{\text{ss}} = \frac{R_i}{l} + \frac{\psi v_i}{\phi l} \qquad (6.15)$$

and the total heat release rate

$$R(T) = R_i + \frac{\psi v_i}{\phi} \tag{6.16}$$

where ϕ is termed the *net branching factor*, ψ represents the summation of the heat-release rates from reactions involving X, and l is a combined heat-loss rate term that is constant provided that the residence time is not varied. $R_i = v_i h_i$.

$R(T)$ will exhibit extrema when ϕ exhibits extrema, namely at

$$\frac{d\phi}{dT} = \frac{d(k_f + k_t - k_b)}{dT} = \frac{1}{RT^2}(E_t k_t - E_b k_b) = 0 \tag{6.17}$$

since k_f is independent of temperature. An extremum exists when $E_t k_t = E_b k_b$ and it will give rise to a maximum in $R(T)$ when the inequality $E_t > E_b$ is satisfied.

The overall results for the heat-release rate when the sequence (6.3′)–(6.6′) is imposed on (6.1)–(6.3) are displayed in Fig. 21. The chain branching via (6.4′) augments the heat-release rate beyond that achieved by the nonchain conversion (6.2) and the parallel initiation of X (6.3). $R(T)$ reaches a maximum and then plunges into a negative temperature-dependent region. $R(T)$ is dominated eventually by (6.2) as the temperature is raised further. At constant residence time, one such curve for $R(T)$ may represent the heat-release rate at a particular reactant concentration in the c.s.t.r. Up to five singularities are possible, as shown by the intersections between R and L (Fig. 21). This additional complexity was anticipated in diagrammatic form independently of Gray and Yang by Vedeneev et al.,[158] and they investigated the nature of the associated singularities.

From the conclusions of Section II, we should expect B and D to be saddle points and, in general, A to be a stable node. E may be an unstable singularity surrounded by a limit cycle, and thus exhibit oscillatory behavior in time, or it may be a stable state (see Section II). The properties of C are discussed below. Criticality is demonstrated by the intersection between R and L' when T_a is raised, but the transition is to state c. High-temperature "ignition" is achieved only when T_a is sufficient to bring about a second tangency condition, as shown by the heat-loss rate L'' (Fig. 21). The two critical jumps may also be brought about by raising the concentration (or pressure) in the reactant.

The characteristic equation for the system (6.13′) and (6.14′) is given by the determinant (2.20), and the nature of singularities is determined by

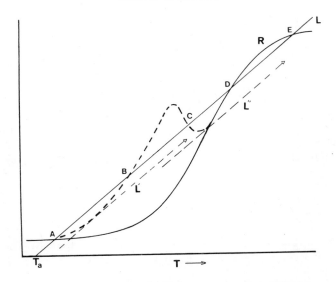

Figure 21. Heat-release rate R as a function of T_a in a nonadiabatic c.s.t.r. The solid curve represents results for first-order exothermic reaction. The dashed curve represents the imposition of complex kinetics involving an overall negative temperature coefficient of the heat-release rate. There is a range of conditions in which five singularities (A–E) can coexist.

the roots λ_1 and λ_2 (Table I). The condition for complex roots, and thus a focus, is

$$(A'+D')^2 - 4(A'D' - B'C') < 0$$

which for the present system [(6.13′) and (6.14′)] takes the form

$$\left[\left(\frac{\partial \dot{T}}{\partial T}\right)_x - \phi\right]^2 + 4\phi\left(\frac{dR}{dT} - l\right) < 0 \qquad (6.18)$$

A focus can exist only at

$$\phi\left(\frac{dR}{dT} - l\right) < 0$$

and since $\phi < 0$ for real concentrations of X, $dR/dT - l < 0$ is a necessary condition. A negative temperature dependence of the heat-release rate, $dR/dT < 0$, favors the existence of a focus, and one may thus exist at C.

For instability, and hence the possibility of a limit cycle surrounding the singularity,

$$A' + D' > 0 \quad \text{or} \quad \left(\frac{\partial \dot{T}}{\partial T}\right)_x + \left(\frac{\partial \dot{x}}{\partial x}\right)_T > 0$$

That is,

$$\frac{dR_i}{dT} + x_{ss}\frac{d\psi}{dT} - l - \phi < 0 \tag{6.19}$$

But for the focus,

$$\frac{dR}{dT} - l < 0$$

or

$$\frac{dR_i}{dT} + \frac{\psi dx_{ss}}{dT} + x_{ss}\frac{d\psi}{dT} - l < 0 \tag{6.20}$$

Thus, if an unstable focus is to exist, $dx_{ss}/dT < 0$ must hold. The existence of a region of negative temperature coefficient in the overall heat-release rate, brought about by a falling concentration of the intermediate propagating species as the temperature is raised, favors the existence of an unstable focus surrounded by a limit cycle. These are sufficient, rather than necessary, conditions. As is exemplified in Fig. 21, we may expect to enter an oscillatory region at C discontinuously by raising the vessel temperature or reactant pressure, and as $dx/dT \rightarrow 0$ (the minimum in dR/dT) oscillatory reaction may give way to a stable state that is a focal singularity.[159] This is in excellent accord with features characterized in the P–T_a ignition diagram (see line AB in Fig. 11). The form of the heat-release rate $R(T)$ shown in Fig. 21 thus implies the existence of two distinct modes of oscillatory behavior. The one at E is characteristic of oscillatory instability resulting from flow interactions in the c.s.t.r., and is limited in amplitude only by the adiabatic temperature excess. The other, at C, is a consequence of the interaction between heat release and complex chemical kinetics and properly deserves the name "thermokinetic oscillation" used by Salnikov[33] and Frank-Kamenetskii[48] or "thermally coupled kinetic oscillations," as suggested by B. F. Gray.

The bounds for the existence of oscillatory cool flames, and the amplitudes and frequencies of their oscillations, cannot be interpreted from the present mode of analysis, nor can the existence of multiple-stage ignitions be inferred from it. A proper interpretation must be based on the analysis of the three-dimensional system of equations representing $\dot{a}$, $\dot{x}$, and $\dot{T}$ [(6.7)–(6.9)]. The mathematical tools required to cope with this situation analytically have yet to be developed. For this reason, Gray and Yang,[38–42] and later Gonda and Gray,[159] adopted the quasi-stationary route to analysis of a two-dimensional model based on a pseudo-zero-order dependence on reactant concentration. Intuitively, this would seem to be a limiting condition for the slow consumption of A with respect to reactions involving X, thus accounting, for example, for the limited lengths of oscillatory trains under closed-vessel conditions. The validity of this analytical technique, and that of the alternative "pseudo-stationary-point analysis" adopted by Halstead, Prothero, and Quinn,[160] has yet to be proved rigorously.

The skeleton framework exploited by Gray and Yang is the simplest unifying relationship between the existence of oscillatory cool flames and a negatively temperature-dependent regime in the overall reaction rate. It reveals how this particular kinetic feature, when thermally coupled, can lead to nonisothermal oscillatory events under nonadiabatic conditions, even in closed vessels.

Identity of the Gray and Yang model with "real" hydrocarbon-oxidation chemistry has yet to be established, even when the skeleton is "dressed" with molecular reactants at each step and additional propagation steps are included. As the reactions in Table VI show, the system includes many variables, and reducing such a scheme to a single kinetic variable by linking, via the stationary-state hypothesis, the concentrations of all other intermediates is destined to yield an expression containing nonlinear dependences on x. The two-variable (T, x) analytical model then loses the tractability exploited by Gray and Yang. While shedding a small area of light on the darkness that covers many-variable, nonlinear problems, the simplification to two variables must be treated with caution; it may impose constraints that are too restrictive, masking, for example, the more elaborate "real" kinetic origins of two-stage and multiple-stage ignitions. Even though the Gray and Yang model predicts a two-stage mode of ignition, this phenomenon and its more complex counterpart are almost certainly the properties of a multidimensional system.

C. Numerical Interpretations of Thermokinetic Phenomena

Computational studies either exploit kinetic schemes that are fully representative of a particular chemical system or they are based on "minimum," generalized kinetic models. The numerical procedure is to integrate the time-dependent differential equations representing the concentration of each species and temperature [e.g., (6.7)–(6.9)]. In each case the validity of the conclusions hangs upon the appropriateness of the kinetic and thermal data; reduced schemes necessarily include rate coefficients that are composites of kinetic parameters for individual elementary processes. The calculations may be applied to either open or closed systems. All numerical models are based on spatially uniform, well-mixed states, and thus quantitative comparisons between numerical and experimental results are restricted to events in the c.s.t.r. Nonisothermal calculations pertaining to closed vessels can do no more than match the classical experimental studies in a superficial qualitative fashion; this cannot be regarded as an adequate test for the validity of a particular kinetic model.

1.   Computations Involving "Full" Kinetic Schemes

The organic oxidation system that has received the most numerical attention is that of acetaldehyde. Despite the richness of its patterns of nonisothermal phenomena, its oxidation can be represented by one of the most simple kinetic schemes among organic substrates, and many of the rate parameters are well established. Griffiths, Skirrow, and Tipper[161] (1968) first applied a degenerate chain-branching scheme to a numerical appraisal of the autocatalysis observed under isothermal, closed-vessel conditions. Building on the Gray and Yang type of skeleton with an elaboration of this particular scheme, Halstead, Prothero, and Quinn[160] (1971) then took a most important step toward the ab initio prediction of oscillatory cool-flame phenomena through chain–thermal interactions. Their achievement was simulating the existence of oscillatory (or quasi-oscillatory) states under closed-vessel conditions. Two-stage and multiple-stage ignitions were not predicted at the start, and the initial 14 elementary reactions were elaborated to a scheme containing 21 reactions to accommodate these phenomena.[162] The results were most encouraging, but not without ambiguity, since these numerical studies predated the experimental investigations of acetaldehyde oxidation in the c.s.t.r. and thus could not be adapted to this more stringent and quantitative experimental test.

Felton, Gray, and Shank[157] took the scheme of 14 reactions used by Halstead et al.[160] and applied it in a numerical analysis of cool-flame oscillatory states in the c.s.t.r. They were able to predict the P–T_a boundaries for the oscillatory realm and to match satisfactorily the periods and amplitudes of the experimentally measured oscillations in the c.s.t.r.[163] By a reduction of the kinetic scheme to three algebraic expressions representing the concentration of the fuel, the concentration of an intermediate organic hydroperoxide (peracetic acid), and the reactant temperature [analogous in form to (6.7)–(6.9)], Felton et al.[157] were able also to investigate some properties of the singularities of the system. Their experimental studies had not been extended to the ignition region and the calculations were not developed to predict facets of it.

The most recent numerical attack on acetaldehyde oxidation[163] has been directed specifically to the prediction of multiple-stage ignition under well-stirred flowing conditions; a kinetic scheme that has proved to be successful for this purpose is given in Table XII. The scheme builds on those schemes used in the earlier work, but incorporates additional reactions of methyl radicals, which previously had been interpreted for simplicity solely as chain-terminating species. The involvement of methyl radicals in chain propagation and branching is a vital link in the evolution of spontaneous ignition. Similarities to the main structural elements of Table VI will be recognized. The results of computations using this kinetic scheme are the first to display repetitive ignition phenomena of any kind and the first to demonstrate the evolution of multiple-stage ignitions in the c.s.t.r. (Fig. 22).

The kinetic scheme is not proposed as the definitive interpretation of acetaldehyde oxidation; indeed, it is almost certainly too simple. Its purpose is to establish a "real" basis from which reductions to the most simple general form may be made and to provide clues to the identities of the radical and molecular intermediates that hold the key to the development of two-stage ignition in nonisothermal conditions. The concept that CH_2O as a precursor to CO and H_2O_2 is crucial to the evolution of hot ignition seems well supported by these calculations.

Perhaps the greatest reward from this numerical study is that it points the way to new experiments: By eliminating time from the solution of the numerical computations, the phase relationships among concentrations of intermediates or among concentrations and temperature may be predicted. The selection of any two variables yields an x–y or x–T phase portrait. When x and y are similarly dependent on a third variable, they generate a single line in the x–y plane. The independent variables x and y that are in phase generate an orbit with an apex at (x_{max}, y_{max}). The independent variables x and y that are out of phase generate a limit cycle. Its existence is due to $dx/dy = 0$ and $dy/dx = 0$ occurring at different coordinates in the x–y plane, corresponding to time-dependent behavior in which $dx/dt \neq 0$ when $dy/dt = 0$, and vice versa. The direction of progress around the cycle reveals the variable that leads in time.

Numerically predicted relationships that emerge from oscillatory four-stage ignition (Fig. 22) based on the kinetic scheme in Table XII are illustrated in Fig. 23–26. The concentrations of carbon monoxide and hydrogen peroxide as products principally of formaldehyde oxidation vary

TABLE XII

Elementary Reaction Scheme from Which Oscillatory, Multiple-Stage Ignitions are Predicted During Oxidation of Acetaldehyde in a c.s.t.r.[a,b]

Initiation

$$CH_3CHO + O_2 \rightarrow CH_3CO + HO_2$$
$$CH_3CO + O_2 \rightarrow CH_3CO_3$$
$$CH_3CO_3 + CH_3CHO \rightarrow CH_3CO_3CH + CH_3CO$$
$$CH_3CO_3H \rightarrow CH_3 + CO_2 + OH$$

Oscillatory Cool-Flame Stage

$$CH_3CO + M \rightarrow CH_3 + CO + M$$
$$CH_3 + O_2 \rightarrow CH_3O_2$$
$$CH_3O_2 \rightarrow CH_3 + O_2$$
$$CH_3 + CH_3 \rightarrow C_2H_6$$
$$CH_3O_2 + CH_3O_2 \rightarrow CH_3O + CH_3O + O_2$$
$$CH_3O + CH_3O \rightarrow CH_3OH + CH_2O$$
$$CH_3O_2 + CH_3CHO \rightarrow CH_3O_2H + CH_3CO$$
$$CH_3O_2H \rightarrow CH_3O + OH$$
$$OH + CH_3CHO \rightarrow CH_3CO + H_2O$$
$$CH_3O + O_2 \rightarrow CH_2O + HO_2$$
$$OH + CH_2O \rightarrow HCO + H_2O$$

Transition to Second-Stage Ignition

$$HCO + M \rightarrow H + CO + M$$
$$HCO + O_2 \rightarrow CO + HO_2$$
$$OH + CO \rightarrow H + CO_2$$
$$H + O_2 \rightarrow OH + O$$
$$H + O_2 + M \rightarrow HO_2 + M$$
$$O + CH_2O \rightarrow CHO + OH$$
$$HO_2 + CH_2O \rightarrow HCO + H_2O_2$$
$$HO_2 + HO_2 \rightarrow H_2O_2 + O_2$$
$$H_2O_2 + M \rightarrow 2OH + M$$
$$OH + H_2O_2 \rightarrow HO_2 + H_2O$$

[a] From Gibson et al.[164]

[b] Although three broad divisions may be recognized, there are strong kinetic interactions among them.

in phase (Fig. 23); they reach different maxima (points 1, 2, and 3 in Fig. 23) at each cycle. The change in concentration of formaldehyde is out of phase with that of carbon monoxide, with formaldehyde leading in time (Fig. 24). $[CH_2O]$ versus T and $[CO]$ versus T relationships are shown in

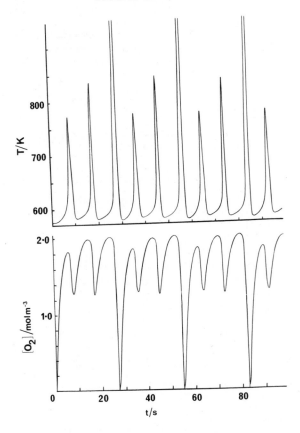

Figure 22. Numerical interpretation of four-stage ignition of $CH_3CHO + O_2$ in a nonadiabatic c.s.t.r. Temperature–time and oxygen concentration–time profiles are displayed. $T_a = 565$ K; $P = 150$ mm Hg; $t_{res} = 3$ s.

Figs. 25 and 26. They reveal that the formaldehyde concentration reaches its maximum before the peak temperature of each cycle is attained. The maximum concentration in carbon monoxide is achieved after the peak temperature in each cool flame; its growth at the third cycle is curtailed by the onset of hot ignition.

Much information is bound up in these phase relationships and their cycle-to-cycle variations. By matching the results of such calculations to direct experimental measurements of the changing concentrations of molecular intermediates, as may be obtained using responsive mass

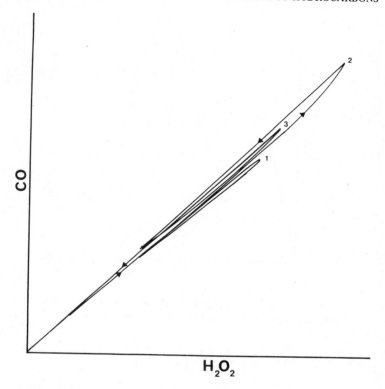

Figure 23. Phase portraits displaying the variation of concentrations of hydrogen peroxide and carbon monoxide during one cycle of a four-stage ignition in acetaldehyde oxidation, determined numerically from the thermally coupled kinetic scheme in Table XII.

spectrometry, the distinctions between "cause" and "effect" may be made with more confidence than has been possible hitherto. Instrumental responses have to be matched or appropriate allowance made in the experimental accumulation of x–T data.

2.   Computations Involving Reduced Kinetic Schemes

Provoked by concern over the limitation imposed on the efficiency of spark-ignition engines by spontaneous ignition and engine knock, Halstead, Kirsch, Prothero, and Quinn[165] sought a model for general applicability to hydrocarbon combustion. Their chemical model for two-stage

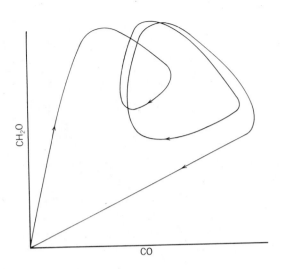

Figure 24. Phase portraits displaying the relationship between the concentrations of formaldehyde and carbon monoxide. Further details as for Fig. 23.

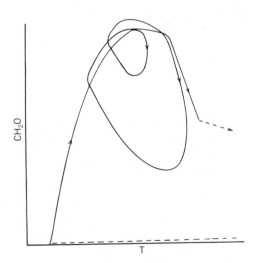

Figure 25. Calculated phase relationship between the concentration of formaldehyde and reactant temperature. Further details as for Fig. 23. The broken lines signify the departure to and return from the (adiabatic) temperature reached in the final stage.

288

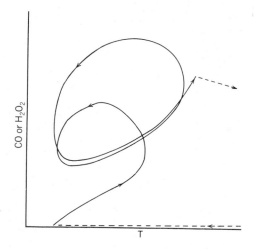

Figure 26.   Calculated phase relationship between the concentration of carbon monoxide and reactant temperature. The phase relationship shown in Fig. 23 establishes that hydrogen peroxide behaves in a fashion qualitatively similar to that shown here for carbon monoxide. Further details as for Fig. 25.

ignition is as follows (see also Fig. 27):

	$\log A\,(\text{s}^{-1}$ or m^3/mol/s)	$E\,(\text{kJ/mol})$	
$RH + O_2 \rightarrow 2R + product$	6	146	(6.21)
$R + O_2 \rightarrow RO_2$	6	0	(6.22)
$RO_2 \rightarrow products\ + OH$	12	63	(6.23)
$RO_2 \rightarrow Q + OH$	18.8	188	(6.24)
$OH + RH \rightarrow R + H_2O$	8	8.4	(6.25)
$RO_2 + RH \rightarrow RO_2H + R$	4	0	(6.26)
$RO_2 + O \rightarrow RO_2H + R$	8	83	(6.27)
$RO_2H \rightarrow 2R + products$	15	167	(6.28)
$R + R \rightarrow inert\ products$	9.3	0	(6.29)
$R + O_2 \rightarrow inert\ products$	6.2	42	(6.30)
$RO_2 + RH \rightarrow H_2O_2 + products + R$	16.9	105	(6.31)
$H_2O_2 \rightarrow 2OH$	18	188	(6.32)

Numerical integration of the conservation equations for eight species and for energy were carried out in a simulation of the high-pressure, closed conditions in an engine. Timescales to ignition were correctly predicted to be on the order of milliseconds; reversion to subatmospheric pressure stretched the simulated induction time more than 1000-fold, as

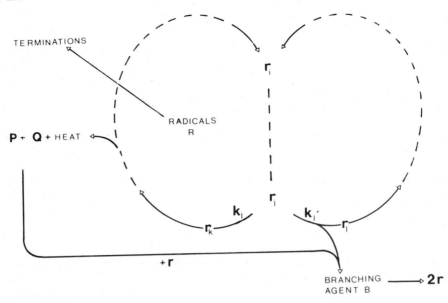

Figure 27. Schematic representation of the interacting thermally coupled kinetic cycles conceived by Halstead et al.[165] to represent spontaneous two-stage ignition of hydrocarbons. r (or r_i, etc.), principal propagating radicals; B, degenerate branching agent; P, final product; Q labile intermediates. The associated conservation equations involve only the species r, B, and Q and the energy-conservation equation.

is observed experimentally. The effect on the induction time of changes of the initial reactant temperature was well matched to experimental measurements in a rapid-compression apparatus. There is a clear manifestation of the negative temperature coefficient of the reaction rate in these results. Interest in engine knock has intensified in recent years, and because of it this work has received justifiable widespread attention as a chemical basis for predicting engine performance.[166]

Superficially, the scheme presented in Table VI has similarities to that adopted by Halstead et al., but there are important differences. First of all, the reactions of Table VI do not represent behavior beyond the realm of stationary reaction or oscillatory cool flames. If ignition is to be expected, the additional reactions of the type discussed in Section V.F and exemplified in Table XII have to be included. Halstead et al. have embraced two-stage ignition in their simplified scheme and, to minimize the numbers of species and steps, they have utilized one intermediate (ROOH) to control the first *and* second stages of ignition. Hydrogen peroxide decomposition alone does not furnish radicals sufficiently rapidly to give a satisfactorily responsive time evolution of the second stage of ignition. To accommodate the additional complexity, a peculiarly high activation energy (188 kJ/mol) is assigned to the formation of the intermediate Q [via (6.24)] to make the generation of ROOH possible [reaction (6.27)] at a second stage. To complement this, the rate-determining control of the first stage of formation of ROOH is thrown onto reaction (6.23) rather than that role being adopted more appropriately by (6.26). The need for this may well have been exacerbated by the neglect of the dissociation of RO_2, and thus the failure to provide the subtle balance between reactions of R and RO_2 that is engendered by the displacement of the equilibrium $R + O_2 \rightleftharpoons RO_2$ as the temperature increases. As discussed in Section V.E, the competitive reaction between R and O_2 [(6.30)] seems to facilitate an important additional route to formaldehyde through alkene oxidation. That formaldehyde can be

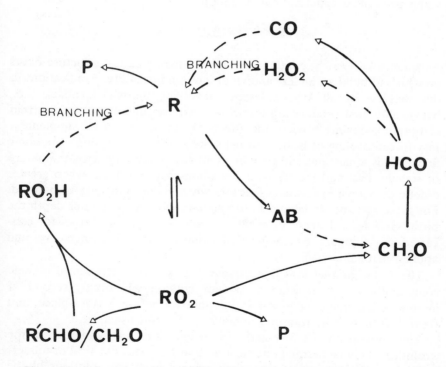

Figure 28. Schematic representation of thermally coupled kinetic cycles representing oscillatory, multiple-stage ignition phenomena of hydrocarbons. These emerge from the schemes displayed in Tables VI and XII. The broken lines signify indirect kinetic routes.

formed via reaction 5a (Table VI) and removed in reaction 6b parallels, in part, the role of Q assigned by Halstead et al., but neither the mechanistic detail nor the kinetic data incorporated by them permit a natural extension to the second, hot stage of ignition along the lines displayed in Table XII. A complementary general kinetic model, incorporating features of Tables VI and XII, may be exemplified by the cycles displayed in Fig. 28.

The requirement of engineers for a chemical model that is sufficiently simple to be incorporated into a full simulation of fluid dynamics and heat transfer in engines may preclude anything more complex than the two interacting branching cycles (Fig. 27) conceived by Halstead and co-workers.[165] However, the ab initio prediction of the effects of fuel structure does not follow readily from such a simple scheme. Despite its chemical pedigree, this model is used to best effect in an empirical mode, namely the derivation of the key elementary parameters by matching of predicted ignition delays for specific control fuels to those measured under prescribed experimental conditions.[167,168]

VII. CONCLUSIONS

A remarkably rich, continually evolving mathematical structure exists on which to build an understanding of ignition instability. Applications to the simplest of all kinetic systems involving thermal feedback (i.e., irreversible, first-order, exothermic systems) are fully developed; current elaborations mainly concern the effects of heat transport via supplementary forced cooling or heating of the system or the imposition of periodic fluctuations. Questions still remain about reaction multiplicity, the nature of singularities, and the types of transitions between them when greater kinetic complexity prevails, especially when chain branching is involved. For these systems we need also to understand in much greater depth the birth, growth, and death of oscillatory states. These are topics for concerted development by combustion chemists, chemical engineers, and applied mathematicians.

The technical questions that surround spontaneous ignition of organic compounds are, first, whether or not ignition will occur; second, if ignition is inevitable, what is the time interval before it takes place; and third, how may it be retarded, inhibited, or extinguished?

The answers to the first and second questions are bound up in the evolution of chain–thermal interactions, and whether analytical or numerical interpretations are sought, they emerge from the solutions of the mass- and energy-conservation equations. The criteria for criticality in the nonadiabatic c.s.t.r. have been developed quantitatively here, and when ignition occurs it may be either oscillatory or a stationary, high-temperature state. The rate of supply (or residence time) is the principal control parameter determining the nature of the upper state. Where there is a sustained reactant supply, but in less rigorously controlled circumstances than those of the c.s.t.r., true periodicity may be replaced by a more erratic, but nevertheless repetitive, ignition; a stabilized flame (the high-temperature state) remains an alternative possibility. When there is no replenishment of the reactant, explosion or ignition can occur only once and cannot be sustained.

Inhibition of spontaneous ignition is a major topic in itself. Nevertheless, broad principles emerge from the foregoing discussion. Control may be brought about by physical or chemical means, either by modification of heat-dissipation rates or by interference with the kinetics of reaction. We may infer that at low temperatures the rate of development of combustion of organic compounds is relatively slow (for example, during the induction time and even during the first stage of two-stage ignition). Characteristic chemical times are compatible with the characteristic times for heat dissipation, so physical control is a viable prospect.

On a chemical front, the retardation or inhibition of oxidation processes by lead alkyls as additives to gasoline is very familiar; they are believed to act via oxidation to lead oxide in finely dispersed particulate form in the engine cylinder, so providing during the induction time an extensive surface at which free radicals are lost and reaction chains terminated.[169]

In the second stage of two-stage ignition, or in single-stage ignition initiated at temperatures above ~ 850 K, the rate of kinetic development far outstrips the ability of the system to dissipate heat. Adiabatic temperature changes occur in consequence, driven by the very strong chemical autocatalysis via H, O, and OH atoms and radicals. The only hope for retardation or inhibition is thus to modify the chain reaction by competitive removal of the principal propagating species. Organohalogen compounds are effective for this purpose and they are believed to play a part in reactions of the type[104,170]

$$CF_3Br + \begin{cases} H \\ O \end{cases} \rightarrow CF_3 + \begin{cases} HBr \\ BrO \end{cases}$$

$$CF_3 + \begin{cases} H & \rightarrow CF_3H \\ O \\ OH & \rightarrow CF_2O + \begin{cases} F \\ HF \end{cases} \end{cases}$$

The halogens are reduced eventually to F_2 and Br_2 via multiple consecutive or competitive elementary steps.

How hydrocarbon structure may affect the evolution of spontaneous ignition has been exemplified in the present text with reference to the prototype isomeric structures $CH_3CH_2CH_2CH_3$ and $CH_3CH(CH_3)CH_3$. At temperatures below ~ 850 K the distinctions emerge principally through competitive and consecutive steps of the following kind (numbered as in Table VI)

$$R + O_2 \rightarrow AB + HO_2 \qquad\qquad 2$$

$$R + O_2 \rightleftharpoons RO_2 \qquad\qquad 3/4$$

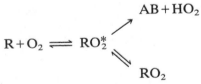

$$\begin{array}{ll} & R'CHO + OH + P_1 \qquad 5a \\ RO_2 \longrightarrow QOOH & \\ & AO + OH + P_2 \qquad 5b \end{array}$$

$$RO_2 + R'CHO \rightarrow RO_2H + R'CO \qquad\qquad 6b$$

$$RO_2H \rightarrow RO + OH \qquad\qquad 8$$

The predominant modes are controlled by the nature of the alkyl moiety R. At the present time (mid-1984) this is an acceptable basis and may serve as the most convenient qualitative expression of such interactions. However, extremely important questions are being raised concerning the mechanisms of reactions 2 and 3/4 and the rate constants assigned to them. There is evidence, at least for ethyl and propyl radicals at temperatures up to 850 K, to suggest that the route to the conjugate alkene occurs via[171]

$$\begin{array}{ll} & AB + HO_2 \\ R + O_2 \rightleftharpoons RO_2^* & \\ & RO_2 \end{array}$$

Interpretations of the reactions of larger alkyl radicals, similarly modified, may have quantitative repercussions with respect to the prediction of spontaneous ignition of alkanes at low temperatures. The qualitative features that give rise to a negative temperature coefficient of the reaction rate and the principles of oscillatory cool-flame phenomena and complex ignitions would still pertain.

Even at rather higher temperatures ($T > 1000$ K), the ignition-delay times for normal and isobutane differ. These distinctions must arise from different routes for degradation of the hydrocarbon structure. A detailed interpretation is outside the scope of the present text; it is the subject of current numerical investigations.[172]

To these comparisons and contrasts of effects arising from different isomeric structural forms must be added the effects of (1) unsaturated linkages in alkenes, conjugated dienes, or acetylenes; (2) cyclic structures; and (3) aromaticity. Far less attention has been given in the past to compounds in these classes than to the saturated chain structures. However, they are the subject of current interest, especially in relation to the safe and efficient use of gasoline and diesel fuel and of extracts from coal. The state of present knowledge is documented in the proceedings of the Nineteenth and Twentieth International Symposia on Combustion (1982 and 1984, respectively).

The following points may be made in conclusion.

1. Hydrocarbons of all types ignite spontaneously above ~750 K; although heat release plays a part at these temperatures, feedback is dominated by H- and O-atom branching reactions. That different reactivities are observed (e.g., as measured by the critical reactant pressure in a closed or flow vessel or by the ignition-delay interval) reflects an influence of reactant structure. The destructions of different types of structures (i.e., straight or branched chains, aliphatic or aromatic rings) do not occur equally readily, even at these temperatures.

2. Above 1000 K, pyrolysis of the hydrocarbon reactant predominates over the oxidative degradation that is associated with lower temperatures. Distinctions between different substrates thus become less marked.

3. Spontaneous ignition brought about at temperatures below ~750 K is due to free-radical oxidation involving facile H-atom transfer by intermolecular and intramolecular routes, as described in Section V. Chain branching and heat release occur in consequence. The dependence of the ease of ignition on carbon-chain (or side-chain) lengths and on the numbers of primary, secondary, or tertiary C–H bonds is reflected in the "knock ratings" of hydrocarbon fuels.[173]

4. Hydrocarbons precluded from routes to low-temperature thermal ignition because they cannot either offer labile H atoms or undergo internal rearrangements with ease include methane, ethane, ethene, benzene, and toluene. Limited reactivity is also conferred on propane, because of its short carbon backbone and its high proportion of primary C–H bonds. Propene oxidation may also be regarded as atypical, since resonance stabilization is conferred on the allyl radical. Among higher

hydrocarbons, neopentene $[(CH_3)_4C]$ is unusual; it offers only primary C–H bonds and there is no conjugate alkene of the neopentyl radical. Aliphatic cyclic hydrocarbons are generally less reactive than their open-chain counterparts; cyclobutane would seem to be an exception to this rule.[174]

5.   To predict the occurrence of spontaneous ignition, it is necessary to establish quantitatively not only detailed kinetic mechanisms and associated rate constants and exothermicities, but also modes of heat dissipation to the surroundings. Progress in the quantitative interpretation of ignition of specific organic substrates has been made within the last decade by recourse to the use of well-stirred flow reactors, for which heat-dissipation rates can be characterized and controlled most easily.

6.   To understand spontaneously ignition in less rigorously (and simply) defined circumstances, it is essential to simplify the kinetic interpretations. A current objective is to seek kinetic generalizations that nonetheless allow for the differing reactivity of each substrate; guidelines are established in this review.

7.   There is no doubt that numerical studies have contributed to the present state of understanding and may be expected to do so to a still greater degree through intelligent-computer interpretation of kinetics and of complex heat- and mass-transport phenomena. However, parallel evolution of analytical approaches to nonlinear dynamics is essential, even if simplifications and reductions to limited numbers of dimensions are a prerequisite. "Hit or miss' probing of parameter space by numerical methods is a grossly uneconomical procedure, and the import of numerical output cannot be appreciated without fundamental insight into the implications of coupling between the nonlinear characteristics of the system.

APPENDIX: LIST OF SYMBOLS

B	Dimensionless adiabatic temperature excess
B^*	Maximum dimensionless stationary temperature in the non adiabatic c.s.t.r. [see (2.41)]
c_p (J/mol K)	Heat capacity of reactants
Da	Damkohler number ($= t_{res}/t_{ch}$)
E_i (J/mol)	Activation energy of ith reaction
f (m^3/s)	Volumeric flow rate
h_i (J/mol)	Exothermicity of ith reaction
ΔH (J/mol)	Molar enthalpy change
K_p (mm Hg)$^{-1}$	Equilibrium constant for $R + O_2 \rightleftharpoons RO_2$
k_i	Rate constant of ith reaction

$L(\theta)$, l (W/K)	Heat-loss-rate term
P (mm Hg)	Total pressure of reactants
$P(i)$ (mm Hg)	Partial pressure of ith component
q (J/mol)	Molar exothermicity
$R(\theta)$ (W)	Heat-release-rate term
R_i (W)	Heat-release rate from initiating process
R (J/mol K)	Gas constant
S (m^2)	Vessel surface area
T (K)	Reactant temperature
T_0 (K)	Reactant inlet temperature
T_a (K)	Vessel-wall temperature
T_{ad} (K)	Adiabatic temperature
T_* (K)	Normalized reference temperature for nonadiabatic c.s.t.r.
t (s)	Time
t_{ch} (s)	Characteristic chemical time at T_* ($\equiv$ reciprocal of pseudo-first-order rate constant)
t_{res} (s)	Mean residence time in the c.s.t.r.
t_N (s)	Characteristic Newtonian heat-transfer time
V (m^3)	Reactor volume
v_i (mol/s)	Rate of initiating process
x, [X] (mol/m^3)	Concentration of species X
γ	Fractional extent of conversion ($= 1 - x/x_0$)
ε	Reciprocal of reduced activation energy ($= RT_*/E$)
θ	Dimensionless temperature excess [$= (T - T_*)E/RT_*^2$]
θ_{ign}, θ_-	Dimensionless critical temperature excess for ignition
θ_{ext}, θ_+	Dimensionless critical temperature excess for extinction
λ	Root of characteristic equation for singularities in two-dimensional phase plane
σ (mol/m^3)	Molar density
χ (W/m^2 K)	Heat-transfer coefficient
ϕ (s^{-1})	Net branching factor ($= k_t - k_b$)
ψ (W/mol)	Summation of molar heat-release rates

Acknowledgments

It is my pleasure to thank Professor P. Gray for many enjoyable and stimulating years of collaborative research. Long-standing associations with Professor B. F. Gray, Professor P. G. Lignola, and Dr. S. M. Hasko and the support of research, technical, and secretarial staffs of the University of Leeds are gratefully acknowledged. Research at Leeds contributing to this review has been funded by SERC, UKAEA and British Gas. Travel grants have been provided by NATO, the EEC, the University of Leeds and The Combustion Institute, British Section.

References

1. T. Ahmad and G. M. Faeth, in *Seventeenth Symp. (Int.) on Combust.*, The Combustion Institute, Pittsburgh, 1978, p. 1149.

2. P. C. Bowes, *Self-Heating: Evaluation and Controlling the Hazards*, Her Majesty's Stationery Office, London, 1984.

3. G. Dixon-Lewis and I. G. Shepherd, in *Fifteenth Symp. (Int.) on Combust*, The Combustion Institute, Pittsburgh, 1974, p. 1483.

4. B. P. Mullins and S. S. Penner, *Explosions, Detonations, Flammability and Ignition*, Pergamon, London, 1959.

5. R. A. Schmitz, in *Chemical Reaction Engineering Reviews* (Adv. in Chem. Ser. 148), H. M. Hurlburt, Ed., American Chemical Society, Washington, D.C., 1975, p. 156.

6. A. G. Gaydon and H. G. Wolfhard, *Flames, Their Structure, Radiation and Temperature*, Chapman and Hall, London, 1970.

7. J. Warnatz, *Combust. Sci. Technol.* **34,** 177 (1983); *Ber. Bunsen Phys. Chem.* **87,** 1008 (1983).

8. C. K. Westbrook and F. L. Dryer, *Progress in Energy and Combustion Science*, **10,** 1 (1984).

9. C. K. Westbrook, F. L. Dryer, and K. P. Schug, in *Nineteenth Symp. (Int.) on Combust.*, The Combustion Institute, Pittsburgh, 1982, p. 153.

10. G. J. Minkoff and C. F. H. Tipper, *Chemistry of Combustion Reactions*, Butterworths, London, 1962.

11. V. Ya. Shtern, *The Gas-Phase Oxidation of Hydrocarbons* (Engl. Trans. M. F. Mullins), Pergamon, London, 1964.

12. J. N. Bradley, *Flame and Combustion Phenomena*, Methuen, London, 1969.

13. R. T. Pollard, in *Comprehensive Chemical Kinetics*, Vol. 17, C. H. Bamford and C. F. H. Tipper, Eds., Elsevier, Amsterdam, 1977, p. 249.

14. J. A. Barnard, in *Comprehensive Chemical Kinetics*, Vol. 17, C. H. Bamford and C. F. H. Tipper, Eds., Elsevier, Amsterdam, 1977, p. 441.

15. N. N. Semenov, *Chemical Kinetics and Chain Reactions*, Oxford Univ. Press, Oxford, 1935.

16. C. N. Hinshelwood, *The Kinetics of Chemical Change*, Oxford Univ. Press, Oxford, 1942.

17. A. B. Nalbandyan and A. A. Mantashyan, *Elementary Processes in Slow Gas Phase Reactions*, Armenian Academy of Sciences, Erevan, 1975 (in Russian).

18. M. Prettre, P. Dumanois, and P. Laffitte, *Comptes Rendus Acad. Sci.* **191,** 329, 414 (1930).

19. R. Ben-Aim and M. Lucquin, *Oxidation and Combustion Reviews*, Vol. 1, C. F. H. Tipper, Ed., Elsevier, Amsterdam, 1965.

20. P. G. Ashmore, F. S. Dainton, and T. M. Sugden, Eds., *Photochemistry and Reaction Kinetics*, Cambridge Univ. Press, Cambridge, 1967.

21. R. N. Pease, *Equilibrium and Kinetics of Gas Reactions*, Princeton Univ. Press, Princeton, 1942.

22. D. Indritz, H. A. Rabitz, and F. W. Williams, *J. Phys. Chem* **81,** 2526 (1977).

23. C. Venkat, K. Brezinsky, and I. Glassman, *Nineteenth Symp. (Int.) on Combust.*, The Combustion Institute, Pittsburgh, 1982, p. 143.

24. J. R. Smith, R. M. Green, C. K. Westbrook, and W. J. Pitz, *Twentieth Symp. (Int.) on Combust.*, The Combustion Institute, Pittsburgh, in press.
25. B. Lewis and G. von Elbe, *Combustion, Flames and Explosions of Gases*, Academic, New York, 1961.
26. S. W. Benson, *The Foundations of Chemical Kinetics*, McGraw-Hill, New York, 1960.
27. M. F. R. Mulcahy, *Gas Kinetics*, Nelson, London 1973.
28. H. Davy, *Phil. Trans. Roy. Soc. London* **107**, 77 (1817).
29. W. H. Perkin, *J. Chem. Soc.* **41**, 363 (1882).
30. J. Bardwell and C. N. Hinshelwood, *Proc. Roy Soc. A* **205**, 375 (1951).
31. L. R. Sochet, J. P. Sawerysyn, and M. Lucquin, in *Oxidation of Organic Compounds—II* (Adv. in Chem. Ser. 76), F. R. Mayo, Ed., American Chemical Society, Washington, D. C., 1968, p. 111.
32. F. E. Malherbe and A. D. Walsh, *Trans. Faraday Soc.* **46**, 824 (1950).
33. I. E. Salnikov, *Zh. Fiz. Khim.* **23**, 258 (1949).
34. A. G. Merzhanov and V. G. Abramov, *Chem. Eng. Sci.* **32**, 475 (1977).
35. Ya. B. Zel'dovich, *Zh. Tekh. Fiz.* **11**, 493 (1941).
36. Ya. B. Zel'dovich and Y. A. Zysin, *Zh. Tekh. Fiz.* **11**, 501 (1941).
37. R. Aris and N. R. Amundson, *Chem. Eng. Sci.* **7**, 121 (1958).
38. B. F. Gray and C. H. Yang, *J. Phys. Chem.* **69**, 2747 (1965).
39. C. H. Yang and B. F. Gray, in *Eleventh Symp. (Int.) on Combust.*, The Combustion Institute, Pittsburgh, 1967, p. 1099.
40. C. H. Yang and B. F. Gray, *Trans. Faraday Soc.* **65**, 1614 (1969); *J. Phys. Chem.* **73**, 3395 (1969).
41. C. H. Yang, *J. Phys. Chem.* **73**, 3407 (1969).
42. B. F. Gray, *Trans. Faraday Soc.* **65**, 1603, 2133 (1969); *ibid.*, in *Reaction Kinetics*, Vol. 1, P. G. Ashmore, Ed., The Chemical Society, London, 1975, p. 309.
43. G. Nicolis and F. Baras, Eds., *Chemical Instabilities*, Reidel, Dordrecht, The Netherlands, 1983.
44. R. J. Field and M. Burger, Eds., *Oscillations and Traveling Waves in Chemical Systems*, Wiley, New York, 1985.
45. J. P. Longwell, E. E. Frost, and M. A. Weiss, *Ind. Eng. Chem.* **45**, 1629 (1953).
46. O. Bilous and N. R. Amundson, *AIChE J.* **1**, 513 (1955).
47. A. Uppal, W. H. Ray, and A. B. Poore, *Chem. Eng. Sci.* **29**, 967 (1974); *Chem. Eng. Sci.* **31**, 205 (1976).
48. D. A. Frank-Kamenetskii, *Diffusion and Heat Transfer in Chemical Kinetics* (Engl. Trans. J. P. Appleton), Plenum, New York, 1969.
49. D. R. Jenkins, V. S. Yumlu, and D. B. Spalding, *Eleventh Symp. (Int.) on Combust.*, The Combustion Institute, Pittsburgh, 1967, p. 779.
50. R. A. Schmitz, R. R. Bautz, W. H. Ray, and A. Uppal, *AIChE J.* **25**, 289 (1979).
51. A. H. Heemskerk and J. M. H. Fortuijn, *Chem. Eng. Sci.* **38**, 1261 (1983).
52. S. F. Bush, *Proc. Roy. Soc. A* **309**, 1 (1969).
53. P. Gray, J. F. Griffiths, S. M. Hasko, and J. R. Mullins, *Int. Symp. on Chem. Reaction Eng.* **8**, 101 (1984).

54. J. F. Griffiths and H. J. Singh, *J. Chem. Soc. Faraday Trans. I* **78,** 747 (1982).

55. A. A. Andronov, A. A. Vitt, and S. E. Khaikin, *Theory of Oscillators* (Engl. Trans. F. Immirzi), Pergamon, London, 1966.

56. M. E. Levy, A. E. Cerkanowicz, and R. F. McAlevy III, in *AIAA 7th Aerospace Sciences Meeting*, American Institute of Aeronautics and Astronautics, New York, 1969, Paper 69–88.

57. T. Furusawa and H. Nishimura, *J. Chem. Eng. Sci. Prog. (Japan)* **1,** 180 (1967).

58. V. Balakotaiah and D. Luss, *Chem. Eng. Commun.* **13,** 111 (1981).

59. V. Caprio, A. Insola, and P. G. Lignola, (a) in *Sixteenth Symp. (Int.) on Combust.*, The Combustion Institute, Pittsburgh, 1977, p. 1155; (b) *Combust. Flame* **43,** 23 (1981); (c) *Archivum Combustionis* **3,** 27 (1983).

60. P. Gray, J. F. Griffiths, S. M. Hasko, and P. G. Lignola, (a) *Proc. Roy. Soc. A* **374,** 313 (1981); (b) *Combust. Flame* **43,** 175 (1981).

61. B. F. Gray, J. F. Griffiths, and J. C. Jones, *Fuel* **63,** 43 (1984).

62. B. D. Hazzard, N. D. Kazarinov, and Y.-H. Wan, *Theory and Applications of Hopf Bifurcation*, Cambridge Univ. Press, Cambridge, 1981.

63. H. T. Davies, *Introduction to Non-linear Differential and Integral Equations*, Dover, New York, 1962.

64. J. F. Griffiths and K. Hasegawa, *Combust. Flame* **45,** 53 (1982).

65. A. J. Lotka, *J. Am. Chem. Soc.* **43,** 1595 (1920).

66. I. Prigogine and R. Lefever, *J. Chem. Phys.* **48,** 1965 (1968).

67. R. J. Field and R. M. Noyes, *J. Chem. Phys.* **60,** 1877 (1974).

68. I. R. Epstein, *J. Phys. Chem.* **88,** 187 (1984).

69. P. Gray and S. K. Scott, *J. Phys. Chem.* **87,** 1835 (1983).

70. P. Gray and S. K. Scott, *Ber. Bunsen Phys. Chem.* **87,** 379 (1983).

71. P. Gray and S. K. Scott, *Chemical Instabilities* G. Nicolis and F. Baras, Eds., Reidel, Dordrecht, The Netherlands, 1983, p. 69.

72. P. Gray and S. K. Scott, *Chem. Eng. Sci.* **38,** 29 (1983).

73. S. K. Scott, *Chem. Eng. Sci.* **38,** 1701 (1983).

74. P. Gray and S. K. Scott, *J. Phys. Chem,* **89,** 21 (1985).

75. P. Gray, J. F. Griffiths, and S. M. Hasko, *Proc. Roy. Soc. A* **396,** 227 (1984).

76. P. G. Lignola, V. Caprio, A. Insola, and G. Mondini, *Ber. Bunsen Phys. Chem.* **84,** 369 (1980).

77. B. F. Gray and P. G. Felton, *Combust. Flame* **23,** 295 (1974).

78. V. Caprio, A. Insola, and P. G. Lignola, in *First Specialists' Meeting of Combust. Inst.*, The Combustion Institute French Section, Bordeaux, 1981, p. 378.

79. P. G. Lignola, E. Reverchon, and R. Piro, in *Twentieth Symp. (Int.) on Combust.*, The Combustion Institute, Pittsburgh, in press.

80. P. G. Lignola and E. Reverchon, *Combust. Flame*, in press.

81. B. F. Gray and J. C. Jones, *Combust. Flame* **57,** 3 (1984).

82. J. F. Griffiths and S. M. Hasko, to be published.

83. J. P. Longwell and M. A. Weiss, *Ind. Eng. Chem. (Ind.)* **47,** 1634 (1955).

84. Ya. G. Gervart and D. A. Frank-Kamenetskii, *Izv. Akad. Nauk, SSSR*, 210 (1942).

85. J. Dutton, *A Study of Oscillatory Processes in a Continuous System*, Ph.D. Thesis, London University, London, 1968.

86. B. F. Gray and J. F. Griffiths, *Chem. Commun.*, 1391 (1969).
87. B. F. Gray, P. Gray, and J. F. Griffiths, in *Thirteenth Symp. (Int.) on Combust.*, The Combustion Institute, Pittsburgh, 1971, p. 239.
88. K. G. Williams, J. E. Johnson, and H. W. Carhart, in *Seventh Symp. (Int.) on Combust.*, The Combustion Institute, Pittsburgh, 1959, p. 592.
89. J. N. Bradley, G. A. Jones, G. Skirrow, and C. F. H. Tipper, in *Tenth Symp. (Int.) on Combust.*, The Combustion Institute, Pittsburgh, 1965, p. 139.
90. F. W. Williams, D. Indritz, and R. S. Sheinson, *Combust. Sci. Technol.* **11**, 67 (1975).
91. J. C. Pope, F. J. Dykstra, and G. Edgar, *J. Am. Chem Soc.* **51**, 1875, 2203, 2213 (1929).
92. R. J. Foresti, *Fifth Symp. (Int.) on Combust.*, The Combustion Institute, Pittsburgh, 1959, p. 582.
93. W. G. Agnew and J. T. Agnew, *Tenth Symp. (Int.) on Combust.*, The Combustion Institute, Pittsburgh, 1965, p. 123.
94. L. R. Cairnie, A. J. Harrison, and R. Summers, in *First Specialists' Meeting (Int.) of Combust. Inst.*, The Combustion Institute French Section, Paris, 1981, p. 366.
95. D. F. Cooke and A. Williams, *Combust Flame* **24**, 245 (1975).
96. J. F. Griffiths and S. M. Hasko, *Proc. Roy. Soc. A* **393**, 371 (1984).
97. A. Fish, W. W. Haskell, and I. A. Read, *Proc. Roy. Soc. A* **323**, 261 (1969).
98. A. S. Sokolik, *Self Ignition, Flame and Detonation in Gases*, Israel Program for Scientific Translations, Jerusalem, 1963.
99. J. F. Griffiths, G. Skirrow, and C. F. H. Tipper, *Combust. Flame* **13**, 195 (1969).
100. J. G. Calvert and J. N. Pitts, *Photochemistry*, Wiley, New York, 1966.
101. A. G. Gaydon, *The Spectroscopy of Flames*, Chapman and Hall, London, 1974.
102. R. W. Sheinson and F. W. Williams, *Combust. Flame* **21**, 221 (1973).
103. J. C. Biordi, C. P. Lazzara, and J. F. Papp, in *Sixteenth Symp. (Int.) on Combust.*, The Combustion Institute, Pittsburgh, 1976, p. 1097.
104. H. Y. Safieh, J. Vandorren, and P. J. van Tiggelen, in *Nineteenth Symp. (Int) on Combust.*, The Combustion Institute, Pittsburgh, 1982, p. 117.
105. E. A. Poladyan, G. K. Grigoryan, L. A. Kachatryan, and A. A. Mantashyan, *Kinetics and Catalysis* **17**, 265 (1976) (Engl. Trans.).
106. A. B. Nalbandyan, E. A. Oganessyan, I. A. Vardanyan, and J. F. Griffiths, *J. Chem. Soc. Faraday Trans. I* **71**, 1203 (1975).
107. M. Carlier and L. R. Sochet, *Combust. Flame* **25**, 309 (1975).
108. P. G. Ashmore, B. J. Tyler, and T. A. R. Wesley, in *Eleventh Symp. (Int.) on Combust.*, The Combustion Institute, Pittsburgh, 1967, p. 1133.
109. D. H. Fine, P. Gray, and R. Mackinven, in *Twelfth Symp. (Int.) on Combust.*, The Combustion Institute, Pittsburgh, 1969, p. 545.
110. R. Hughes and R. F. Simmons, in *Twelfth Symp. (Int.) on Combust.*, The Combustion Institute, Pittsburgh, 1969, p. 449.
111. J. A. Barnard and A. Watts, in *Twelfth Symp. (Int.) on Combust.*, The Combustion Institute, Pittsburgh, 1969, p. 365.
112. P. Gray, J. F. Griffiths, and S. K. Scott, *Proc. Roy. Soc. A*, **397**, 21 (1985).
113. J. F. Griffiths, unpublished results.
114. J. Chamboux and M. Lucquin, *J. Chim. Phys.* **59**, 797 (1962).

115. P. G. Felton, M.Sc. Dissertation, Univ. of Leeds, Leeds, 1971.

116. G. A. Luckett and R. T. Pollard, *Combust. Flame* **21,** 265 (1973).

117. A. J. Brown, N. Burt, G. A. Luckett, and R. T. Pollard, in *Symposium on the Mechanisms of Hydrocarbon Reactions*, Eds. F. Marta and D. Kallo, Elsevier, Amsterdam, 1975.

118. C. T. Bowman, in *Fifteenth Symp. (Int.) on Combust.*, The Combustion Institute, Pittsburgh, 1974, p. 869.

119. G. A. Luckett, unpublished results.

120. J. F. Griffiths, unpublished results.

121. J. W. Falconer and J. H. Knox, *Proc. Roy. Soc. A* **250,** 453 (1959).

122. T. Berry, C. F. Cullis, and D. L. Trimm, *Proc. Roy. Soc. A* **316,** 377 (1970).

123. J. A. Barnard and A. W. Brench, *Combust. Sci. Tech.* **15,** 243 (1977).

124. J. G. Atherton, A. J. Brown, G. A. Luckett, and R. T. Pollard, in *Fourteenth Symp. (Int.) on Combust.*, The Combustion Institute, Pittsburgh, 1973, p. 513.

125. A. Fish, *Combust. Flame* **13,** 23 (1969).

126. D. D. Drysdale and R. G. W. Norrish, *Proc. Roy. Soc. A* **308,** 305 (1969).

127. C. F. Cullis, A. Fish, M. Saeed, and D. L. Trimm, *Proc. Roy. Soc. A* **289,** 402 (1966).

128. A. P. Zeelenberg and H. W. de Bruijn, *Combust. Flame* **9,** 281 (1965).

129. A. Fish, *Proc. Roy. Soc. A* **298,** 204 (1967).

130. P. Barat, C. F. Cullis, and R. T. Pollard, *Proc. Roy. Soc. A* **325,** 469 (1971).

131. A. Fish and J. P. Wilson, in *Thirteenth Symp. (Int.) on Combust.*, The Combustion Institute, Pittsburgh, 1971, p. 229.

132. C. J. Luck, A. R. Burgess, D. H. Desty, D. M. Whitehead, and G. Pratley, in *Fourteenth Symp. (Int.) on Combust.*, The Combustion Institute, Pittsburgh, 1973, p. 501.

133. P. Barat, C. F. Cullis, and R. T. Pollard, *Proc. Roy. Soc. A* **329,** 443 (1972).

134. C. F. Cullis, M. M. Hirschler, and R. L. Rogers, *Proc. Roy. Soc. A* **382,** 429 (1982).

135. D. J. M. Ray, R. Ruiz Diaz, and D. J. Waddington, in *Fourteenth Symp. (Int.) on Combustion.*, The Combustion Institute, Pittsburgh, 1973, p. 259.

136. R. R. Baldwin, D. E. Hopkins, A. C. Norris, and R. W. Walker, *Combust. Flame* **15,** 33 (1970).

137. R. R. Baldwin, C. J. Everett, D. E. Hopkins, and R. W. Walker, *Trans. Faraday Soc.* **66,** 189 (1970).

138. R. R. Baker, R. R. Baldwin, and R. W. Walker, (a) *Trans. Faraday Soc.* **66,** 2812 (1970); (b) *J. Chem. Soc. Faraday Trans. I* **74,** 2229 (1978).

139. R. R. Baker, R. R. Baldwin, A. R. Fuller, and R. W. Walker, *Trans. Faraday Soc.* **71,** 736 (1975).

140. R. R. Baldwin and R. W. Walker, *J. Chem. Soc. Faraday Trans. I* **75,** 140 (1979).

141. B. H. Bonner and C. F. H. Tipper, *Combust. Flame* **9,** 387 (1965).

142. T. Mill, F. Mayo, H. Richardson, K. Irwin, and D. L. Allara, *J. Am. Chem. Soc.* **94,** 6802 (1972).

143. D. M. Brown and A. Fish, *Proc. Roy. Soc. A* **308,** 547 (1969).

144. R. W. Walker, in *Reaction Kinetics*, Vol. 1, P. G. Ashmore, Ed., The Chemical Society, London, 1975, p. 161.

145. R. W. Walker, in *Gas Kinetics and Energy Transfer*, Vol. 2, P. G. Ashmore and R. P. Donovan, Eds., The Chemical Society, London, 1976, p. 296.

146. S. W. Benson, *J. Am. Chem. Soc.* **87,** 972 (1965).

147. A. Fish, *Quart. Rev.* **XVIII,** 243 (1964); *Oxidation of Organic Compounds—II* (Adv. in Chem. Ser. 76), F. R. Mayo, Ed., American Chemical Society, Washington, D.C., 1968, p. 69.

148. T. Berry, C. F. Cullis, M. Saeed, and D. L. Trimm, in *Oxidation of Organic Compounds—II* (Adv. in Chem. Ser. 76), F. R. Mayo, Ed., American Chemical Society, Washington, D.C., 1968, p. 86.

149. S. W. Benson, in *Oxidation of Organic Compounds—II* (Adv. in Chem. Ser. 76), F. R. Mayo, Ed., American Chemical Society, Washington, D.C., 1968, p. 143; *Prog. Energy Combust. Sci* **7,** 125 (1981).

150. J. H. Knox, in *Oxidation of Organic Compounds—II* (Adv. in Chem Ser. 76), F. R. Mayo, Ed., American Chemical Society, Washington, D.C., 1968, p. 1.

151. R. R. Baldwin, M. W. M. Hisham, and R. W. Walker, *J. Chem. Soc. Faraday Trans. I* **78,** 1615 (1982).

152. D. H. Slater and J. G. Calvert, in *Oxidation of Organic Compounds—II* (Adv. Chem. Ser. 76), F. R. Mayo, Ed., American Chemical Society, Washington, D.C., 1968, p. 58.

153. R. R. Baldwin and R. W. Walker, in *Eighteenth Symp. (Int.) on Combust.,* The Combustion Institute, Pittsburgh, 1981, p. 819.

154. D. Malcolm, Ph.D. Thesis, Hull University, Hull, 1982.

155. J. R. Bond, P. Gray, and J. F. Griffiths, in *Seventeenth Symp. (Int.) on Combust.,* The Combustion Institute, Pittsburgh, 1978, p. 811.

156. I. E. Salnikov and B. V. Vol'ter, *Dokl. Akad. Nauk. SSSR* **152,** 171 (1963).

157. P. G. Felton, B. F. Gray, and N. Shank, *Combust. Flame* **27,** 363 (1976).

158. V. I. Vedeneev, U. M. Gershenson, and O. M. Sarkisov, *Armenian J. Chem.* **20,** 968 (1967).

159. I. Gonda and B. F. Gray, *Proc. Roy. Soc. A* **389,** 133 (1983).

160. M. P. Halstead, A. Prothero, and C. P. Quinn, *Proc. Roy. Soc. A* **322,** 377 (1971).

161. J. F. Griffiths, G. Skirrow, and C. F. H. Tipper, *Combust. Flame* **12,** 360 (1968).

162. M. P. Halstead, A. Prothero, and C. P. Quinn, *Combust. Flame* **20,** 211 (1973).

163. B. F. Gray, P. G. Felton, and N. Shank, in *Second Eur. Combust. Sympos.,* R. Delbourgo Ed., The Combustion Institute French Section, Orleans, 1975, p. 103.

164. C. Gibson, P. Gray, J. F. Griffiths, and S. M. Hasko, in *Twentieth Symp. (Int.) on Combust.,* The Combustion Institute, Pittsburgh, in press.

165. M. P. Halstead, L. J. Kirsch, A. Prothero, and C. P. Quinn, *Proc. Roy. Soc. A* **346,** 515 (1975).

166. B. Natarajan, M.Sc. Dissertation, Princeton Univ., Princeton, 1983.

167. M. P. Halstead, L. J. Kirsch, and C. P. Quinn, *Combust. Flame* **30,** 45 (1977).

168. L. J. Kirsch and C. P. Quinn, in *Sixteenth Symp. (Int.) on Combust.,* The Combustion Institute, Pittsburgh, 1976, p. 233.

169. A. D. Walsh, in *Low Temperature Oxidation*, W. Jost, Ed., Gordon & Breach, New York, 1965, p. 329.

170. C. K. Westbrook, in *Nineteenth Symp. (Int.) on Combust.,* The Combustion Institute, Pittsburgh, 1982, p. 127.

171. I. R. Slagle, J.-Y. Park, and D. Gutman, in *Twentieth Symp. (Int.) on Combust.*, The Combustion Institute, Pittsburgh, in press.

172. C. K. Westbrook, to be published.

173. H. K. Livingston, *Ind. Eng. Chem.* **43,** 2834 (1951).

174. L. Kuhn, personal communication.

AUTHOR INDEX

Numbers in parentheses are reference numbers and indicate that the author's work is referred to although his name is not mentioned in the text. Numbers in *italics* show the pages on which the complete references are listed.

305

314

Walkafen, G. E., 5(8), 28(8), *43*
Walker, R. W., 259(138, 139), 260(136–140), 264(144, 145), 265(144), 267(151), 268(151), 270(140, 145), 271(153), *302, 303*
Wallenstein, M. B., 119(38), 121(38), 136(38), *197*
Walsh, A. D., 207(32), 293(164), *299, 303*
Walters, E. A., 174(171), *202*
Wan, Y.-H., 226(62), 228(62), *300*
Wanderlingh, F., 28(44), 32(44), 33(44), 34(56), 35(56), *44*
Wang, M. C., 95(38), *108*
Wankenne, H., 125(67, 125), *198*
Warhaftig, A. L., 119(38), 121(38), 136(38), *197*
Warnatz, J., 206(7), 256(7), *298*
Watts, A., 243(111), *301*
Weber, R., 175(173), *202*
Weeks, J. D., 50(13), 81(28), 93(29), 94(30), 95(30), 97(29, 30, 36), 103(30, 36), (31), (35), *108*
Weiss, M., 140(103), 163(157), *199, 201*
Weiss, M. A., 210(45), 237(83), *299, 300*
Weiss, M. J., 126(73), 140(103), *198, 199*
Welge, K. H., 113(9), *196*
Wenzel, J. T., 16(28), *43*
Werner, A. S., 115(17), 148(129), 150(129), 162(156), 182(184), *197, 200, 201*
Wertheim, M. A., 98–100(39), *109*
Wesley, T. A. R., 243(108), *301*
Westbrook, C. K., 206(8, 9), 207(24), 256(8), 272(8, 9), 293(120), 295(172), *298, 299, 303, 304*
Weston, R. E., 136(94), *199*
White, M. G., 117(29), *197*
Whitehead, D. M., 257(132), *302*
Whitehead, J. C., 112(3), *196*
Whitten, G. Z., 119(40), *197*
Wiafe-Akenten, J., 28(44), 32(44), 33(44), *44*
Widon, B., 50(14), *108*
Wigner, E., 126(75), *198*

Willett, G. D., 148(124, 131), 151(124), 152(124), 153(143), 155(147), 156(142), *200, 201*
Williams, A., 240(95), *301*
Williams, F. W., 207(22), 240(90), 242(102), 243(102), *298, 301*
Williams, K. G., 240(88), *301*
Wilson, J. P., 257(131), *302*
Wilson, K. R., 189(195), *202*
Witiak, D. N., 158(151), *201*
Wittig, C., 113(8), *196*
Wolf, R. J., 121(45), 123(45), *198*
Wolff, P. A., 49(9), *107*
Wolfhard, H. G., 206(6), *298*
Wood, D. W., 2(7), *43*
Woodin, R. L., 145(117), 146(117), *200*
Woodward, A., 117(29), *197*
Woodward, W. S., 116(24), *197*

Yamazaki, T., 146(121), 154(121), 183(121), 186(121), *200*
Yang, C. M., 208(38–41), 252(38–41), 275(38–41), 276(38–41), 282(38–41), *299*
Yek, Y., 28(44), 32(44), 33(44), *44*
Yip, S., 23(40), 28(40), 30(48), 36(62), *44*
Yoshida, S., 144(110), *199*
Yumlu, V. S., 215(49), *299*

Zare, R. N., 112(2), 116(25, 26), 117(32), 128(89, 90), 147(26, 89), 163(90), *196, 198, 199*
Zeelenberg, A. P., 257(128), *302*
Zeldovic, Ya. B., 81(27), *108*, 208(35, 36), 210(35, 36), *299*
Zheleznyi, B. V., 5
Zheng, C.-F., 128(90), 165(190), *199*
Zielinska, B. J. A., 49(11), 81(11), 83(11), 87(11), 93(11), 94(30), 95(30), 97(30), 103(30), (35), *108*
Zundel, G., 26(69), 38(69), 40(69), 41(69), *45*
Zwanzig, R., 102(40), *109*

SUBJECT INDEX

Acetaldehyde oxidation:
 in closed vessels, 251–252
 cool-flame multiplicity, 251–252
 full kinetic scheme, 283–287
 oscillatory states, 246–249
 P-T_a ignition diagram, 244–251
 stationary states, 245–246
 in unstirred closed vessel, 249–251
Acetone ion:
 dissociation of isomers, 174–177
 "nonstatistical" decay, 174–177
 potential-energy diagram, 176
Acetylenes, fluorescence from, 128
Adiabatic channel theory, of transition state,
 127–128
Alkane oxidation:
 at high temperatures, 272–274
 initiation phase, 264
 of molecular intermediates, 271–272
 propagation phase, 264–266
 stoichiometry and thermochemistry,
 255–257
Alkene oxidation:
 conjugate formation, 270
 equilibrium phase, 266–267
Alkylperoxy radicals:
 isomerization and decomposition, 267–269
 in low-temperature oxidation, 261–263
Alkyl radicals, in low-temperature oxidation,
 261–263
Allene ion, isomerization $vs.$ dissociation,
 148–150
Ammonia ion:
 angular momentum effects, 126
 KERD, 182–186
 photoelectron spectrum, 183
 polyatomic dissociation rates, 124–125
 rotational predissociation in, 182–186
 translational energy released, 185
Angular momentum, effect on density of
 states, 123–126

Aniline ions:
 dissociation and kinetic shift, 141–144
 energy diagram, 143
 heats of formation, 143
Appearance potential, and kinetic-shift
 problem, 139
Argonne National Laboratory, 28
Aromaticity, of gaseous hydrocarbons, 295
Autocorrelation function, of characteristic
 function, 96
Autoionization:
 in iodobenzene dissociation, 138
 ion preparation by, 189
 Rydberg states, 189
 threshold electron technique, 189
 trifluoro-chloromethane ion, 188–189

Balance equations:
 entropy, 65–71
 interfacial, general form, 60–61
Barycentric velocity, definition of, 58
Belousov-Zhabotinsky reaction, in
 autocatalytic equations, 233
Benzene ion:
 isomerization $vs.$ dissociation, 150–151
 isomers of, 151
 PEPICO studies, 165
 transition state in, 127
Benzonitrile ions:
 dissociation and kinetic shift, 144–145
 entropy of activation, 144
 infrared fluorescence, 145
 transition state, 144
Bifurcation:
 Hopf phenomena, 228–229
 saddle-node, 222
Brillouin light scattering, and water dynamics,
 33–35
Bromobenzene:
 density of vibrational states, 119, 120
 dissociation rates, 133–137

315